Principles of Proteomics

Second Edition

Principles of Proteomics

Second Edition

Richard M. Twyman

Garland Science
Taylor & Francis Group
NEW YORK AND LONDON

Garland Science
Vice President: Denise Schanck
Assistant Editor: David Borrowdale
Production Editor and Layout: Ioana Moldovan
Copy Editor: Sally Huish
Proofreader: Mac Clarke
Illustrations: Oxford Designers & Illustrators
Cover Design: Armen Kojoyian
Indexer: Bill Johncocks

ISBN 978-0-8153-4472-8

Library of Congress Cataloging-in-Publication Data
Twyman, Richard M.
Principles of proteomics / Richard M. Twyman. -- Second edition.
pages cm
Includes bibliographical references.
ISBN 978-0-8153-4472-8 (alk. paper)
1. Proteomics. 2. Proteins. I. Title.
QP551.T94 2014
572'.6--dc23
2013021694

Published by Garland Science, Taylor & Francis Group, LLC,
an informa business,
711 Third Avenue, New York, NY, 10017, USA, and 3 Park Square,
Milton Park, Abingdon, OX14 4RN, UK.

Printed in the United States of America

15 14 13 12 11 10 9 8 7 6 5 4 3 2 1

Visit our website at http://www.garlandscience.com

Front cover image:
Courtesy of Gabriel Mazzucchelli, Mass Spectrometry Laboratory, GIGA Proteomics, University of Liège, Belgium.

About the author:
Richard M. Twyman studied genetics at Newcastle University, where he gained a first-class honors degree, and then obtained his doctorate in molecular biology at Warwick University. After working as a postdoctoral research fellow at the MRC Laboratory of Molecular Biology in Cambridge, he became a full-time scientific writer, initially at the John Innes Centre in Norwich and then as the director of Twyman Research Management Ltd, a company that develops and manages scientific projects and provides assistance with the preparation of scientific manuscripts. He is the author of many science textbooks and is actively involved in many current research projects and lecture courses. He is a visiting professor of biotechnology at the University of Lleida in Spain.

Preface to the second edition

When I wrote the first edition of *Principles of Proteomics* in 2003, it was the first book that had attempted to cover the entire field of proteomics in broad strokes rather than focusing on specialized individual technologies. The first edition was published when proteomics was an emerging discipline, still unsure of its footing although confident in its abilities, with many technology platforms jostling for attention and consideration. Nearly a decade later, writing the second edition has proven a significant challenge. Although proteomics has stabilized, with certain technologies becoming unshakably established and others becoming obsolete, the cutting edge still boasts a rich and diverse source of novel technology platforms seeking to capture the proteome in ever more detail and on a scale barely conceived at the beginning of the millennium. But proteomics has also become increasingly commercialized. It is a billion-dollar industry, with many companies vying for attention, providing technologies, solutions, and contract research to other companies, who are in turn interested in using proteomics to find disease biomarkers, drug targets, vaccine candidates, novel chemical inhibitors, improved enzymes for industrial processes, and products to protect plants, the food chain, and the environment. Keeping up with the pace of change while still being aware of the fundamental aspects of proteomics, the core principles that make it possible in the first place, is a difficult task made more difficult by the dominant position of proprietary technologies, and the explosion in patents relating to proteomic technologies and strategies for processing proteomics data.

Despite the above, we must remember that proteomics is still about the global analysis of proteins. It seeks to achieve what genomics cannot—that is, a complete description of living cells in terms of all their functional components, brought about by the direct analysis of those components rather than the genes that encode them. Proteins offer a rich source of data, including sequences, structures, and biochemical and biological functions, which are influenced by modifications, subcellular localization, and, perhaps most important of all, the interactions among proteins and with other molecules. If genes are the instruction carriers, proteins are the molecules that execute those instructions. Genes are the instruments of change over evolutionary timescales, but proteins are the molecules that define which changes are accepted and which are discarded. It is from proteins that we shall learn how living cells and organisms are built and maintained and what leads to their dysfunction.

Although now firmly established, proteomics is still a difficult subject to penetrate for those not familiar with the terminology and technology, including experts in one area of proteomics venturing into another. There is still a great deal of jargon and many hyphenated acronyms that make sense once explained but otherwise remain mystifying; and there is still a high turnover of methods at the cutting edge, making it difficult to keep up. This situation is exacerbated by the increasing integration of proteomics with other areas of large-scale biology as researchers attempt to model cellular processes by looking not only at the functional components, but also at the information (genes, transcripts) and the outputs (metabolites, phenotypes) and how these are linked into networks and systems.

As I stated in the preface to the first edition, it is my hope that this book will be useful to those who need a broad overview of proteomics and what it has to offer. It is not meant to provide expertise in any particular area: there are plenty of other books that deal with specific technologies and their applications, the processing and archiving of proteomic data, and the integration of proteomics with other disciplines. The aim of this book is to pull together the different proteomics technologies and their applications, and present them in what I hope is a simple, logical, and user-friendly manner. After a brief introductory chapter providing an updated perspective on the history of proteomics since the turn of the millennium, the major proteomics technologies are discussed in more detail: two-dimensional gel electrophoresis, multidimensional liquid chromatography, mass spectrometry, sequence analysis, structural analysis, methods for studying protein interactions and modifications, and the development and applications of protein microarrays. These chapters have been broadened to account for new developments since the first edition, but I have made every effort to keep the material as concise as possible, since the brevity of the first edition was one of its strengths. I have assumed necessarily that the reader has a working knowledge of molecular biology and biochemistry. Each chapter has a short bibliography listing classic papers and useful reviews that will help the interested reader delve deeper into the literature.

The second edition would not have been possible without the help and support of the editorial team at Garland Science, so I extend special thanks to Gina Almond, David Borrowdale, and Ioana Moldovan for their dedication and assistance during the writing and revision process. I would also like to thank friends and colleagues who provided feedback on the first edition or suggestions for the second edition or who pointed out errors and omissions.

As ever, this book is dedicated with love to my parents, Peter and Irene, to my children, Emily and Lucy, and to Hannah, Joshua, and Dylan.

Richard M. Twyman

August 2013

Instructor Resources Website

Accessible from www.garlandscience.com, the Instructor Resource Site requires registration and access is available only to qualified instructors. To access the Instructor Resource Site, please contact your local sales representative or email science@garland.com.

The images in *Principles of Proteomics* are available on the Instructor Resource Site in two convenient formats: PowerPoint® and JPEG, which have been optimized for display. The resources may be browsed by individual chapter or a search engine.

Resources available for other Garland Science titles can be accessed via the Garland Science Website.

Acknowledgments

The author and publisher of *Principles of Proteomics* gratefully acknowledge the contributions of the following reviewers in the development of this book:

Vasco A. de Carvalho Azevedo	Universidade Federal de Minas Gerais, Brazil
Venkatesha Basrur	University of Michigan, USA
Richard Edwards	University of Southampton, UK
Rob Ewing	Case Western Reserve University, USA
Yao-Te Huang	College of Life Sciences, China Medical University, Taiwan
André Klein	Hogeschool Leiden, Netherlands
Sunny Liu	North Carolina State University, USA
Metodi Metodiev	University of Essex, UK
Peter Nilsson	AlbaNova University Center, Sweden
Joanna Rees	University of Cambridge, UK
Dacheng Ren	Syracuse University, USA
Anikó Váradi	The University of the West of England, UK

Contents

CHAPTER 1

The origin and scope of proteomics

1

1.1 INTRODUCTION

Proteomics is the systematic, large-scale analysis of proteins. It is based on the concept of the **proteome** as a complete set of proteins produced by a given cell, tissue, or organism, either as a complete protein catalog or as a list of proteins produced under a defined set of conditions. Proteins are involved in almost every conceivable biological activity, so a comprehensive analysis of the proteins in the cell provides a unique global perspective showing how these molecules interact and cooperate to create and maintain a working biological system. The cell responds to internal and external changes by regulating the level and activity of its proteins, so changes in the proteome, either qualitative or quantitative, provide a snapshot of the cell in action. The proteome is a complex and dynamic entity that can be defined in terms of the sequence, structure, abundance, stability, localization, modification, interaction, and biochemical function of its components, providing a rich and varied source of data. The analysis of these various properties of the proteome requires an equally diverse range of technologies, which are the subject of this book.

This introductory chapter considers the importance of proteomics in the context of large-scale biology, discusses some of the major goals of proteomics, and introduces the major technology platforms. We begin by tracing the origins of proteomics in the genomics revolution of the 1990s and following its evolution from a concept to a mainstream technology with a global market value that is predicted to exceed $6 billion by 2015.

1.2 THE BIRTH OF LARGE-SCALE BIOLOGY AND THE "OMICS" ERA

The overall goal of molecular biology is to determine the functions of genes and their products. This allows them to be linked into pathways and networks that should ultimately lead to a detailed understanding of how biological systems work. Until the turn of the millennium, molecular biology research focused predominantly on the isolation and characterization of individual genes and proteins because there was neither the information nor the technology available for investigations on a global scale. The only way to study biological systems was to break them down into their components, look at these individually, and then attempt to deduce how the system worked as a whole by proposing hypotheses that could be tested in further experiments. This is known as the **reductionist approach**.

The face of biological research began to change in the 1990s as technological breakthroughs made it possible to carry out **large-scale DNA** (deoxyribonucleic acid) **sequencing**. Until this point, the sequences of individual genes and proteins had accumulated slowly and steadily as researchers cataloged individual discoveries. This can be seen from the steady growth in the **INSDC nucleotide sequence databases** from 1980 to 1990, when the total amount of stored sequence data reached 10 million base pairs (**Figure 1.1**). During this time, almost all DNA sequencing was performed manually using the **Sanger chain termination method** (**Box 1.1**). The 1990s saw the advent of automated DNA sequencing, which allowed sequence data to be gathered at an increasing rate and ensured that the databases grew exponentially well into the 2000s. In the early 1990s, much of the new sequence data was represented by expressed sequence tags (ESTs), which are short fragments of DNA obtained by the random sequencing of cDNA (complementary DNA) libraries. In 1995, the first complete cellular genome sequence was published, that of the bacterium *Haemophilus influenzae*. This represented a new paradigm in molecular biology because for the first time the data existed to characterize a complete biological system. Over the next few years, more than 100 further genome sequences were completed, including the human genome, which was essentially finished in 2003. A lot of the data added to the databases after this point was in the form of random genomic clones resulting from **whole-genome shotgun** projects, basically massive collections of sequences covering the entire genome, which were then assembled into contigs using powerful computers. The rate of sequence data accumulation continued to increase in the 2000s, mainly because the throughput of automated Sanger sequencing continued to increase despite the inherent limitations of the underlying technology (**Box 1.2**). This involved the development of capillary sequencing machines that could carry out large numbers of automated reactions in parallel, day and night. To cope with this influx of data, two of the INSDC partners collaborated to launch the **Trace Archive** in 2001, to collect raw data produced at sequencing centers around the world. The amount of data in the archive doubled every 10 months between 2001 and 2006.

In 2005, there was another paradigm shift when the first **next-generation sequencing** methods began to displace the Sanger technique. Several next-generation sequencing technologies now exist based on different underlying principles, but they are united by their ability to yield millions of short DNA sequences in parallel (**Box 1.3**). To give some insight into the pace of change,

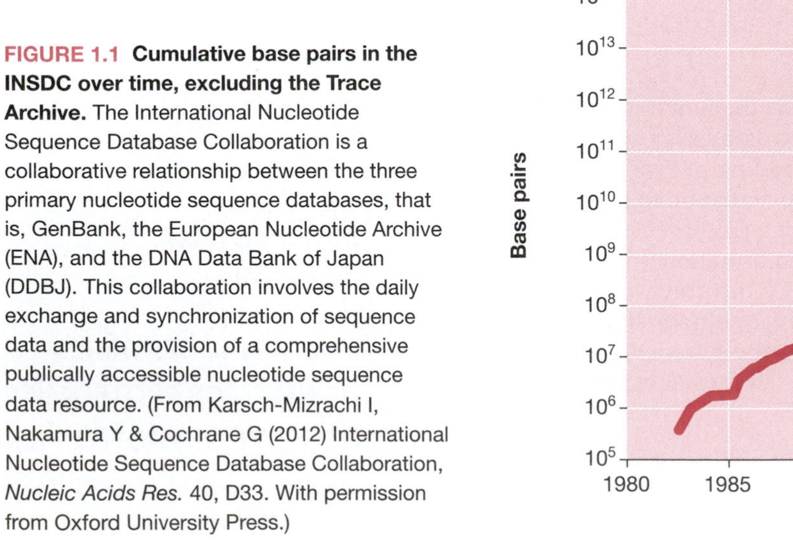

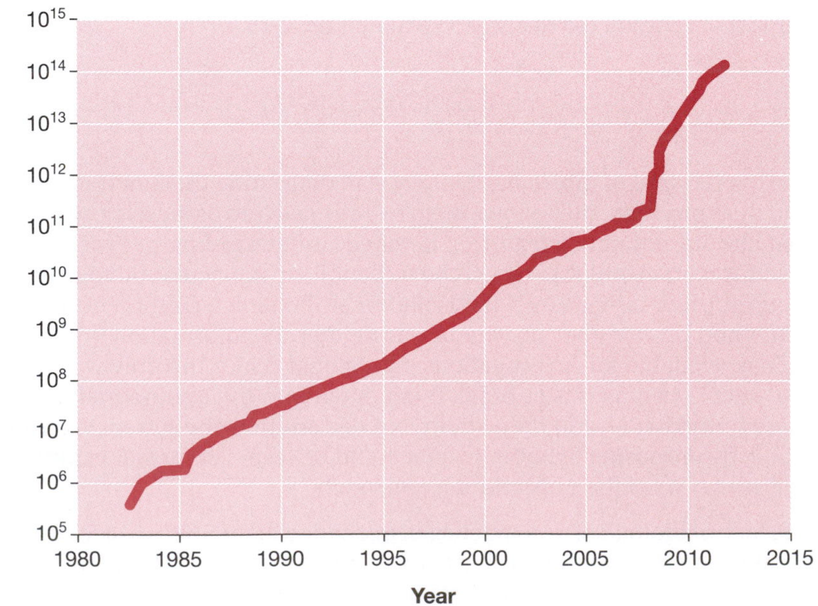

FIGURE 1.1 Cumulative base pairs in the INSDC over time, excluding the Trace Archive. The International Nucleotide Sequence Database Collaboration is a collaborative relationship between the three primary nucleotide sequence databases, that is, GenBank, the European Nucleotide Archive (ENA), and the DNA Data Bank of Japan (DDBJ). This collaboration involves the daily exchange and synchronization of sequence data and the provision of a comprehensive publically accessible nucleotide sequence data resource. (From Karsch-Mizrachi I, Nakamura Y & Cochrane G (2012) International Nucleotide Sequence Database Collaboration, *Nucleic Acids Res.* 40, D33. With permission from Oxford University Press.)

the original Human Genome Project took a decade of work involving a huge consortium of researchers using Sanger sequencing and billions of dollars of funding, but with the advent of next-generation sequencing there are now companies that offer to sequence individual human genomes for less than $100,000 in a few weeks. The first phase of the 1000 Genomes Project was completed in 2010 and this aims to sequence at least 1000 different human genomes every two years, which represents an output of 10 billion bases (three complete human genomes) every 24 hours. The three members of the INSDC began to archive raw next-generation sequence data reads between 2007 and 2008, and in 2009 launched the collaborative **Sequence Read Archive** (**SRA**) to accommodate the additional data, accounting for

BOX 1.1 RELATED TECHNOLOGIES.
Sanger's method for DNA sequencing.

Frederick Sanger's chain termination method for DNA sequencing (also known as the dideoxy method) exploits the ability of DNA polymerases to synthesize complementary strands of varying lengths on a single-stranded DNA template when provided with a short, labeled primer and a mixture of all four standard 2′-deoxynucleoside triphosphates (dNTPs) plus a small amount of a specific, chain-terminating analog (a 2′, 3′-dideoxynucleoside triphosphate, or ddNTP). The normal substrates for DNA synthesis are the four dNTPs representing the nucleosides adenosine (A), cytidine (C), guanosine (G), and thymidine (T). These possess a hydroxyl group at the C3′ position allowing the formation of a phosphodiester bond with the next nucleotide incorporated into the DNA strand. The corresponding ddNTPs lack this hydroxyl group and the strand cannot be elongated once a ddNTP is incorporated, that is, it acts as a chain terminator.

The original Sanger method comprises four parallel reactions, each incorporating the components required for DNA synthesis (the template, a radiolabeled primer, DNA polymerase, and four dNTPs) plus small amount of one of the four corresponding ddNTPs. In each reaction, the ddNTP is incorporated randomly when the template exposes the complementary base, generating a population of DNA molecules with a common 5′ end corresponding to the primer, but a variable 3′ end always representing the same base depending on which analog has been included. The four reaction products are denatured and separated in adjacent lanes by polyacrylamide gel electrophoresis, which has sufficient resolution to separate DNA molecules differing in length by one base. Exposure of the dried gel to X-ray film reveals a ladder of bands, which can be used to read off the sequence.

BOX 1.2 BACKGROUND ELEMENTS.
The limitations of Sanger sequencing.

The Sanger chain termination method for DNA sequencing dominated molecular biology for approximately 25 years (1980–2005). During this time, the throughput of the method increased substantially through cumulative technological improvements, including modifications allowing the use of double-stranded DNA templates, better enzymes, and more sensitive labels. However, the most significant improvement was achieved by switching from the use of radiolabeled primers to the use of four ddNTPs labeled with different fluorophores. This allowed the four reactions to be separated in a single gel lane (because the four sets of products produce different signals) and allowed the sequence to be read automatically by detecting fluorescence in real time during electrophoresis (rather than several days later by autoradiography). This not only increased throughput but also improved sequencing accuracy by reducing the human role in sequence interpretation and providing sufficient capacity to allow both DNA strands to be sequenced a number of times. Even higher throughput was achieved by replacing slab gels (which are labor-intensive and time-consuming) with capillary electrophoresis, which is up to five times faster at separation, reduces artifacts, and involves minimal operator handling. Capillary electrophoresis runs that handle up to 384 reactions simultaneously were the basis of factory-style Sanger sequencing programs that yielded up to 1 million base pairs of sequence each day. The resulting capillary trace data were processed automatically and subjected to rigorous quality control to yield high-quality datasets.

Even so, there are three irreconcilable bottlenecks in Sanger sequencing, namely the requirement to prepare template DNA, then carry out the chain termination reaction, and then separate the products. All these processes take time. These limitations have been addressed by today's "next-generation" sequencing methods, as discussed in Box 1.3.

the sudden surge in deposited sequences from 2009 in Figure 1.1, which is remarkable considering that the *y* axis has a logarithmic scale. Or in other words, the rate of sequence data accumulation is more than exponential at the time of writing. Based on these technological advances, the INSDC

BOX 1.3 RELATED TECHNOLOGIES.
Next-generation sequencing.

The so-called "next-generation" sequencing methods were developed to overcome the inherent limitations of the Sanger chain termination method, namely the need for template preparation and the time taken to complete the chain termination reaction and product separation. The bottleneck caused by template preparation was initially addressed by combining DNA sequencing with the polymerase chain reaction (cycle sequencing), which in its most extreme form can sequence uncloned source DNA directly. More recently, this has been superseded by the use of single-molecule templates immobilized either on a solid substrate or within an oil droplet, which can be sequenced directly or amplified *in situ* (emulsion PCR). The bottleneck caused by chain termination and product separation has been addressed by sequencing DNA in real time and increasing the throughput by extensive miniaturization as discussed below. Many of these methods have now been adopted as the basis of RNA profiling as well as DNA sequencing, as discussed in Box 1.4. They all produce short sequence reads (50–100 bp) but in huge amounts, allowing sequences to be assembled by analyzing overlaps and quality to be tested by sequencing the same DNA segment many times.

454 sequencing
This platform is a high-throughput form of pyrosequencing, in which the incorporation of a nucleotide into DNA is recorded in real time by detecting the release of pyrophosphate. As DNA polymerase moves along the template, each of the four nucleoside triphosphates is fed sequentially into the reaction and then removed. When one of the nucleotides is incorporated, the released pyrophosphate is detected as a flash of light. Multiplexing is achieved by constraining individual sequencing reactions onto microbeads where the template has been amplified by emulsion PCR. The beads are channeled into wells on a picotiter plate, which allows between one and two million reactions to be monitored in parallel.

Illumina/Solexa sequencing
This is based on reversible chain termination, that is, a chain-terminating nucleotide analog is incorporated but can then be cleaved and removed so chain extension can resume after a pause, allowing the fluorescent label to be detected. This method is therefore the closest conceptually to the original Sanger method. The Illumina/Solexa platform involves solid-phase *in situ* template amplification on a glass slide followed by sequencing with four-color blocked reversible terminators that are detected by total internal reflection fluorescence (TIRF) imaging using two lasers. A similar platform known as HeliScope uses non-amplified single-molecule templates.

SOLiD sequencing
This platform is based on the detection of ligation products. Sequencing by ligation involves the "interrogation" of a primed, single-stranded DNA template with a short degenerate oligonucleotide probe containing one or two discriminating bases identified by a specific fluorescent label. If the discriminatory bases match the template immediately adjacent to the primer then the oligonucleotide will anneal and can be ligated to the primer. Otherwise, ligation will not be possible and the probe will be washed away. The sequence adjacent to the primer can therefore be determined by fluorescence detection after washing.

DNA nanoball sequencing
This method uses rolling circle replication to amplify small fragments of genomic DNA into DNA nanoballs, which are then characterized using sequencing by ligation. This platform is offered by CompleteGenomics.

HeliScope sequencing
HeliScope sequencing uses DNA fragments with added poly(A) tail adapters attached to the surface of a flow cell prior to extension-based sequencing achieved by the cyclic addition of individual fluorescently labeled nucleotides prior to washing and signal detection by fluorescence imaging.

SMRT sequencing
Single-molecule real-time sequencing is a sequencing-by-synthesis approach using zero-mode wave guides (small wells containing immobilized DNA polymerase) and fluorescently labeled nucleotides in solution. The wells are constructed so that only fluorescence signals at the base of the well can be detected, allowing the detection of detached fluorescent labels as the corresponding nucleotide is incorporated into the DNA strand.

Emerging methods
Several additional sequencing technologies are considered promising but have yet to reach mainstream development because of technical limitations. These include nanopore sequencing in which a single DNA strand is drawn through a narrow portal and the sequence is determined by measuring the variable but base-specific differences in charge across the pore; ion semiconductor sequencing, which is based on the detection of hydrogen ions that are released during polymerization; and sequencing by hybridization. Although the latter has not been developed into a commercial platform, it is a good example of a next-generation technology because it does not rely on DNA synthesis and therefore does not involve the detection of reaction products. It is the only method that provides instant sequence readout capability, albeit only for short sequences at the current time. The basis of sequencing by hybridization is the annealing of a labeled DNA probe (the sequence to be determined) to an oligonucleotide chip containing arrays of every possible oligonucleotide of a certain length (for example, all possible octanucleotides = 65,536 sequences). The probe will only hybridize to complementary octanucleotides, which should allow the sequence to be reconstructed as a series of overlapping complementary eight-nucleotide fragments.

databases surpassed 100 billion base pairs of DNA in 2009 and reached 100 trillion base pairs in 2011. The ability to produce such massive amounts of sequence data with ever decreasing effort and expense means that it is now considered straightforward to sequence an entire genome as a first step toward characterizing an organism.

The large-scale sequencing projects ushered in the **genomics** era, which led in time to the concept of "**omics**" as a term for genomics and its derivatives, as discussed in the following section. This effectively removed the information bottleneck in accessing the genome and brought about the realization that biological systems, although large and very complex, are ultimately finite. In the 1990s, the idea formed that it might be possible to study biological systems in a global or holistic manner if sufficient amounts of data could be collected and analyzed, simply by cataloging and enumerating the components. However, although the technology for genome sequencing had advanced rapidly, the technology for studying the functions of the newly discovered genes lagged far behind. The databases became clogged with anonymous sequences and gene fragments, and the problem was exacerbated by the unexpectedly large number of new genes found even in well-characterized organisms. As an example, consider the yeast *Saccharomyces cerevisiae*, which was thought to be one of the best-characterized model organisms prior to the completion of its genome-sequencing project in 1996. Over 2000 genes had been characterized in traditional experiments and it was thought that genome sequencing would identify at most a few hundred more. Scientists got a shock when they found the yeast genome contained more than 6000 potential genes, nearly a third of which were unrelated to any previously identified sequence (**Figure 1.2**). Even today, nearly a quarter of the predicted open reading frames in the *S. cerevisiae* genome remain either unconfirmed or without functional annotations.

There are several related terms that describe questionable or unconfirmed sequences. A sequence is described as **unconfirmed** or **questionable** when there is only marginal evidence that it represents a gene. It may be short or may lack certain aspects of a gene while possessing others, suggesting it could be a gene remnant or fragment (that is, a pseudogene) even if it shows homology to known genes. On the other hand, an **orphan gene** has been shown to function as a gene (for example, expression may have been demonstrated) but the sequence is unrelated to any other known gene, that is, it is not a member of a known gene family. This precludes functional annotation by homology but not by independent means, so an orphan gene may not necessarily lack a functional annotation. Several related genes may be grouped into an "orphan family", although this is an oxymoron and a novel family designation is preferred. Finally, a **hypothetical protein** is a protein that is predicted to exist based on the existence of a gene sequence, but direct proof at the protein level does not exist. A hypothetical protein may be the product of an unconfirmed sequence, an orphan gene, or a well-known gene family. Hypothetical proteins can often be promoted to extant proteins by using proteogenomics for the analysis of genomes (Chapter 5).

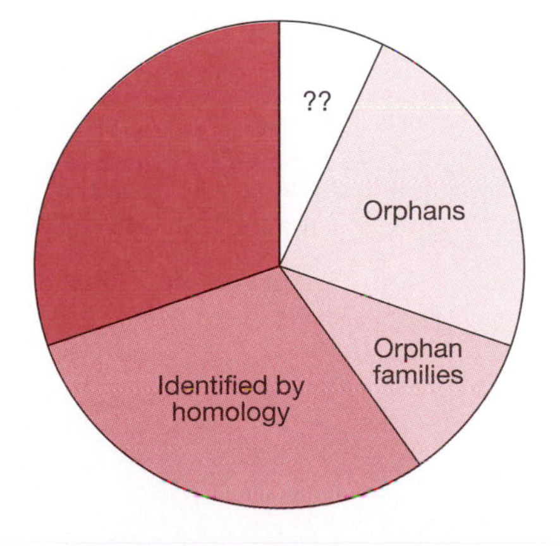

FIGURE 1.2 Distribution of yeast genes by annotation status in the aftermath of the *Saccharomyces cerevisiae* genome project. (?? shows questionable open reading frames)

The availability of masses of anonymous sequence data for hundreds of different organisms has precipitated a number of fundamental changes in the way research is conducted in the molecular life sciences. Traditionally, gene function had been studied by moving from phenotype to gene, an approach sometimes called **forward genetics**. An observed mutant phenotype (or in some cases a purified protein) was used as the starting point to map and identify the corresponding gene, and this led to the functional analysis of that gene and its product. The opposite approach, sometimes termed **reverse genetics**, is to take an uncharacterized gene sequence and modify it to see the effect on phenotype. As more uncharacterized sequences have accumulated in databases, the focus of research has shifted from forward to reverse genetics. Similarly, most research prior to 1995 was **hypothesis-driven**, in that the researcher put forward a hypothesis to explain a given observation, and then designed experiments to prove or disprove it. The genomics revolution instigated a progressive change toward what can arguably be called **discovery-driven** research, in which the components of the system under investigation are collected irrespective of any hypothesis about how they might work.

The final paradigm shift concerns the sheer volume of data generated in current experiments. Whereas in the past researchers have focused on individual gene products and generated rather small amounts of data, now the trend is toward the analysis of many genes and their products and the generation of enormous datasets that must be mined for salient information using computers. Advances in genomics have thus forced parallel advances in **bioinformatics**, the computer-aided handling, analysis, extraction, storage, and presentation of biological data.

1.3 THE GENOME, TRANSCRIPTOME, PROTEOME, AND METABOLOME

As large-scale biology has progressively supplanted reductionist experiments, so it has been necessary to re-evaluate the central dogma of molecular biology, which states that a gene is transcribed into RNA (ribonucleic acid) and then translated into protein (**Figure 1.3a**). It has already been necessary to tinker with the dogma to account for new discoveries such as reverse transcription, but large-scale biology has forced a reappraisal of the dogma based on scale. The new paradigm is that the **genome** (all the genes in the organism) gives rise to the **transcriptome** [the complete set of mRNA (messenger RNA) in any given cell], which is then translated to produce the **proteome** (the complete collection of proteins in any given cell) (Figure 1.3b). The proteome is largely responsible for the complete set of chemical compounds found in a cell or organism, which constitutes the **metabolome**. The metabolome is intricately involved in the regulation of the genome, transcriptome, and proteome, thus completing the biological system. By harnessing all this information simultaneously and using it to study and model living organisms, we have now entered the era of systems biology (**Box 1.4**).

The genome differs from the transcriptome and proteome in two important ways. First, the genome has a defined and limited information content because it is a linear sequence of nucleotides. The transcriptome and proteome are much more complex than the genome because a single gene can produce many different mRNAs and proteins. Different transcripts can be generated by alternative splicing, alternative promoter or polyadenylation site usage, and special processing strategies such as RNA editing. Different proteins can be generated by the alternative use of start and stop codons and the proteins synthesized from these mRNAs can be modified in various different ways during or after translation. Some types of modification, such as glycosylation, tend to be permanent. Others, such as phosphorylation, are transient and are often used to regulate protein activity and/or interactions

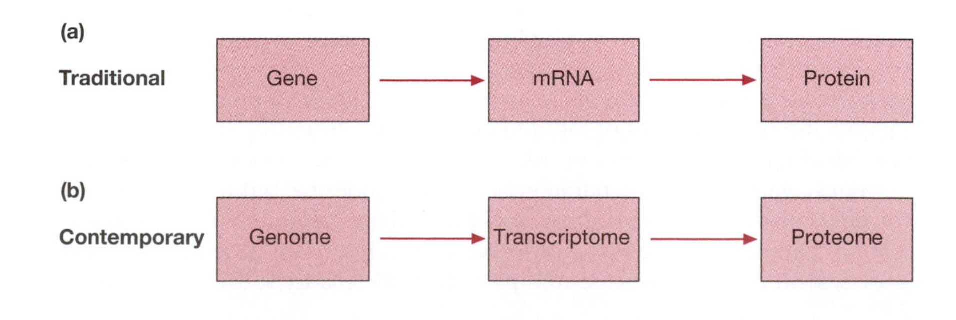

FIGURE 1.3 The new paradigm in molecular biology—the focus on single genes and their products has been replaced by global analysis.

BOX 1.4 RELATED TECHNOLOGIES.
Beyond proteomics—metabolomics and systems biology.

Proteomics can be regarded as the global analysis of the final stage of the expression of biological information stored in DNA, resulting in the production of functional molecules—proteins—that carry out a diverse range of activities in the cell (see Box 1.8 for more information on the functions of proteins). However, the end products of cellular processes orchestrated by proteins—for example, in their capacity as enzymes, receptors, transporters, and components of signaling pathways—are the small molecules making up the metabolic profile of the cell, the complete set of which is defined as the metabolome. Metabolomics is thus the global study of metabolites, completing the chain of information from DNA through RNA and protein to the biochemical output of the cell or organism. In many ways, the metabolome can be regarded as an even better snapshot of the functioning cell or organism than the proteome, because it provides an instant readout of physiological status in real time, hence the widespread use of specific metabolites as diagnostic markers.

Like the proteome, the metabolome is dynamic and full of diverse structures that are impossible to analyze with any single method. Depending on the properties of different classes of metabolites, they may be separated by gas chromatography (for volatiles), HPLC (high-performance liquid chromatography; for nonvolatiles), and capillary electrophoresis (for charged molecules) (also see Chapter 2). These techniques can be coupled to various forms of mass spectrometry (discussed in detail in Chapter 3) for detection and identification according to their fragmentation patterns or to nuclear magnetic resonance spectroscopy (Chapter 6).

with other molecules. The same protein can be modified in many different ways, giving rise to innumerable variants. For example, about 70% of human proteins have the potential to be glycosylated and the glycan chains can have many different structures. Often there are several glycosylation sites on the same protein, and different glycan chains can be added to each site. The largest known number of glycosylation sites on a single polypeptide is greater than 20, giving the potential for millions of possible glycoforms. Over 400 different types of post-translational modification have been documented, creating a massive source of proteome diversity. Therefore, although it is estimated that the human genome contains between 20,000 and 25,000 genes, it is likely that the proteome catalog comprises more than a million proteins when post-translational modification is taken into account. The human gene number was initially estimated at 50,000–100,000 based on EST data. This number has been progressively revised downwards following the sequencing and annotation of the human genome, but even with all this information to hand there is still no precise answer. Part of the problem is that different approaches to defining genes give different answers. For example, Ensembl release 67.37 indicates there are 20,115 genes whereas UniProt defines 20,231 genes (see also Box 5.1). Only by increasing diversity at the transcriptome and proteome levels, can the biological complexity of humans be explained compared with nematodes (~18,000 genes), fruit flies (~12,000 genes), and yeast (~6000 genes).

The other major difference between the genome and the transcriptome and proteome is that the genome is a static information resource that, with few exceptions, remains the same regardless of cell type or environmental conditions. In contrast, both the transcriptome and proteome are dynamic entities, whose content can fluctuate dramatically under different conditions due to the regulation of transcription, RNA processing, RNA stability, protein synthesis, protein modification, and protein stability. The transcriptome and proteome vary qualitatively (the type of mRNAs and proteins that are present) and also quantitatively (the levels of different mRNAs and proteins fluctuate over time and in response to internal and external stimuli). Again, much of the increase in biological complexity between simple organisms, such as yeast, and complex organisms, such as mammals and higher plants, is generated at the levels of the transcriptome and proteome.

1.4 FUNCTIONAL GENOMICS

The complete genome sequences that are now available for a large number of important organisms provide potential access to every single gene and therefore pave the way for large-scale functional analysis, an approach known as **functional genomics**. However, even complete gene catalogs provide at best a list of components, and no more explain how a biological system works than a list of mechanical parts explains how to drive a car. Before we can begin to understand how these components build a bacterial cell, a mouse, an apple tree, or a human being, we must understand not only what they do as individual entities, but also how they interact and cooperate with each other. Because the genome is only a blueprint, functional relationships among genes can only be inferred. Direct evidence must be gathered by studying the behavior of gene products at the levels of the transcriptome and proteome. The need for such analysis has encouraged the development of novel technologies that allow large numbers of mRNA and protein molecules to be studied simultaneously.

Transcriptomics is the systematic, global analysis of mRNA

Because the genomics revolution was based on technological advances in large-scale DNA cloning and sequencing, it made good sense to put these technologies to work in the functional analysis of genes. The first functional genomics methods were therefore based on DNA sequencing, and were used to study mRNA expression profiles on a global scale. This gave rise to the field now known as transcriptomics. The expression profile of a gene can reveal much about its role in the cell and can also help to identify functional links to other genes. For example, the expression of many genes is restricted to specific cells or developing structures suggesting that the genes have particular functions in those places (such as insulin, which is expressed solely in pancreatic β-cells). Other genes are expressed in response to external stimuli. For example, they might be switched on or switched off in cells exposed to endogenous signals such as growth factors, or environmental molecules such as DNA-damaging chemicals. Genes with similar expression profiles are likely to be involved in similar processes, and demonstrating that an uncharacterized gene has a similar expression profile to a gene whose function is already known may allow the first gene to be functionally annotated on the basis of "guilt by association." Furthermore, mutating one gene may affect the expression profiles of others, helping to link those genes into functional pathways and networks.

The first transcriptomics technologies were based on a concept now known as **census sequencing**, which refers to the collection and counting of short representative cDNA sequences (**tags**) that are sufficient to identify the corresponding mRNAs. The number of times a given sequence appears is indicative of the relative abundance of that mRNA in the source tissue. In the original method, clones were randomly picked from cDNA libraries and 200–300 bp sequences known as **expressed sequence tags** (**ESTs**) were generated using Sanger's chain termination method. This was an expensive and laborious way to compare mRNA levels within a given sample and it was difficult to compare mRNA levels between samples without carefully prepared comparable cDNA libraries. Alternative methods were therefore devised involving either the rapid quantitative representation of mRNA abundance using techniques such as **differential display PCR** (polymerase chain reaction) or the acquisition of very short **sequence tags** (9–15 bp), many of which could be analyzed at the same time, for example, **serial analysis of gene expression** (**SAGE**) and **massively parallel signature sequencing** (**MPSS**). These tag-based techniques were more reliable than large-scale cDNA sequencing but were complex to realize. The advent of next-generation sequencing methods (Box 1.3) has made it possible to collect millions of longer sequence tags (~50 bp) rapidly and inexpensively, rendering techniques such as SAGE largely redundant. These new methods (collectively known as **RNA-Seq**) are widely used today. The principles of census sequencing techniques are outlined briefly in **Box 1.5**.

The major alternative transcriptomics technology is based on DNA microarrays, which are miniature devices onto which many different DNA sequences are immobilized in the form of a grid. There are two major types, one made by the mechanical spotting of DNA molecules onto a coated glass slide and one produced by *in situ* oligonucleotide synthesis (the latter are also known as **oligonucleotide chips**). Although manufactured in completely different ways, the principles of mRNA analysis are much the same for each device. Expression analysis is based on **multiplex hybridization** using a complex population of labeled DNA or RNA molecules (**Figure 1.4**

BOX 1.5 RELATED TECHNOLOGIES.

Sequence sampling and display techniques for the global analysis of gene expression.

Sampling of cDNA libraries

Randomly picked clones are sequenced and searched against databases to identify the corresponding genes. The frequency with which each sequence is represented provides a rough guide to the relative abundances of different mRNAs in the original sample. This is an expensive and labor-intensive approach, particularly if several cDNA libraries need to be compared.

Analysis of EST databases

Expressed sequence tags are signatures generated by the single-pass sequencing of random cDNA clones. If EST data are available for a given library, the abundance of different transcripts can be estimated by determining the representation of each sequence in the database. This is a rapid approach, advantageous because it can be carried out entirely *in silico*, but it relies on the availability of EST data for relevant samples.

Differential display PCR

This is a display method that was devised for the rapid identification of cDNA sequences that are differentially expressed across two or more samples. The method has insufficient resolution to cope with the entire transcriptome in one experiment, so populations of labeled cDNA fragments are generated by RT-PCR (reverse transcriptase polymerase chain reaction) using one oligo-dT primer and one arbitrary primer, producing pools of cDNA fragments representing subfractions of the transcriptome. The equivalent amplification products from two biological samples (that is, products amplified using the same primer combination) are then run side by side on a sequencing gel, and differentially expressed cDNAs are revealed by quantitative differences in band intensities. This technique homes in on differentially expressed genes but false positives are common and other methods must be used to confirm the predicted expression profiles.

Serial analysis of gene expression (SAGE)

In this technique, very short sequence tags representing many cDNAs are joined together in a concatemer, which is sequenced. The tags may be as short as 9–15 bp but this is still adequate to resolve individual cDNA sequences, allowing them to be counted. The method is complex but it essentially involves cleaving a cDNA population with a frequent-cutter restriction enzyme and capturing the poly(A) tail and short exposed fragment. Ligation to a linker containing the recognition site for a type IIS restriction enzyme (which cuts a few base pairs downstream) then generates a sequence tag of defined length. Pairs of linker tags are ligated and the linkers are used as primer annealing sites to amplify the paired tags by PCR. The linkers are then released and the paired tags ligated to form large concatemers for sequencing and counting. SAGE is much more efficient than standard cDNA sampling because 50–100 tags can be counted for each sequencing reaction.

Massively parallel signature sequencing (MPSS)

Like SAGE, the MPSS technique involves the collection of short sequence tags from many cDNAs. However, unlike SAGE (where individual tags are cloned in series for identification by conventional sequencing), MPSS relies on the parallel analysis of thousands of cDNAs attached to microbeads in a flow cell by progressive sequence decoding. As with SAGE, a type IIS restriction enzyme is used to generate the sequence data, but whereas only a single SAGE tag is produced for each cDNA, in MPSS the enzyme is used to expose sequential four-base overhangs on each cDNA, which are "decoded" by a matching adaptor oligonucleotides identified by specific fluorophores. The cDNAs are progressively digested and decoded in four-nucleotide "bites."

RNA-Seq

Sequence sampling comes full circle with RNA-Seq, which is basically the same as the original cDNA sampling method except here the mRNA is reverse-transcribed from source and the resulting cDNA population is randomly sequenced "deeply," that is, millions of short sequences are obtained, using the next-generation sequencing methods described in Box 1.3. This provides statistically highly reliable data about the relative abundance of different mRNA species in a sample and is suitable for direct comparisons between samples.

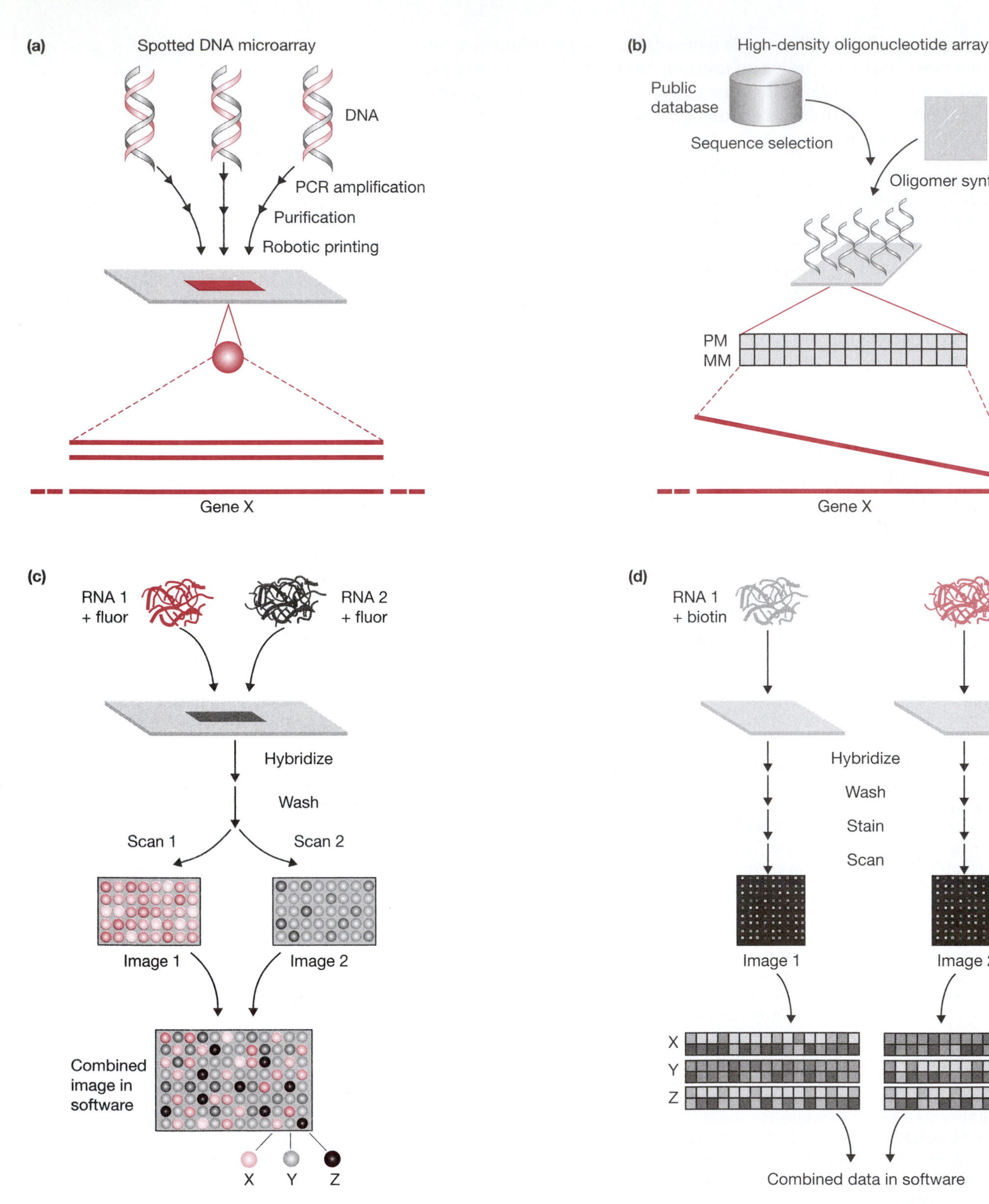

and color plates). For both devices, a population of mRNA molecules from a particular source is reverse-transcribed en masse to form a representative complex cDNA population. In the case of spotted microarrays, a fluorophore-conjugated nucleotide is included in the reaction mix so that the cDNA population is universally labeled. In the case of oligonucleotide chips, the unlabeled cDNA is converted into a labeled cRNA (complementary

FIGURE 1.4 Expression analysis with DNA microarrays. (a) Spotted microarrays are produced by the robotic printing of amplified cDNA molecules onto glass slides. Each spot or feature corresponds to a contiguous gene fragment of several hundred base pairs or more. (b) High-density oligonucleotide chips are manufactured using a process of light-directed combinatorial chemical synthesis to produce thousands of different sequences in a highly ordered array on a small glass chip. Genes are represented by 15–20 different oligonucleotide pairs (PM, perfectly matched; MM, mismatched) on the array. (c) On spotted arrays, comparative expression assays are usually carried out by differentially labeling two mRNA or cDNA samples with different fluorophores. These are hybridized to features on the glass slide and then scanned to detect both fluorophores independently. Shaded dots labeled X, Y, and Z at the bottom of the image correspond to transcripts present at increased levels in sample 1 (X), increased levels in sample 2 (Y), and similar levels in samples 1 and 2 (Z). (d) On Affymetrix GeneChips, biotinylated cRNA is hybridized to the array and stained with a fluorophore conjugated to avidin. The signal is detected by laser scanning. Sets of paired oligonucleotides for hypothetical genes present at increased levels in sample 1 (X), increased levels in sample 2 (Y), and similar levels in samples 1 and 2 (Z) are shown. See also color plates section for color clarity. (From Harrington CA, Rosenow C & Retief J (2000) *Curr. Opin. Microbiol.* 3, 285. With permission from Elsevier.)

RNA) population by the incorporation of biotin, which is later detected with fluorophore-conjugated avidin. The complex population of labeled nucleic acids is then applied to the array and allowed to hybridize. Each individual feature or spot on the array contains 10^6–10^9 copies of the same DNA sequence, and is therefore unlikely to be completely saturated in the hybridization reaction. Under these conditions, the intensity of the signal at each address on the array is proportional to the relative abundance of that particular cDNA or cRNA in the mixture, which in turn reflects the abundance of the corresponding mRNA in the original source population. Therefore, the relative levels of thousands of different transcripts can be monitored in one experiment. Comparisons between samples may be achieved by hybridizing labeled cDNA or cRNA prepared from each of the samples to identical microarrays, or by using different fluorophores to label different cDNA populations and scanning at different emission wavelengths to compare the two signals at each address (Figure 1.4 and color plates). Microarrays have been widely used to infer global trends in gene expression under different conditions (for example, normal tissue versus tumor tissue, responses of cell cultures to different media additives), but the experiments must be carefully controlled and the data analyzed using approved statistical methods to ensure reproducibility.

Large-scale mutagenesis and interference can also determine the functions of genes on a global scale

One of the most straightforward ways to establish the function of a gene is to mutate it and observe the resulting phenotype. Mutations have been at the forefront of biological research since the beginning of the twentieth century, but only in the 1990s did it become practical to generate **comprehensive mutant libraries**, that is, collections of organisms with systematically produced mutations affecting every gene in the genome. Like transcriptomics, such developments relied on prior advances in large-scale clone preparation and sequencing.

Mutagenesis strategies can be divided into two approaches. The first is genome-wide mutagenesis by **homologous recombination**, which involves the deliberate and systematic inactivation of each gene in the genome through replacement with a DNA cassette containing a nonfunctional sequence (**Figure 1.5**). This form of gene replacement is termed **gene knockout** and produces **null mutations** that cause complete **loss-of-function** phenotypes, although, due to genetic redundancy, some null mutations have no apparent effect. The knockout approach can be used on a genome-wide scale only where the organism in question has a fully sequenced genome and is amenable to homologous recombination. Thus far, systematic homologous recombination has been restricted to the relatively small genomes of yeast

FIGURE 1.5 Large-scale mutagenesis by gene knockout in yeast. This has been achieved by systematically replacing each endogenous gene (*gray bar*) with a nonfunctional sequence or marker (*red bar*) inserted within a homology cassette. Recombination occurs at the homologous flanking regions (X) leading to the replacement of the functional endogenous gene with its nonfunctional counterpart.

and bacteria because individual mutagenesis cassettes are required for every gene, although there is also a community-based effort to generate a comprehensive library of conditional knockout mutants in mouse embryonic stem (ES) cells. Homologous recombination can also be achieved in the fruit fly *Drosophila melanogaster* and in other model organisms, but genome-wide gene knockout projects for these organisms have yet to be implemented. Homologous recombination can also be used for a related method known as **knock-in** where new sequences are introduced at specific loci. In the context of proteomics this is particularly relevant because it provides the means to generate comprehensive proteomic libraries with proteins expressed as fusions to markers that facilitate protein localization or purification, a subject we consider in more detail in Chapter 7.

The second mutagenesis approach is genomewide **random mutagenesis**. This is actually one of the oldest and most established techniques in genetics, but the genomics revolution has changed the way it is used. Originally, populations were mutagenized by irradiation or exposure to mutagenic chemicals, and large-scale screens were performed to identify mutants impaired for a particular biological process of interest. More recently, insertional mutagenesis has become the method of choice. This involves the random integration of a DNA cassette which generates a mutant phenotype when it interrupts a gene. Whereas irradiation and chemical mutagenesis tend to introduce point mutations that need to be laboriously mapped, **insertional mutagenesis** can be achieved with a specific DNA sequence that can be identified by hybridization or PCR, allowing the direct cloning of flanking sequences and hence the rapid identification of the interrupted gene. Although not as systematic as homologous recombination, insertional mutagenesis is applicable in a wider range of organisms because it does not require a completed genome sequence and it is much easier to perform. Insertional mutant libraries have therefore been produced in many species, including bacteria, yeast, the nematode *Caenorhabditis elegans*, the fruit fly, the mouse, the zebrafish, and many plants. Some of these libraries have been generated using modified **transposons**, which have a natural tendency to become mobile in the genome, others using artificial cassettes, and in plants a popular strategy is to use the T-DNA (transfer DNA) from *Agrobacterium tumefaciens*, which naturally integrates into the genome of many plant species. Like homologous recombination cassettes, insertional mutagenesis constructs can be designed to collect information about the gene in addition to generating a mutant phenotype (**Box 1.6**). Furthermore, transcriptional and translational fusions can be used to monitor the expression of the interrupted gene and localize the protein, whereas the inclusion of a strong, outward-facing promoter can activate genes adjacent to the insertion site generating strong, gain-of-function phenotypes caused by overexpression or ectopic expression. An example of a highly modified insertional construct used in yeast is shown in **Figure 1.6**.

Mutagenesis strictly requires the alteration of a DNA sequence, but functional analysis can also be achieved by interrupting gene expression. There are many ways to achieve this, but the one that has become most popular for genome-scale analysis is **RNA interference** (**RNAi**), which results in the rapid degradation of specific mRNAs and the generation of a loss-of-function **phenocopy** of a mutant phenotype by abolishing gene expression (also

BOX 1.6 ALTERNATIVE APPLICATIONS.
Advanced insertional elements for functional genomics.

As discussed in the main text, both homologous recombination constructs and insertional mutagenesis cassettes primarily generate data by mutation, allowing the analysis of the loss-of-function phenotype. However, by innovative construct design, much more information can be generated about gene function at the mRNA and protein levels. Insertional cassettes are therefore a key element of proteomic analysis.

Gene traps

The **gene trap** is an insertion element that contains a reporter gene downstream of a splice acceptor site. The reporter gene encodes a product that can be detected and visualized using a simple assay, for example, the *lacZ* gene encodes the enzyme β-galactosidase, which converts the colorless substrate X-gal into a dark blue product, and the *gfp* gene encodes green fluorescent protein, which emits light under appropriate illumination. If the gene trap integrates within the transcription unit of an endogenous gene, the splice acceptor site causes the reporter gene to be recognized as an exon, allowing it to be incorporated into a transcriptional fusion product. Because this fusion transcript is expressed under the control of the interrupted gene's promoter, the expression pattern revealed by the reporter gene is often identical to that of the interrupted endogenous gene. Early gene trap vectors depended on in-frame insertion, but the incorporation of internal ribosome entry sites, which allow independent translation of the reporter gene, circumvent this limitation.

Enhancer traps

The **enhancer trap** is an insertion construct in which the reporter gene lies downstream of a minimal promoter. Under normal circumstances, the promoter is too weak to activate the reporter gene, which is therefore not expressed. However, if the construct integrates in the vicinity of an endogenous enhancer, the marker is activated and reports the expression profile driven by the enhancer.

Activation traps

The **activation trap** is an insertion construct containing a strong, outward-facing promoter. If the element integrates adjacent to an endogenous gene, that gene will be activated by the promoter. Unlike other insertion vectors, which cause loss of function by interrupting genes, an activation tag causes gain of function through overexpression or ectopic expression.

Protein localization traps

Protein localization traps are insertion constructs or homologous cassettes that identify particular classes of protein based on their localization in the cell. For example, a construct has been described in which the reporter gene is expressed as a fusion to the transmembrane domain of the CD4 type I protein. If this inserts into a gene encoding a secreted product, the resulting fusion protein contains a signal peptide and is inserted into the membrane of the endoplasmic reticulum (ER) in the correct orientation to maintain β-galactosidase activity. However, if the construct inserts into a different type of gene, the fusion product is inserted into the ER membrane in the opposite orientation and β-galactosidase activity is lost.

Purification tags

Both homologous recombination cassettes and insertional constructs can also be used to introduce epitopes so that the resulting fusion protein can be localized by immunohistochemistry or purified by immunoaffinity chromatography. One of the most innovative uses of this approach is to trap protein complexes and identify interaction partners, a form of co-immunopurification that is discussed in more detail in Chapter 7.

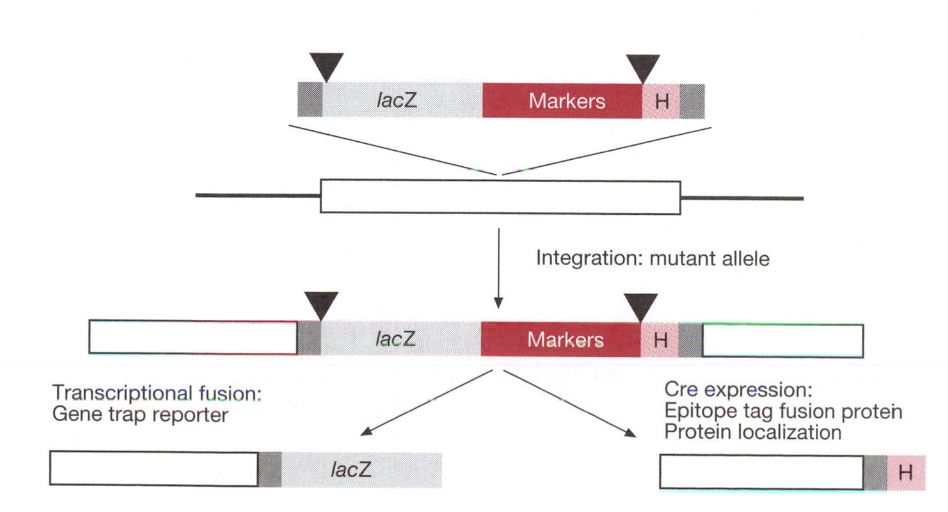

FIGURE 1.6 Multifunctional *E. coli* Tn*3* cassette used for random mutagenesis in yeast. The cassette comprises Tn*3* components (*dark gray*), *lacZ* (*light gray*), selectable markers (*red*) and an epitope tag such as His_6 (*pink*, H). The *lacZ* gene and markers are flanked by *lox*P sites (*black triangles*). Integration generates a mutant allele, which may or may not reveal a mutant phenotype. The presence of the *lacZ* gene at the 5′ end of the construct allows transcriptional fusions to be generated, so the insert can be used as a reporter construct to reveal the normal expression profile of the interrupted gene. If Cre recombinase is provided, the *lacZ* gene and markers are deleted, leaving the endogenous gene joined to the epitope tag, allowing protein localization to be studied.

known as **gene silencing**). The effect is triggered by double-stranded RNA (dsRNA) and appears to result from the induction of a ubiquitous defense mechanism that protects cells from viruses, which often use a dsRNA replicative intermediate. The defense mechanism destroys the inducing dsRNA molecule and targets any single-stranded RNA in the cell with the same sequence (Box 1.7). In the context of functional genomics, RNAi is useful because the introduction of a dsRNA molecule homologous to an endogenous gene results in the rapid destruction of any corresponding mRNA and hence the potent silencing of that gene at the post-transcriptional level. The ease with which RNAi can be initiated has allowed large-scale RNAi programs to be carried out, most notably in the nematode worm *C. elegans*, where the phenomenon was discovered. These experiments involved the synthesis of thousands of dsRNA molecules and their systematic administration to worms by microinjection, soaking, or feeding. A screen was carried out in which nearly 17,000 bacterial strains were generated and fed to worms, each strain expressing a different dsRNA, representing 86% of the genes in the *C. elegans* genome.

BOX 1.7 BACKGROUND ELEMENTS.
How does RNA interference work?

The mechanism of RNA interference is complex, but involves the conversion of a dsRNA molecule into duplexes, usually about 21 bp in length with short 3′ overhangs, by a dsRNA-specific endonuclease called **Dicer** (Figure 1). The short duplexes are known as **small interfering RNAs** (**siRNAs**). In addition to introducing dsRNA, RNAi can be induced by expressing sense and antisense transgenes, hairpin constructs, or introducing synthetic siRNAs directly. This is required in mammalian cells, where longer dsRNAs induce nonspecific silencing. Each siRNA is unwound into two single-stranded components: the passenger strand, which is degraded, and the guide strand, which is incorporated into a sequence-specific RNA endonuclease known as the **RNA-induced silencing complex** (**RISC**). When the guide strand base-pairs with a complementary mRNA, a protein component of the RISC known as Argonaute cleaves the mRNA, resulting in its rapid and efficient degradation. RNAi is a systemic phenomenon in some species because siRNAs appear to replicate and move between cells, spreading the effect throughout the organism. RNAi-induced silencing is transient in mammals and *Drosophila*, but in nematodes and plants it is heritable. In plants, this reflects cross-talk between the RNAi pathway and gene-specific DNA methylation.

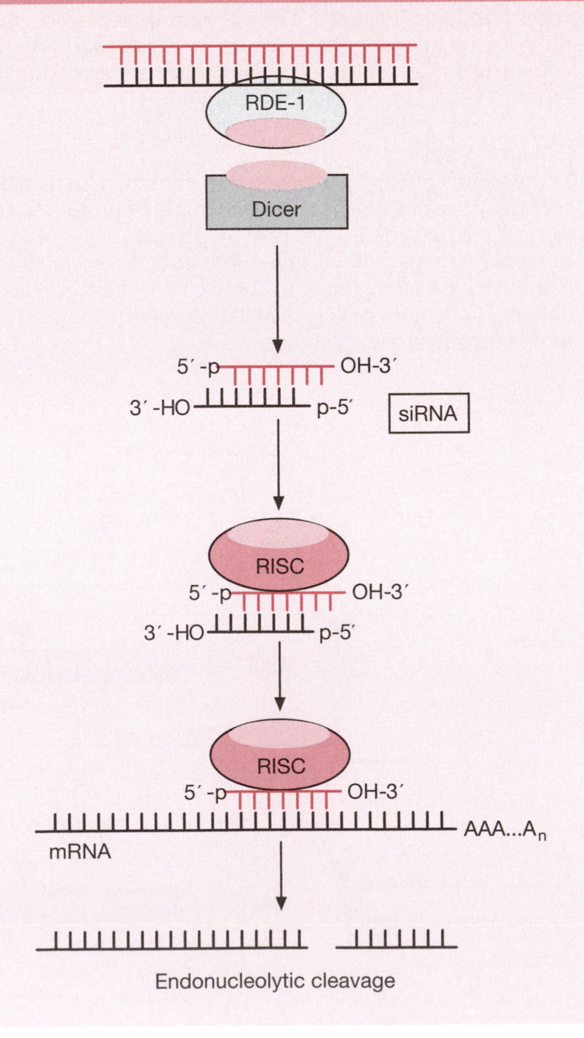

FIGURE 1 The mechanism of RNA interference. Double-stranded RNA (dsRNA) is recognized by the protein RDE-1, which recruits a nuclease known as Dicer. This cleaves the dsRNA into short fragments, 21–23 bp in length with two-base overhangs. The fragments are known as small interfering RNAs (siRNAs). The siRNA is incorporated into the RNA-induced silencing complex (RISC). The siRNA serves as guide for RISC and, upon perfect base pairing, the target mRNA is cleaved in the middle of the duplex formed with the siRNA. (From Voinnet O (2002) *Curr. Opin. Plant Biol.* 5, 444. With permission from Elsevier.)

RNAi is not suitable for all species, particularly zebrafish and the model frog *Xenopus laevis,* where nonspecific gene silencing effects are commonplace (a similar phenomenon occurring in mammals can be overcome by inducing RNAi with synthetic siRNA because the nonspecific effect is triggered by long dsRNA molecules; see Box 1.7). In these species, oligonucleotides are used instead as gene-specific inhibitors and they work by blocking splicing, protein synthesis, or regulatory interactions (depending on the design and position within the mRNA). Natural oligonucleotides are unstable, so chemically modified **morpholino antisense oligonucleotides** are preferred because these are resistant to degradation. Genome-wide morpholino oligo collections are not yet available, although mid-scale screening services for both zebrafish and *Xenopus* have been described. The genome-wide analysis of gene function using mutagenesis and interference methods is sometimes termed phenomics.

1.5 THE NEED FOR PROTEOMICS

Transcriptomics and phenomics have risen quickly to dominate functional genomics because they are based on high-throughput clone generation and sequencing, two of the technology platforms that saw rapid development in the genomics revolution. But what do they really tell us about the working of biological systems? Nucleic acids are undoubtedly important molecules, but they are only information carriers. Therefore, the analysis of genes (by mutation) or of mRNA (by RNA interference or transcriptomics) can only tell us about protein function indirectly. Proteins are the actual functional molecules of the cell (**Box 1.8**). They are responsible for almost all the biochemical activity of the cell through interactions with each other and with a diverse spectrum of other molecules. In this sense, they are functionally the most relevant components of biological systems and a true understanding of such systems can only come from the direct study of proteins. The importance of proteomics can be summarized as follows:

- The function of a protein depends on its structure and interactions, neither of which can be predicted accurately based on sequence information alone. Only by looking at the structure and interactions of the protein directly can definitive functional information be obtained.

- Mutations and RNA interference are coarse tools for large-scale functional analysis. If the structure and function of a protein is already understood in detail, precise mutations can be introduced to investigate its function further. However, for the large-scale analysis of gene function, the typical strategy is to completely inactivate each gene (resulting in the absence of the protein) or to overexpress it (resulting in overabundance or ectopic activity). In each case, the resulting phenotype may not be informative. For example, the loss of many proteins is lethal, and although this tells us the protein is essential, it does not tell us what the protein actually does. Random mutagenesis can produce informative mutations serendipitously, but there is no systematic way to achieve this. Some proteins have multiple functions in different times and/or places, or have multiple domains with different functions, and these cannot be separated by a blanket mutagenesis approach.

- The abundance of a given transcript may not reflect the abundance of the corresponding protein. Transcriptome analysis tells us the relative abundance of different transcripts in the cell, and from this we infer the abundance of the corresponding protein. However, the two may not be related, because of post-transcriptional gene regulation. Not all the mRNAs in the cell are translated, so the transcriptome may include gene products that are not found in the proteome. Similarly, rates of protein synthesis and protein turnover differ among transcripts, and therefore the abundance of a transcript does not necessarily correspond to the abundance of the encoded protein. The transcriptome may not accurately represent the proteome either qualitatively or quantitatively.

- Much protein diversity is generated post-transcriptionally. Many genes, particularly in eukaryote systems, give rise to multiple transcripts by alternative splicing. These transcripts often produce proteins with different functions. Mutations, acting at the gene level, may therefore abolish the functions of several proteins at once. Splice variants are represented by different transcripts, so it should be possible to distinguish them by RNA interference and transcriptome analysis, but some transcripts give

BOX 1.8 BACKGROUND ELEMENTS. The central importance of proteins.

The term protein was coined in 1838 by the Swedish chemist Jöns Jacob Berzelius to describe a particular class of macromolecule, abundant in living organisms, and made up of linear chains of **amino acids**. The term is derived from the Greek word *proteios* meaning "of the first order" and was chosen to convey the central importance of proteins in the human body. As our knowledge of this class of macromolecules has grown, this definition seems all the more appropriate. We have discovered that proteins are vital components of almost every biological system in every living organism. There are thousands of different proteins in even the simplest of cells and they form the basis of every conceivable biological function.

Most of the biochemical reactions in living cells are catalyzed by proteins called **enzymes**, which bind their substrates with great specificity and increase the reaction rates millions or billions of times. Several thousand enzymes have been cataloged. Some catalyze very simple reactions, such as phosphorylation or dephosphorylation, while others orchestrate complex and intricate processes such as DNA replication and transcription. Proteins can also transport or store other molecules. Examples include ion channels (which allow ions to pass across otherwise impermeable membranes), ferritin (which stores iron in a bioavailable form), hemoglobin (which transports oxygen), and the component proteins of larger structures such as nuclear pores and plasmodesmata.

Other proteins have a structural or mechanical role. All eukaryotic cells possess a cytoskeleton comprising three types of protein filament: microtubules made of tubulin, microfilaments made of actin, and intermediate filaments made of specialized proteins such as keratin. Unlike enzymes and storage proteins, which tend to be globular in structure and soluble in aqueous solvents, the cytoskeletal proteins are fibrous and can link into bundles and networks. Such proteins not only provide mechanical support to the cell, but they can also control intracellular transport, cell shape, and cell motility. For example, microtubule networks help to separate chromosomes during mitosis and to transport vesicles and other organelles from site to site within the cell. They also form the core structures of cilia and flagella. Actin filaments form contractile units in association with proteins of the myosin family. This actin–myosin interaction provides muscle cells with their immense contractile power. In other cells, actin filaments have a more general role in facilitating cell movement and changing cell shape, for example by forming a contractile ring during cell division. In multicellular organisms, further structural proteins are deposited in the extracellular matrix, which consists of protein fibers embedded in a complex gel of carbohydrates. Such proteins, which include collagen, elastin, and laminin, contribute to the mechanical properties of tissues. Cell adhesion proteins, such as cadherins and integrins, help to stick cells together and to their substrates.

Another important role for proteins is communication and regulation. Most cells bristle with receptors for various molecules allowing them to respond to changes in the environment. These receptors are specialized proteins that either span the membrane, with domains poking out each side, or are tethered to it. In some cases, the ligands that bind to these receptors are also proteins: many hormones are proteins (for example, growth hormone and insulin), as are most developmental regulators, growth factors, and cytokines. In this way, a protein secreted by one cell can bind to a receptor on the outside of another and influence its behavior. Inside the cell, further proteins are involved in signal transduction, the process by which a signal arriving at the surface of the cell mediates a specific effect inside. Often, the ultimate effect is to change the pattern of gene expression in the responding cell by influencing the activity of regulatory molecules called transcription factors, which are also proteins. Other proteins are required for mRNA processing, translation, protein sorting in the cell, and secretion. More specialized examples of proteins involved in communication include the light-sensitive protein rhodopsin, which is required for light perception in the retinal rod cells of the eye, and the voltage-gated ion channels required for the transmission of nerve impulses along axons.

A final category of proteins encompasses those involved in "species interactions," that is, attack, defense, and cooperation. All pathogenic microorganisms produce proteins that interact with the proteins of their host to enable infection and reproduction. For example, viruses have proteins that allow them to bind to the cell surface and facilitate entry, and some may have further proteins that interact with the machinery that controls cell division and protein synthesis, hijacking these processes for their own needs. Bacterial toxins, such as the cholera, tetanus, and diphtheria toxins, are proteins. And the molecules we use to protect ourselves against invaders—for example, antibodies, complement, and so forth—are also proteins.

rise to multiple proteins whose individual functions cannot be studied other than at the protein level.

- Protein activity often depends on post-translational modifications, which are not predictable from the level of the corresponding transcript. Many proteins are present in the cell as inert molecules that need to be activated by processes such as proteolytic cleavage or phosphorylation. In cases where variations in the abundance of a specific post-translational variant are significant, this means that only proteomics provides the information necessary to establish the function of a particular protein.

- The function of a protein often depends on its localization. Although there are some examples of mRNA localization in the cell, particularly in early development, most trafficking of gene products occurs at the protein level. The activity of a protein often depends on its location, and many proteins are shuttled between compartments (for example, the cytosol and the nucleus) as a form of regulation. The abundance of a given protein in the cell as a whole may therefore tell only part of the story. In some cases, it is the distribution of a protein rather than its absolute abundance that is important.

- Some biological samples do not contain nucleic acids. One practical reason for studying the proteome rather than the genome or transcriptome is that many important samples do not contain nucleic acids. Most body fluids, including serum, cerebrospinal fluid, and urine, fall into this category, but the protein levels in such fluids are often important determinants of disease progression (for example, proteins shed into the urine can be used to follow the progress of bladder cancer). Although nucleic acids are present in fixed biological specimens, they are often degraded or cross-linked beyond use, and protein analysis provides the only feasible means to study such material. It has also recently been shown that proteins may be better preserved than nucleic acids in ancient biological specimens, such as Neanderthal bones.

- Proteins are the most therapeutically relevant molecules in the body. Although there has been recent success in the development of drugs (particularly antivirals) that target nucleic acids, most therapeutic targets are proteins and this is likely to remain so for the foreseeable future. Proteins also represent useful biomarkers and may be therapeutic in their own right (see Chapter 10).

1.6 THE SCOPE OF PROTEOMICS

Proteins are diverse molecules that can be studied using a range of different methods depending on which properties are targeted, for example, physicochemical properties, on/off expression, abundance, sequence, structure, modification, localization, interaction with other molecules, transport, stability, and biochemical, cellular, and organism-level biological function. Proteomics can therefore be divided into several major but overlapping branches, which embrace these different contexts and help to integrate proteomic data into a comprehensive understanding of biological systems. The rest of this book looks at these individual components of proteomics and the associated technologies.

Protein identification and quantitation are the most fundamental aspects of proteomic analysis

A typical proteomic analysis involves the separation of complex protein mixtures (Chapter 2), the identification of individual components (Chapter 3), and their systematic quantitative analysis, often including comparative analysis across related samples (Chapter 4). The major forms of data gathered in this approach are protein identification, the presence/absence of particular proteins in particular samples (on/off expression), and protein abundance, the latter often described as **protein expression profiling** or **expression proteomics** (**Figure 1.7**). Methods for the separation of protein mixtures based on two-dimensional gel electrophoresis (2DGE) were first developed in the 1970s and even at this time it was envisaged that databases could be created to catalog the proteins in different cells and look for differences representing alternative states, such as healthy and diseased tissue. Many of the statistical analysis methods now associated with microarray analysis (such as clustering algorithms and multivariate statistics) were developed originally in the context of 2DGE protein analysis, but technical limitations initially prevented reproducible separation and the identification of separated proteins. The major breakthrough in expression proteomics was made in the early 1990s when **mass spectrometry** techniques were adapted for protein identification, and algorithms were designed for database searching using mass spectrometry data. Today, thousands of proteins can be separated, identified, quantified, cataloged, and compared to reveal the proteins that are differentially expressed among different samples and to characterize post-translational modifications. The key technologies in expression proteomics are 2DGE and multidimensional liquid chromatography (MDLC) for protein separation, mass spectrometry and informatics for protein identification and image analysis, and mass spectrometry informatics for protein quantitation. The application of the above techniques in the analysis of post-translational modifications is considered in Chapter 8. An emerging trend in expression proteomics that overlaps considerably with functional proteomics is the use of protein microarrays for ultra-high-throughput analysis (Chapter 9).

Important functional data can be gained from sequence and structural analysis

Although proteomics as we understand it today would not have been possible without advances in DNA sequencing, it is worth remembering that the first protein sequence (insulin, 51 amino acids, completed in 1956) was determined 10 years before the first RNA sequence (a yeast tRNA, 77 bases, completed in 1966) and 13 years before the first DNA sequence (the *Escherichia coli lac* operator in 1969). Until DNA sequencing became routine

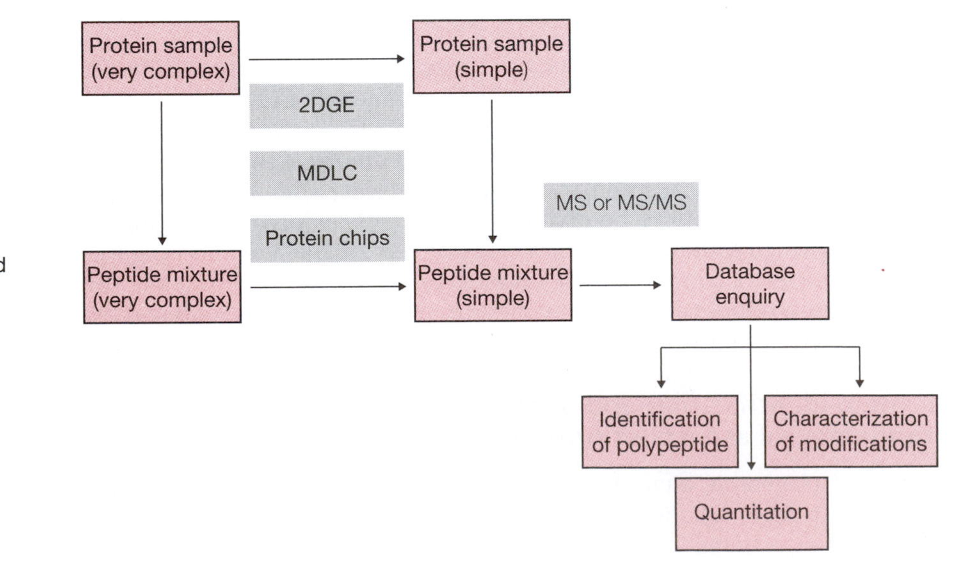

FIGURE 1.7 Expression proteomics is concerned with protein identification and qualitative analysis. This figure shows the aims of expression proteomics and major technology platforms used. See Chapters 2–4 and 8–9 for further information. 2DGE, two-dimensional gel electrophoresis; MS, mass spectrometry; MS/MS, tandem mass spectrometry; MDLC, multidimensional liquid chromatography.

in the 1980s, it was usually the protein sequence that was determined first, allowing the design of probes that could be used to isolate the corresponding cDNA or genomic sequence from a DNA library. Protein sequencing by Edman degradation (see Chapter 3) often provided a crucial link between the activity of a protein and the genetic basis of a particular phenotype, and it was not until the mid-1980s that it first became commonplace to predict protein sequences from genes rather than to use protein sequences for gene isolation.

The increasing numbers of stored protein and nucleic acid sequences, and the recognition that functionally related proteins often had similar sequences, catalyzed the development of statistical techniques for sequence comparison that underlie many of the core bioinformatics methods used in proteomics today (Chapter 5). Nucleic acid sequences are stored in three **primary sequence databases** (GenBank, ENA, and DDBJ) and are exchanged daily under the umbrella of the INSDC. These databases also contain protein sequences that have been translated from DNA sequences, and these protein sequences are archived in a separate collaborative database called UniProt, which was formed in 2002 from the formerly separate but overlapping protein databases Swiss-Prot, TrEMBL, and PIR. One of the major breakthroughs in proteomics was the development of algorithms that allowed the sequence databases to be searched using mass spectrometry data (Chapter 3), which means that proteins identified in large-scale experiments can be identified and linked to related proteins and their corresponding genes that have already been deposited. Protein sequences often allow the nature of specific post-translational modifications to be predicted (Chapter 8). Furthermore, similar sequences give rise to similar structures, which infer similar interactions and biochemical functions. The study of three-dimensional protein structures therefore also provides important additional data to determine the way proteins function and interact.

Structural proteomics is underpinned by technologies such as X-ray diffraction and nuclear magnetic resonance spectroscopy, and has given rise to another branch of bioinformatics concerned with the storage, presentation, comparison, and prediction of structural data (Chapter 6). The Protein Data Bank is the primary protein structure database (http://www.rscb.org) and is the major depository for publically accessible protein structure information, managed by four participating databases that comprise the Worldwide Protein Data Bank (wwPDB). At the time of writing, the database contains more than 80,000 structures. Technological developments in structural proteomics have centered on increasing the throughput of structural determination and the initiation of projects for the systematic analysis of protein structures representing the entire proteome.

Interaction proteomics and activity-based proteomics can help to link proteins into functional networks

Interaction proteomics (**interactomics**) considers the genetic and physical interactions among proteins as well as interactions between proteins, nucleic acids, and small molecules. The analysis of protein interactions can provide information not only about the function of individual proteins but also about how proteins are linked in complexes, pathways, and networks. Interaction proteomics relies on many different technology platforms to provide diverse information, and is closely linked with **activity-based proteomics** (the direct, large-scale analysis of protein functions) and technologies to predict and determine protein localization (**Figure 1.8**). Conceptually, the most ambitious aspect of interaction proteomics is the creation of **proteome linkage maps** based on binary interactions between individual proteins and higher-order interactions determined by the systematic analysis of **protein complexes**. Key technologies in this area include

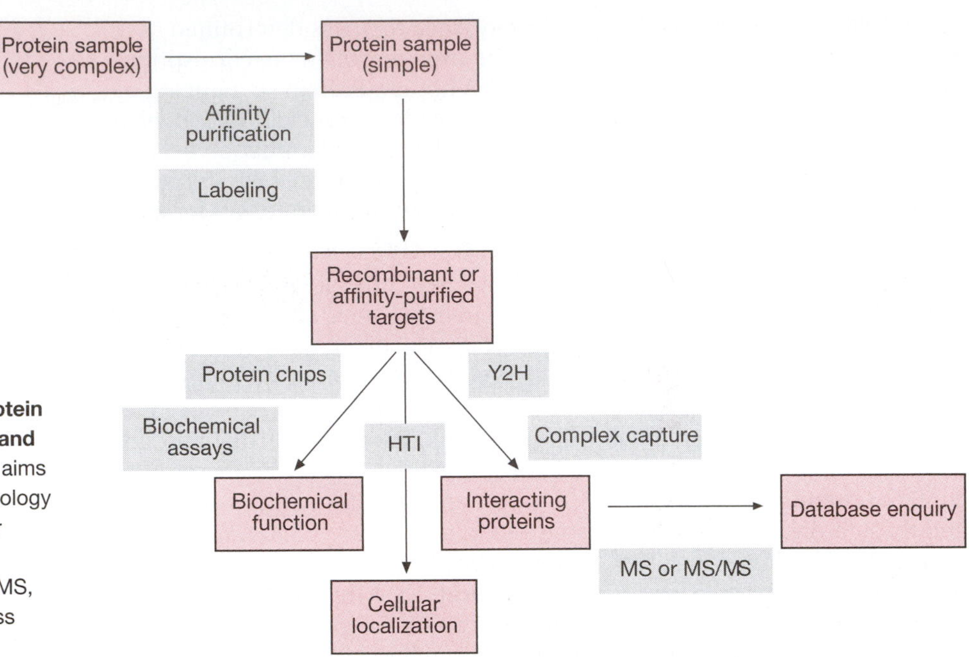

FIGURE 1.8 Functional proteomics is concerned with the investigation of protein interactions and biochemical, cellular, and system functions. This figure shows the aims of functional proteomics and major technology platforms used. See Chapters 7 and 9 for further information. HTI, high-throughput imaging; Y2H, yeast two-hybrid system; MS, mass spectrometry, MS/MS, tandem mass spectrometry.

two-hybrid/protein complementation assays for binary interactions, affinity purification/mass spectrometry for the analysis of protein complexes, and the analysis of restricted subsets of the proteome (**sub-proteomics**) such as **organellar proteomics** and **membrane proteomics** (Chapter 7). Interactions between proteins and nucleic acids underlie many important processes, including gene regulation, whereas protein interactions with small molecules are of interest because they govern important biological processes such as enzyme–substrate, receptor–ligand, and drug–target behavior. These types of interactions are often investigated using both biochemical assays and structural analysis methods such as X-ray diffraction. The characterization of protein interactions with small molecules can play an important role in drug development.

1.7 CURRENT CHALLENGES IN PROTEOMICS

One factor that underpinned the success of the early large-scale DNA sequencing projects was the adoption of a common technology platform and a consistent data format to ensure that data could be shared unambiguously across the world. Proteomics differs from the above in two key areas. First, the types of data generated in proteomics experiments are diverse (reflecting the different properties of proteins that are investigated) and second, there are several competing or complementary technologies for each type of data, with different data formats. DNA sequencing was almost entirely based on the Sanger method until approximately 2005 and the raw datasets were almost entirely sequence traces from capillary electrophoresis that could be converted into curated sequences using a globally accepted informatics approach. In contrast, proteomics datasets can be generated by 2DGE or MDLC, proteins can be identified using a range of mass-spectrometry-based informatics approaches, and the data produced by each platform are distinct. Hurdles must be overcome at every stage of analysis, from sample preparation through to database management, but particularly the integration of diverse data sets (**Figure 1.9**). The complexity of proteomics datasets also means that rigorous standards are required to ensure reproducibility and unambiguous interpretation. This has been enacted by the **HUPO**

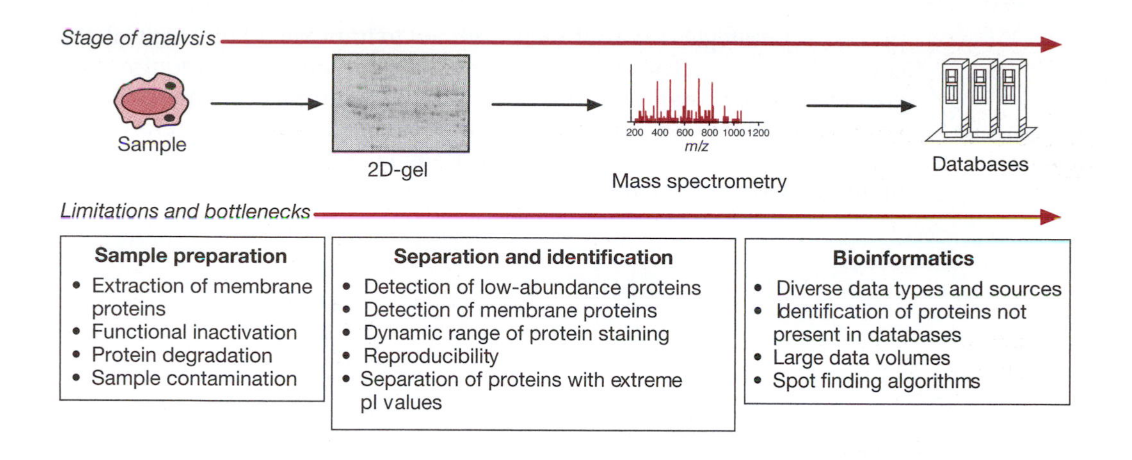

FIGURE 1.9 Challenges facing proteomics with current technology.

Proteomics Standards Initiative as the **minimum information about a proteomics experiment** (**MIAPE**), which is discussed in **Box 1.9**.

Another key challenge in proteomics is the lack of an amplification method equivalent to the polymerase chain reaction for nucleic acids, which means that scarce proteins are difficult to detect and the quality of proteomic data relies on the sensitivity of the technology. Sensitivity is especially important considering the dynamic range of protein abundances in typical biological samples, which has been estimated at 10^5 for tissues and up to 10^9 for body fluids such as serum. Some of the major proteomics technologies also suffer from high rates of false-positive and false-negative results, for example, the yeast two-hybrid system for the detection of binary protein interactions (Chapter 7). Although proteomics methods, instruments, and data

BOX 1.9 BACKGROUND ELEMENTS.
Minimal Information About a Proteomics Experiment (MIAPE).

The advent of functional genomics saw an explosion in research based on large-scale biology, resulting in the publication of increasingly large and complex datasets as well as interpretations based on them. The first functional genomics platform to experience this transition to the mainstream was the use of microarrays to study gene expression profiles, and this led in 2001 to the development of the **Minimum Information About a Microarray Experiment** (**MIAME**) initiative as an attempt to standardize the way both data (the experimental results) and metadata (data about the experimental setup) were presented. The initiative was supported by scientific journal editors and database curators to ensure that microarray data met strict quality standards before they could be accepted for publication, helping to ensure reproducibility and the unambiguous interpretation of experimental results. Part of the MIAME initiative was also the development of common data exchange formats (initially MAGE-ML, an XML-based microarray and gene expression data **markup language**, which is now being superseded by the simpler MAGE-TAB) and knowledge models (MGED-Ontology).

The success of MIAME and MAGE led to similar initiatives in other areas of large-scale biology. The ongoing task of developing a Minimal Information About a Proteomics Experiment (MIAPE) standard is being handled by the **Proteomics Standards Initiative** (**PSI**) a working group of the Human Proteome Organization (HUPO). As above, this involves the development of data/metadata standards, a mark-up language for data exchange, and ontologies for consistent annotation. Unlike MIAME, the MIAPE initiative has to account for a number of different technologies, and hence there are separate modules focusing on gel electrophoresis, gel image informatics, chromatographic separations, mass spectrometry, mass spectrometry informatics, and capillary electrophoresis, plus additional working groups looking at quantitative mass spectrometry, protein modifications, and protein interactions (**Minimal Information About a Molecular Interaction Experiment**, **MIMIx**, and Minimum Information About a Protein Affinity Reagent, MIAPAR). The standards are regularly updated on the HUPO PSI Website, which can be accessed at: http://www.psidev.info/groups. Since 2007, minimum standards initiatives have been grouped under the umbrella of the **Minimum Information About a Biomedical or Biological Investigation** (**MIBBI**), which also shows early-stage MI projects before publication, for example MIAPepAE (Minimum Information About a Peptide Array Experiment), which will become relevant as the methods discussed in Chapter 9 become more commonplace.

processing strategies are continually refined to improve quality control, the comprehensive analysis of complex biological systems will require even greater sensitivity and resolution so that experiments produce reproducible datasets. In this context, it is unlikely that proteomics will completely replace hypothesis-driven research, and well-designed and well-executed experiments are required to confirm proteomic data.

FURTHER READING

Altelaar AF, Munoz J & Heck AJ (2013) Next-generation proteomics: towards an integrative view of proteome dynamics. *Nat. Rev. Genet.* 14, 35–48.

Anderson NG & Anderson NL (1998) Proteome and proteomics: new technologies, new concepts, new words. *Electrophoresis* 19, 1853–1861.

Blackstock W & Weir M (1999) Proteomics: quantitative and physical mapping of cellular proteins. *Trends Biotechnol.* 17, 121–127.

Blow N (2009) Systems biology: untangling the protein web. *Nature* 460, 415–418.

Chuang HY, Hofree M & Ideker T (2010) A decade of systems biology. *Annu. Rev. Cell Dev. Biol.* 26, 721–744.

Coelho PSR, Kumar A & Snyder M (2000) Genome-wide mutant collections: toolboxes for functional genomics. *Curr. Opin. Microbiol.* 3, 309–315.

Dufva M (2009) Introduction to microarray technology. *Methods Mol. Biol.* 529, 1–22.

Gygi SP, Rochon Y, Franza BR & Aebersold R (1999) Correlation between protein and mRNA abundance in yeast. *Mol. Cell Biol.* 19, 1720–1730.

Hardy S, Legagneux V, Audic Y & Paillard L (2010) Reverse genetics in eukaryotes. *Biol. Cell* 102, 561–580.

Johnson CH & Gonzalez FJ (2012) Challenges and opportunities of metabolomics. *J. Cell Physiol.* 227, 2975–2981.

Kamath RS, Fraser AG, Dong Y et al. (2003) Systematic functional analysis of the *Caenorhabditis elegans* genome using RNAi. *Nature* 421, 231–237.

Lee KH (2001) Proteomics: a technology-driven and technology-limited discovery science. *Trends Biotechnol.* 19, 217–222.

Marioni JC, Mason CE, Mane SM et al. (2008) RNA-seq: an assessment of technical reproducibility and comparison with gene expression arrays. *Genome Res.* 18, 1509–1517.

Martin JA & Wang Z (2011) Next-generation transcriptome assembly. *Nat. Rev. Genet.* 12, 671–682.

Martzen MR, McCraith SM, Spinelli SL et al. (1999) A biochemical genomics approach for identifying genes by the activity of their products. *Science* 286, 1153–1155.

Merriman B, Ion Torrent R&D Team & Rothberg JM (2012) Progress in ion torrent semiconductor chip based sequencing. *Electrophoresis* 33, 3397–3417.

Metzker ML (2010) Sequencing technologies—the next generation. *Nat. Rev. Genet.* 11, 31–46.

Morozova O, Hirst M & Marra MA (2009) Applications of new sequencing technologies for transcriptome analysis. *Annu. Rev. Genomics Hum. Genet.* 10, 135–151.

Nakamura Y, Cochrane G, Karsch-Mizrachi I & International Nucleotide Sequence Database Collaboration (2013) The International Nucleotide Sequence Database Collaboration. *Nucleic Acids Res.* 41(Database issue), D21–D24.

Ozsolak F & Milos PM (2011) RNA sequencing: advances, challenges and opportunities. *Nat. Rev. Genet.* 12, 87–98.

Pandey A & Mann M (2000) Proteomics to study genes and genomes. *Nature* 405, 837–846.

Patterson SD & Aebersold RH (2003) Proteomics: the first decade and beyond. *Nat. Genet.* 33 (Suppl), 311–323.

Shendure J & Lieberman Aiden E (2012) The expanding scope of DNA sequencing. *Nat. Biotechnol.* 30, 1084–1094.

Simon R (2008) Microarray-based expression profiling and informatics. *Curr. Opin. Biotechnol.* 19, 26–29.

Thompson JF & Oliver JS (2012) Mapping and sequencing DNA using nanopores and nanodetectors. *Electrophoresis* 33, 3429–3436.

Tyers M & Mann M (2003) From genomics to proteomics. *Nature* 422, 193–197.

Velculescu VE, Zhang L, Vogelstein B & Kinzler KW (1995) Serial analysis of gene expression. *Science* 270, 484–487.

Wang Z, Gerstein M & Snyder M (2009) RNA-Seq: a revolutionary tool for transcriptomics. *Nat. Rev. Genet.* 10, 57–63.

CHAPTER 2

Strategies for protein separation

2

2.1 INTRODUCTION

The analysis of proteins, whether on a small or a large scale, requires methods for the separation of protein mixtures into their individual components. Protein separation methods can be placed on a sliding scale from fully selective to fully nonselective. Selective methods aim to isolate individual proteins from a mixture, usually by exploiting specific properties like their binding specificity. Such affinity-based methods are discussed in more detail in Chapter 7. In this chapter, we focus on nonselective separation methods, which aim to take a complex protein mixture and fractionate it in an unbiased manner so that all the individual proteins, or at least a substantial subfraction, are available for further analysis. Such methods lie at the heart of proteomics and exploit the general properties of proteins such as their mass and charge.

In proteomics, protein separation technology is pushed to its limits. The ultimate goal is to resolve all the individual proteins in a cell or other sample. As stated in Chapter 1, in a eukaryotic cell this may represent tens or hundreds of thousands of different proteins when post-translational modifications are taken into consideration. These proteins are chemically diverse and thus it is difficult to devise a separation method that will represent all proteins equally. Even after many years of development, the most sophisticated proteomics separation methods result in the underrepresentation of certain protein classes and are therefore at least partially selective.

Whether protein separation is selective, partially selective or nonselective, it is important to remember that the underlying principle is always the exploitation of physical and chemical differences between proteins that cause them to behave differently in particular environments. These physical and chemical differences are determined by the number, type, and order of amino acids in the protein, and by the presence and type of any post-translational modifications.

2.2 GENERAL PRINCIPLES OF PROTEIN SEPARATION IN PROTEOMICS

Many techniques can be used to separate complex protein mixtures in what at least approaches a nonselective fashion, but not all of these techniques are suitable for proteomics. One major requirement is high resolution. The separation technique should produce fractions that are simple mixtures of proteins or peptides (**Box 2.1**), which essentially rules out one-dimensional techniques (that is, those that exploit a single chemical or physical property as the basis for separation) because they do not provide sufficient resolving power. All the techniques discussed in this chapter are therefore multidimensional, that is, two or more different fractionation principles are employed one after another. The degree to which the multiple techniques achieve separation under different principles is described as **orthogonality**. The other major requirement in proteomics is high throughput. The separation technique should resolve all the proteins in one experiment and should ideally be easy to automate. The most suitable methods for automation are those that rely on differential rates of migration to produce fractions that can be displayed or collected, a process generally described as **separative transport**. A final requirement is that the fractionation procedure should be compatible with downstream analysis by mass spectrometry, as this is the major technology platform for high-throughput protein identification (Chapter 3). The two groups of techniques that have come to dominate proteomics are **two-dimensional gel electrophoresis** (**2DGE**) and **multidimensional liquid chromatography** (**MDLC**), the latter occasionally combined with further separation techniques such as one-dimensional gel electrophoresis, capillary electrophoresis, or chromatofocusing.

BOX 2.1 BACKGROUND ELEMENTS.
The need for peptides in proteomics.

The typical workflow of a proteomics experiment involves the separation of proteins and/or peptides followed by the identification of either selected proteins or, in the true sense of proteomics, all the proteins in the original sample, by mass spectrometry. We discuss the use of mass spectrometry in proteomics in more detail in Chapter 3, but it is clear from the technologies discussed in this chapter that many of the techniques for protein separation are actually applied to **peptides** derived from proteins by **digestion with proteolytic enzymes**, not the proteins themselves. The reason for this is simple: mass spectrometry has an upper limit on mass detection, which means the majority of full proteins cannot be detected. The cutoff is approximately 50 kDa for standard methods, which represents a protein of fewer than 500 amino acids, and proteins larger than this are excluded from analysis. Some recent specialized methods that have been used to push the boundaries of detection are discussed in Chapter 3.

In most cases, it is necessary to reduce the size of the analyzed molecules by fragmentation while retaining the ability to identify the parental molecule. Proteins are therefore digested with trypsin or another enzyme that breaks each protein molecule into fragments whose sizes can be predicted from sequence data (based on the well-characterized cleavage sites), allowing them to be assigned to their parent proteins using appropriate database searching algorithms (Chapter 3). In 2DGE, the proteins are digested after separation, so each protein spot is converted into a collection of peptides representing the protein(s) in the spot. In multidimensional chromatography methods, it is much more common to digest the proteins into peptides at source and then separate the peptides, as the fractions can be fed automatically into the mass spectrometer, an approach known as **bottom-up proteomics**. The main challenge with this approach is that the mixture of proteins, already highly complex, becomes even more complex when digested into peptides because each protein yields 20–50 peptides depending on its size. The higher resolving power of multidimensional chromatography techniques can be used to prevent the mass spectrometer being overwhelmed by large numbers of peptides per fraction. However, only a small number of peptides need to be positively identified to determine the presence of the parent protein. Therefore, another way to address the complexity problem is to reduce the number of peptides in the sample by affinity enrichment techniques that preserve as far as possible protein representation (p. 36). Another alternative, analogous to the 2DGE workflow, is to select whole proteins for analysis and then generate fragments by collision rather than digestion, an approach known as **top-down proteomics**. The relative merits of these different methods for protein identification are discussed in more detail in Chapter 3.

2.3 PRINCIPLES OF TWO-DIMENSIONAL GEL ELECTROPHORESIS

Electrophoresis separates proteins by mass and charge

Any charged molecule in solution will migrate in an applied electric field, a phenomenon known as **electrophoresis**. The rate of migration depends on the strength of the electric field and the **charge density** of the molecule, that is, the ratio of charge to mass. Because dissolved proteins carry a net charge, a protein mixture in solution can in theory be fractionated by electrophoresis since different proteins, with different charge densities, migrate toward the appropriate electrodes at different rates. In practice, however, effective separation is never achieved because all the proteins are initially distributed throughout the solution.

The answer to this problem is to load the protein mixture in one place within the electrophoresis buffer, allowing proteins with different charge densities to migrate as discrete zones. However, there are many practical reasons why standard electrophoresis is not carried out in free solution. One is that any disturbance to the solution will disrupt the electrophoresis zones. Even if extreme precautions are taken to avoid shocks and vibrations from outside, electrophoresis generates heat and the convection currents within the buffer will disperse the zones quite effectively. Another reason is that the narrow protein zones generated by electrophoresis are broadened by diffusion, which acts quickly to homogenize the protein mixture once the electric field is removed (**Figure 2.1**). These effects are minimized if electrophoresis is carried out in very narrow vessels (**capillary electrophoresis**, **Box 2.2**)

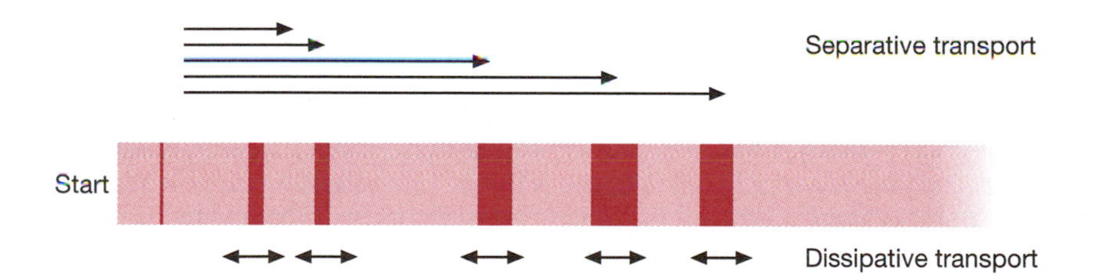

FIGURE 2.1 Separative and dissipative transport during zone electrophoresis.

BOX 2.2 ALTERNATIVE APPLICATIONS.
Capillary electrophoresis in proteomics.

Capillary electrophoresis (CE) is similar in principle to gel electrophoresis but it is carried out in glass tubes that are typically about 50 μm in diameter and up to 1 m in length. The tubes may or may not be filled with gel, but the presence of gel facilitates sieving of the proteins or peptides and enhances size-dependent separation, a variant known as **capillary gel electrophoresis** (CGE). The thin tubes are efficient at dissipating heat, allowing the use of strong electric fields. This means the separations are rapid and can be monitored in real time rather than at the experimental end point. The key benefit of capillary electrophoresis in proteomics is therefore that it allows the separative principles of electrophoresis (that is, the propensity to separate charged and polar molecules efficiently) to be coupled with orthogonal chromatography methods (typically RP-HPLC) to rapidly generate highly resolved fractions that can be fed automatically into a mass spectrometer. The narrow diameter of the capillary tubes requires small sample volumes. Therefore, another application of capillary electrophoresis in proteomics is the separation of peptides in relatively simple mixtures, such as the tryptic digests of purified proteins or spots excised from two-dimensional gels. The miniaturization of capillary electrophoresis can be taken further with the development of chip-based separation devices based on microfluidics technology, which we consider briefly in Chapter 9.

and/or within a stabilizing matrix such as paper or gel, since the latter also allows the separated proteins to be fixed in place once the procedure is complete. Polyacrylamide gels are favored because they facilitate protein separation by sieving, and gels with different pore sizes can be produced easily and reproducibly by varying the concentration of acrylamide in the polymerization mixture, allowing the preferential separation of proteins with a particular range of molecular masses (p. 29). **Polyacrylamide gel electrophoresis** (**PAGE**) is therefore one of the most widely used protein separation techniques in molecular biology.

As discussed above, all high-resolution protein fractionation methods employ multidimensional separation processes that exploit different properties of proteins for separation in each dimension. Although PAGE separates proteins according to both charge and mass, exploiting both these principles in the same dimension still results in a low-resolution separation because the separation zones overlap extensively. It was appreciated as early as the 1970s that the most efficient way to fractionate protein mixtures would be to apply these principles one after the other in orthogonal dimensions. It has therefore been necessary to devise modifications of gel electrophoresis that achieve separation on the basis of charge alone and on the basis of mass alone. These modified techniques are applied consecutively in 2DGE.

Isoelectric focusing separates proteins by charge irrespective of mass

The first-dimension separation in 2DGE is usually **isoelectric focusing** (**IEF**), in which proteins are separated on the basis of their net charge irrespective of their mass. The underlying principle is that electrophoresis is carried out in a pH gradient, allowing each protein to migrate to its **isoelectric point**, that is, the pH value (**pI**) at which the protein has no net charge (**Figure 2.2**). In standard electrophoresis, there is no pH gradient because the electrophoresis buffer has a uniform pH. Therefore, the charge density of each protein remains the same during electrophoresis and, in time, each protein reaches either the anode or the cathode. In the case of IEF, the charge density of each protein decreases as it moves along the pH gradient toward its isoelectric point. When the isoelectric point is reached, the charge density of the protein is zero and it can no longer migrate in the applied electric field. Diffusion still acts against this tendency to focus at a single position in the

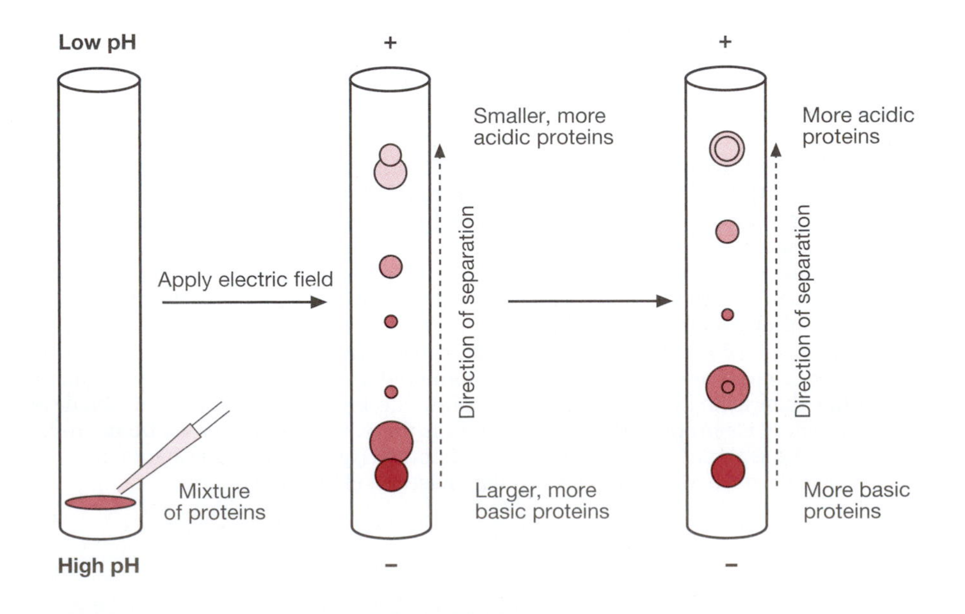

FIGURE 2.2 The principle of isoelectric focusing. A mixture of proteins is loaded at the basic end of a gel that has a pH gradient. An electric field is applied and the proteins separate according to their charge, each protein focusing at a position where the surrounding pH value is such that the protein has zero net charge, that is, its isoelectric point (pI). Larger proteins will move more slowly through the gel, but with sufficient time will catch up with small proteins of equal charge. The circles represent proteins, with shading to indicate protein pI values and diameters representing molecular mass.

gel, but a protein diffusing away from its isoelectric point becomes charged and therefore moves back toward its focus. Proteins with different pI values, as determined by the number and type of acidic and basic amino acid residues they contain, therefore focus at different positions in the pH gradient. Although there may be an initial sieving effect that also separates the proteins on the basis of their size, running the gel for a suitably long duration ensures that all proteins "catch up," reaching their isoelectric points and achieving size-independent separation.

The pH gradient in an IEF gel can be established in two ways. The original method was to use **synthetic carrier ampholytes**, which are collections of small amphoteric molecules with pI values corresponding to a given pH range (**Figure 2.3a**). Initially, there is no pH gradient in the gel because all the ampholytes are evenly distributed; the pH of the electrophoresis buffer is the average of that of the ampholyte molecules. When the electric field is applied, however, the ampholytes themselves are subject to electrophoresis. The most acidic ampholyte moves toward the anode, the most basic ampholyte moves toward the cathode, and all the other ampholytes establish intermediate zones according to their pI values. Once this stacking process is completed, the system has reached an equilibrium characterized by a continuous pH gradient. Proteins, which migrate much more slowly than the ampholyte molecules, then begin to move toward their isoelectric points in the gel. The proteins can be added to the gel before the electric field is applied or after a period of **pre-focusing**.

There are several problems with the use of ampholytes that lead to poor reproducibility in 2DGE experiments. One of the most serious limitations is **cathodic drift**, where the ampholytes themselves migrate to the cathode due to a phenomenon called **electro-osmotic flow** (bulk solvent movement toward the cathode). This results in pH gradient instability as basic ampholytes are progressively lost from the system. In practical terms, it is difficult to maintain a pH gradient that extends far beyond pH 7–8 over prolonged gel runs, resulting in the loss of many basic proteins unless the migration profile and timing are very carefully controlled.

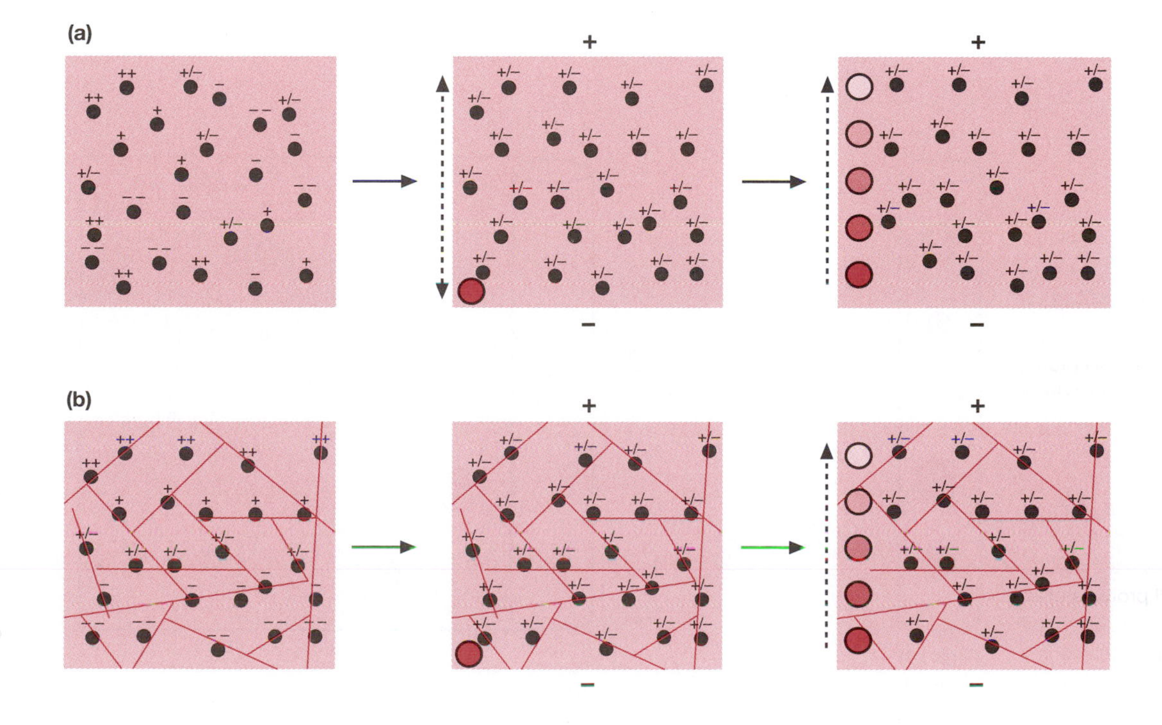

FIGURE 2.3 Different ways of forming a pH gradient for isoelectric focusing. (a) With ampholytes, the buffering molecules are free to diffuse and initially are distributed evenly so there is no pH gradient. When an electric field is applied, the ampholytes establish a pH gradient and become charge-neutral. This leads to separation of proteins according to pI values. (b) In the case of an immobilized pH gradient, the buffering molecules are attached to the polyacrylamide gel matrix. No movement of the buffering molecules occurs when the electric field is applied, but proteins are separated. Dotted arrows show the direction of separation. Shading in circles indicates protein pI values.

One way in which the problem of cathodic drift has been addressed is a procedure known as **non-equilibrium pH gradient electrophoresis** (**NEpHGE**). In this method, the protein sample is applied to the acidic end of the gel (rather than the basic end, which is generally the case for standard isoelectric focusing) so that all the proteins are positively charged at the beginning of the gel run. If run to completion, the basic ampholytes and the basic proteins would still run off the end of the gel, but the essential principle of NEpHGE is that the gel is not run for long enough to allow the system to reach equilibrium. Rather, the proteins are separated in a rapidly forming pH gradient that never becomes stable. The main drawback is that conditions of separation are difficult to reproduce.

Both these problems—cathodic drift and poor reproducibility—have been addressed by the development of **immobilized pH gradient** (**IPG**) gels, in which the buffering groups are attached to the polyacrylamide matrix of the gel (Figure 2.3b). This is now the standard approach in proteomics, where reproducibility is a key issue. The IPG is established using a collection of non-amphoteric molecules called **Immobilines** that contain a weakly acidic or basic buffering group at one end and an acrylic double bond to facilitate the immobilization reaction at the other. The gel is run in the normal way, but the pH gradient exists before the electric field is applied, and remains stable even when the gel is run for a long time. When the sample is loaded, the proteins migrate to their isoelectric points as in conventional isoelectric focusing. Carrier ampholytes may also be added to the IPG gel buffer as these are thought to increase protein solubility and prevent nonproductive interactions between proteins and the Immobiline reagents that can lead to precipitation and artifacts, particularly in the basic region of the gel.

SDS-PAGE separates proteins by mass irrespective of charge

The second-dimension separation in 2DGE is generally carried out by standard **SDS-PAGE** (**sodium dodecylsulfate polyacrylamide gel electrophoresis**) and separates the proteins according to molecular mass irrespective of charge (**Figure 2.4**). The basis of the technique is the exposure of proteins to the anionic detergent **sodium dodecylsulfate** (**SDS**), which denatures the proteins and binds stoichiometrically to the polypeptide backbone (1.4 g SDS per g polypeptide), thereby imparting a uniform

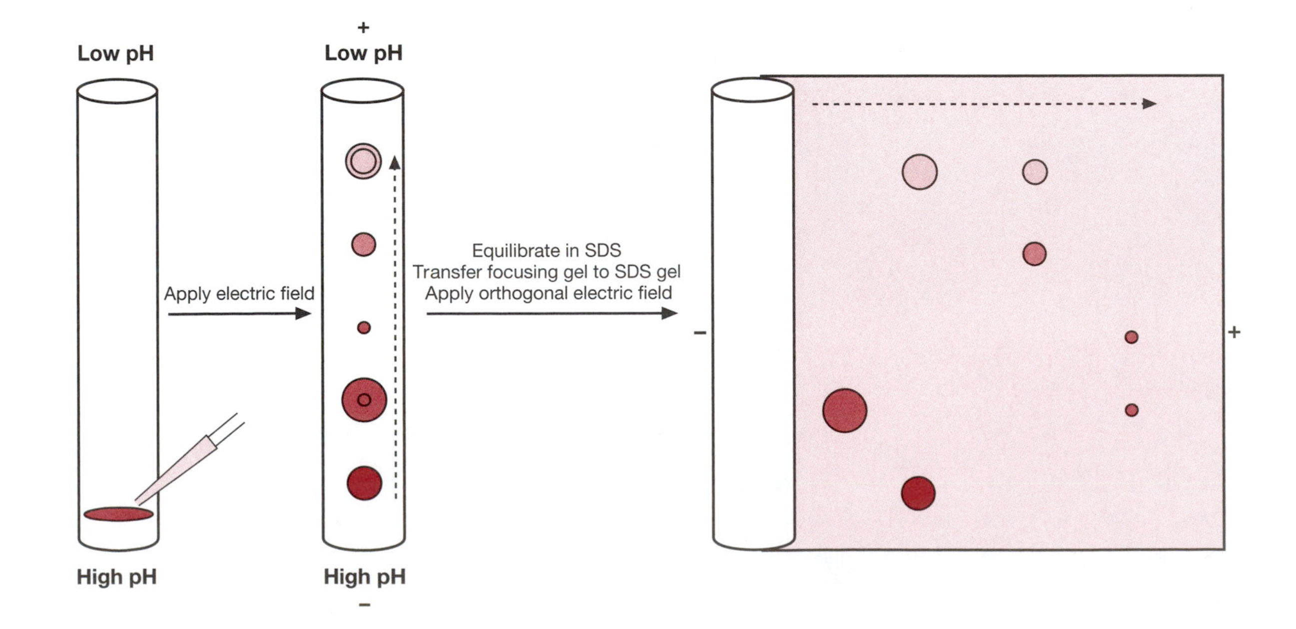

FIGURE 2.4 Two-dimensional electrophoresis using a tube gel for isoelectric focusing and a slab gel for SDS-PAGE. The proteins are separated in the first dimension on the basis of charge and in the second dimension on the basis of molecular mass. The circles represent proteins, with shading to indicate protein pI values and diameters representing molecular mass. The dotted line shows the direction of separation.

negative charge. The presence of tens or hundreds of SDS molecules on each polypeptide dwarfs any intrinsic charge carried by the proteins themselves, and stoichiometric binding means that larger proteins bind more SDS than smaller proteins. This has two important consequences that ensure separation on the basis of mass alone. First, all protein–SDS complexes have essentially the same charge density, and second, the relative differences in mass between proteins are maintained in the protein–SDS complexes.

The gel enhances the size-dependent separation by sieving the proteins as they migrate. The sieving effect depends on the pore size of the gel, which is in turn dependent on the gel concentration (total concentration of monomer as a percentage of the gel volume before it is cast, denoted by **%T**). In the case of polyacrylamide gels, the monomer is made up of the gelling agent acrylamide and the cross-linking agent bis-acrylamide. The pore size also depends to a certain extent on the proportion of the monomer, by mass, represented by bis-acrylamide (denoted by **%C**). Generally, as %T increases, the pore size decreases because more of the gelling agent is present per unit volume of the gel. In standard gels (where %T $\leq$ 15%), the minimum pore size is achieved when %C is approximately 5%. Below this value, there are fewer cross-links and the minimum pore size is larger. Above this value, the acrylamide molecules become over-linked and form dense bundles interspersed with large cavities, and the minimum pore size again becomes larger. Therefore, by holding the amount of bis-acrylamide at 5%, the pore size of the gel can be effectively controlled by varying the total concentration of the monomer. The optimum value for %C—that is, the value required to achieve minimum pore size—increases above 5% when %T > 15%. Gels can be cast with %T values ranging from 1% to more than 30%. Gels with concentrations lower than about 3% are required for the sieving of very large proteins ($M_r \geq 10^6$; M_r, relative molecular mass) but are fragile, and are generally stabilized by the inclusion of agarose (which does not sieve the proteins but provides a support matrix). Gels with concentrations over 30% can sieve very small proteins ($M_r = 10^3$). The mass of the proteins in the sample can be estimated by including, in one of the lanes of the gel, a series of protein markers whose masses are known.

2.4 THE APPLICATION OF 2DGE IN PROTEOMICS

The four major advantages of 2DGE are robustness, reproducibility, visualization, and compatibility with downstream microanalysis

Two-dimensional gel electrophoresis is the oldest of the proteomic separation methods, which means it has been investigated thoroughly in many laboratories, providing a unique insight into its strengths and weaknesses. The influence of different parameters on experimental variability, even between laboratories, has been examined in detail, and standard operating procedures can eliminate most operational variations leaving sample preparation and image analysis as greater sources of error than the separation process itself. Whereas the chromatography-based separations discussed later in the chapter are often accepted without the need for repetition, the poor reproducibility of early 2DGE experiments has encouraged researchers to run parallel gels comprising four or more biological replicates per experiment to control for sources of variation, and this has been adopted by many journals as a requirement for publication. Paradoxically, 2DGE experiments are therefore highly reproducible due to their reputation for being difficult to reproduce.

The visual aspect of 2DGE and the compatibility with microanalysis techniques such as blotting followed by protein detection with antibodies is also advantageous, as this can help to identify proteins where other techniques fail. This is particularly relevant in the case of proteins that carry post-translational modifications, which often alter the pI of a protein and generate linear "trails" on two-dimensional gels that indicate the different spots belong to a common parent molecule. As discussed in Chapter 8, the use of modification-specific antibodies can help to identify the modified variants, particularly where samples differ in the extent of modification rather than the overall abundance of a protein. The identification of degraded proteins is another context in which the visual aspect of 2DGE is more useful than the unbiased sampling achieved by chromatography-based methods.

The four major limitations of 2DGE are resolution, sensitivity, representation, and compatibility with automated protein analysis

Multidimensional electrophoretic protein separation techniques have been practiced since the 1950s, but the origin of the 2DGE method now used in proteomics is more recent. Protocols involving sequential isoelectric focusing and SDS-PAGE were developed in the mid-1970s, but the first proteomic analysis using the now standard approach of loading the sample at one end of a thin IEF gel before transferring the separated proteins to a slab gel for the second separation was published in a landmark 1975 paper by Patrick O'Farrell (see Further Reading). In this study, proteins from the bacterium *Escherichia coli* were separated by isoelectric focusing in a tube gel, that is, a gel cast in a thin tube. When the IEF run was complete, the tube was cracked open and the proteins exposed to SDS by immersion of the gel in an SDS solution. The tube gel was then attached to a SDS-PAGE slab gel—that is, a flat gel cast between two plates—and the focused proteins were separated in the orthogonal dimension on the basis of size. After nonselective staining, the result was a two-dimensional protein profile in which approximately 1000 individual fractions were distributed over the gel as a series of spots, representing approximately 20% of the *E. coli* proteome (**Figure 2.5**).

The basic procedure for 2DGE has changed little since this time, although the rather cumbersome tube gels (which were fragile and subject to nonlinear deformation) have been largely replaced by IPG **strip gels**, which are easier to handle and give more reproducible separations. However, proteomics takes the power of 2DGE to its limits and a number of operational problems have been identified in terms of resolution, sensitivity, representation, and compatibility with downstream mass spectrometry. These limitations are discussed in more detail below together with strategies that have been used to overcome them.

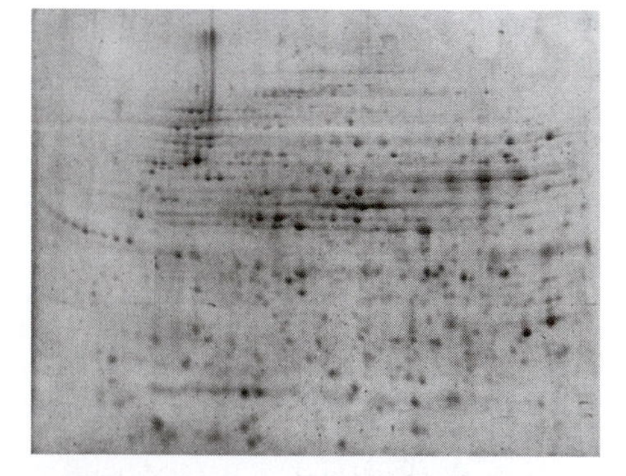

FIGURE 2.5 Separation of 240 μg of *E. coli* proteins by 2DGE, pH range 4–7 and mass range 10–120 kDa. The gel was stained with ruthenium II tris (bathophenanthroline disulfonate) and scanned with a Fuji FLA-3000 laser scanner, with blue SHGlaser. (Courtesy of Michael Lieber, raytest GmbH; see http://www.raytest.com)

The resolution of 2DGE can be improved with giant gels, zoom gels, and modified gradients, or by pre-fractionating the sample

Contemporary standard 2DGE systems, which are based on first-dimension isoelectric focusing using IPG strips followed by second-dimension SDS-PAGE, are capable of resolving approximately 2500 protein spots on a routine basis. However, the proteome of a complex eukaryotic cell may be more than an order of magnitude larger than this. Even in a simple eukaryotic system such as yeast, where alternative splicing and post-translational protein modifications are the exception rather than the rule, individual protein spots on standard two-dimensional gels may comprise several different co-migrating proteins, which can make certain types of downstream analysis more complex due to the presence of multiple proteins in the same spot. It is accepted that 2DGE performs better with lower-complexity systems, although several innovations have been introduced to simplify the analysis of complex proteomes.

The resolution of 2DGE depends on the separation length in both dimensions, and can thus be increased if very large-format gels are used. For example, IEF tube gels and IPG strips >30 cm in length have been used to achieve maximal separation in the first dimension, in combination with very large SDS slab gels that also provide a separation distance of >30 cm. Although such gels can be difficult to handle, they can resolve up to 10,000 protein spots. Another way to increase the resolution of 2DGE is to use multiple IEF gels, each with a narrow pH range. These are known as **zoom gels**. Following second-dimension SDS-PAGE and image analysis, the images of the separate zoom gels can be stitched together by computer to produce a composite of the entire proteome (**Figure 2.6**). In one demonstration, the use of six zoom gels allowed the separation of more than 3000 *E. coli* proteins representing approximately 70% of the proteome. The combination of long separation distances and narrow pH ranges can be used to maximize the resolution of such gels. Alternatively, to increase the resolution of proteins within a particular pH range, gels with **nonlinear pH gradients** can be produced. This is achieved simply by increasing the spacing between the appropriate Immobiline reagents, and is often used to "flatten" the pH gradient between pH 4 and 7, which accounts for the majority of proteins in the proteome (**Figure 2.7**).

Finally, resolution can be increased by various forms of pre-fractionation prior to electrophoresis, to simplify the protein mixture that is being analyzed. This can be achieved, for example, by focusing on a particular sub-proteome (for example, by isolating specific organelles), or by solvent extraction, sucrose density centrifugation, or the affinity-based enrichment or depletion of particular proteins. Pre-fractionation is critical for the successful use of zoom gels because proteins representing pI values outside the nominal pH range tend to precipitate in concentrated zones at the electrodes and distort the focusing of the remaining proteins due to osmotic effects.

The sensitivity of 2DGE depends on the visualization of minor protein spots, which can be masked by abundant proteins

The proteins in a cell differ in abundance over four to six orders of magnitude, with most of the total protein content represented by a relatively small number of abundant or superabundant proteins. For example, it is possible to detect about 4000 of the 5000 verified gene products from the yeast *Saccharomyces cerevisiae*, but 50% of the proteome represents the output of just 130 genes, and 75% of the proteome represents the output of the 400

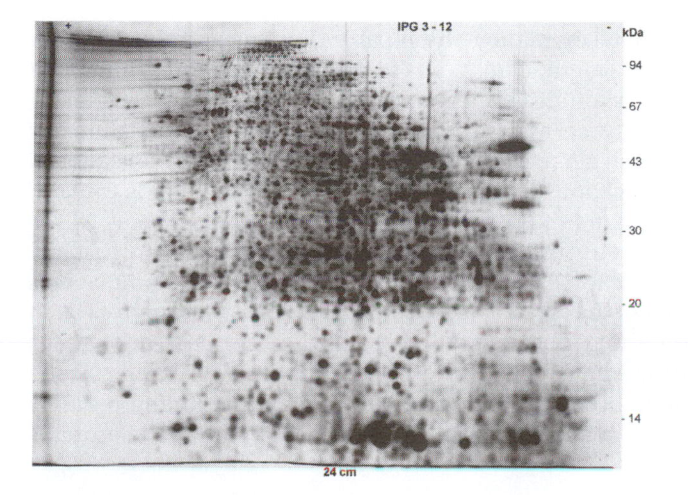

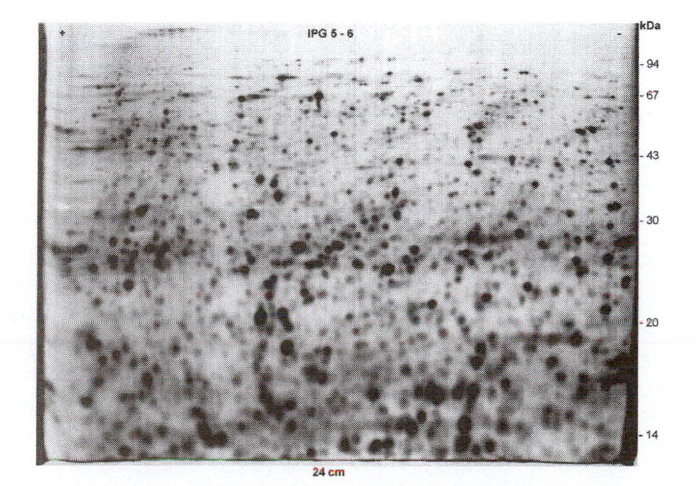

FIGURE 2.6 Both images represent mouse liver proteins separated by two-dimensional gel electrophoresis and silver stained to reveal individual protein spots. The left image is a wide pH range gel (pH 3–12) whereas the right image is a narrow pH range gel, which zooms proteins in the pH 5–6 range. Note that in the wider range gel, most proteins are clustered in the middle, reflecting the fact that most proteins have pI values in the 4–7 range. (Courtesy of the Swiss Institute of Bioinformatics, Geneva, Switzerland.)

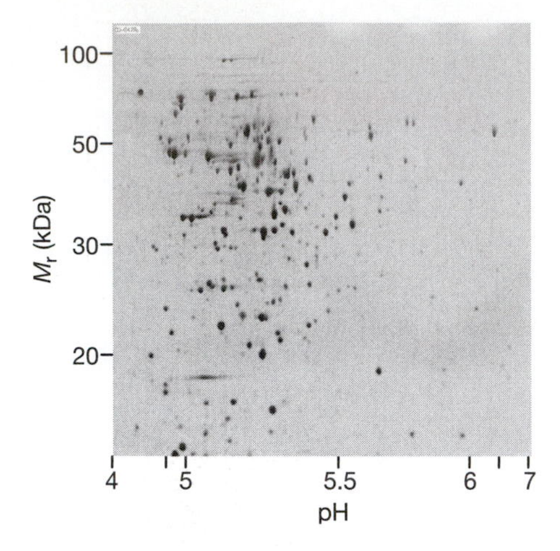

FIGURE 2.7 Higher-resolution separation can be achieved by flattening the pH gradient between pH 4 and pH 7, which accounts for the majority of proteins. (Courtesy of the Swiss Institute of Bioinformatics, Geneva, Switzerland.)

genes with the highest expression levels. Perversely, the least abundant (and therefore most difficult to detect) proteins are often those with the most interesting and enlightening functions, for example transcription factors, signaling proteins, and other regulators. In human body fluids, the dynamic range of protein concentration values may be even higher (nine orders of magnitude) and one extreme example is human serum, where the 12 most abundant proteins account for 95% of the proteome.

The sensitivity issues that affect 2DGE fall into two categories. The first is the difficulty in visually detecting the rarest proteins. Protein identification methods based on mass spectrometry are sensitive enough to work with a few picomoles of protein, but this requires the corresponding protein spots to be isolated and submitted for analysis. However, 2DGE is unique among proteomic methods in that the data submitted for analysis by mass spectrometry are acquired visually, and the identification of protein spots depends on the sensitivity of staining, which is limited to the low nanogram range (Chapter 4). If the entire gel were divided into a grid and each segment submitted for unbiased MS analysis, it might be possible to detect some of the least abundant proteins, but typical experiments involve the selection of "interesting" spots, usually those showing some form of differential expression between samples. There is no gel staining method that can adequately cover the entire dynamic range of proteins in even a simple prokaryotic cell.

The second problem is caused by the prevalence of abundant proteins. As stated above, the most abundant proteins in the cell tend to account for the vast majority of the proteome, and these tend to generate large spots that mask smaller ones. In principle, the detection of rare proteins could be achieved by loading more of the total protein onto the gel, but, in practice all this achieves is the formation of larger spots representing abundant proteins, which obscure more of the gel area. Some sensitivity problems can therefore be addressed by increasing the resolution of two-dimensional gels (see above) since this facilitates better separation of proteins with similar electrophoretic properties. The use of narrow-range IPG gels in combination with pre-fractionation or affinity-depletion of very abundant proteins can also resolve the problems caused by masking, particularly because more sample can be loaded, but this relies on the development of bespoke sample preparation methods catering for specific abundant proteins. More progress has been made in terms of general preparation methods by reducing the dependence on visual protein spot selection using alternative electrophoresis formats (**Box 2.3**).

The representation of hydrophobic proteins is an intractable problem reflecting the buffers required for isoelectric focusing

Proteins are diverse in terms of their chemical and physical properties, so it is impossible to devise a method that leads to the unbiased representation of all proteins on polyacrylamide gels. The most important determining factor is the IEF solubilization step, and for general applications the procedure has not changed very much since it was first developed in 1975. The standard lysis buffer includes a chaotropic agent to disrupt hydrogen bonds (urea or a combination of urea and thiourea), a non-ionic detergent such as NP-40 (definitely not SDS, because this is highly charged!), a reducing agent (usually dithiothreitol or β-mercaptoethanol (although these are charged molecules and they migrate out of the gel during IEF, so non-charged alternatives such as tributylphosphine may be more suitable), and, if desired, ampholytes representing the desired pH range. These conditions are not suitable for the solubilization of membrane proteins and this is why membrane proteins are underrepresented on standard gels. The recovery of membrane proteins

BOX 2.3 ALTERNATIVE APPLICATIONS.
Alternative approaches to 2DGE.

IEF-MS
In this approach, the SDS-PAGE separation step is replaced with MALDI-TOF (matrix-assisted laser desorption/ionization time-of-flight) mass spectrometry. Mass spectrometry is used primarily for protein identification, as discussed in detail in Chapter 3. In this variant application, however, the technique is used simply to list the masses of the different proteins found in each region of the IEF gel, thus providing a virtual two-dimensional separation. The procedure involves soaking the IEF gel in a matrix compound suitable for MALDI analysis (Chapter 3) and scanning the dried gel at close intervals to ionize the proteins and determine their masses. Although rapid and easy to automate, proteins > 50 kDa tend not to be detected, because they are difficult to ionize. Also, there appears to be no easy way to integrate this form of protein separation with conventional downstream MS analysis for protein identification.

SDS-PAGE-MS/MS
This variation of the classical proteomics strategy omits the isoelectric focusing step, and is one solution to the underrepresentation of membrane proteins in conventional 2DGE (which is generally caused by incompatibility with the IEF buffer). In a second application, SDS-PAGE-MS/MS is used to build a list of proteins in a particular sample, but it is not possible to derive any quantitative data from such experiments. The availability of more convenient LC-MS and CE-MS platforms for shotgun proteomics (see Chapter 3) means that SDS-PAGE-MS/MS is now rarely used for this purpose. More often, the technique is used for the analysis of very simple protein mixtures, such as affinity-purified complexes as in the study of protein interactions (see Chapter 7). When the protein mixture is simple, it can be presumed that each SDS-PAGE band contains only one protein, so this is broadly equivalent to the resolution of more complex mixtures using orthogonal separations. The most straightforward example is **in-gel LC-MS (GeLC-MS)**, in which samples are separated by one-dimensional SDS-PAGE, the gel lane is divided into slices, and each slice is separately reduced, alkylated, and digested with trypsin prior to LC-MS or LC-MS/MS analysis to determine the peptides present in each region of the gel.

Native PAGE
Native PAGE involves the same principles as conventional PAGE except the proteins are not denatured during extraction, thus preserving native structures. Native PAGE is therefore useful for the separation of intact multimeric proteins (for example antibodies, which have four separate chains joined by disulfide bonds) and also for protein complexes, a topic we explore again in Chapter 7. Whereas native PAGE is intended to separate proteins by mass, a specialized variant known as **QPNC-PAGE (quantitative preparative native continuous polyacrylamide gel electrophoresis)** can resolve native proteins by isoelectric point and is particularly useful for the isolation of intact metalloproteins.

can be increased by choosing stronger detergents, such as CHAPS, and by selectively enriching the initial sample for membrane proteins, for example by preparing membrane fractions, but this has not resulted in a generally applicable solution. The only methods that have satisfactorily addressed the underrepresentation of membrane proteins are those that abandon the IEF step all together, such as combined one-dimensional PAGE and liquid chromatography (GeLC-MS, see Box 2.3) or separation methods based on chromatography alone (see below).

Other classes of proteins that are traditionally very difficult to separate by standard 2DGE include highly basic histones, other chromatin proteins, and ribosomal proteins. Special separation methods have been devised in these cases. For example, a 2DGE approach that has been widely used for the separation of histones involves a first separation carried out on an acid–urea gel (which separates the proteins on the basis of size) and a second-dimension separation carried out on an acid–urea–Triton gel. Triton is a detergent that binds to histones depending on their degree of hydrophobicity; thus, the more hydrophobic histones have a reduced mobility. A specific problem with nuclear proteins is their tendency to aggregate under normal electrophoresis conditions, and modified buffers are required to avoid this. Interestingly, 2DGE has proven to be a much more satisfactory method for the separation of nucleolar proteins than liquid chromatography, presumably because the proteins are more soluble and resolvable under standard IEF conditions.

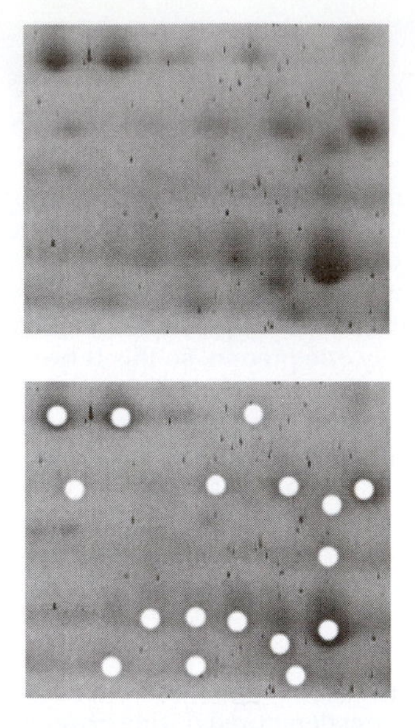

FIGURE 2.8 Section of a silver-stained two-dimensional gel before and after processing with spot excision robot using a 2 mm plastic picking tip.

Downstream mass spectrometry requires spot analysis and picking

The data produced by 2DGE experiments are visual in nature, so downstream analysis involves capturing the images from stained two-dimensional gels and then isolating particular spots for further processing and mass spectrometry. This process is difficult to automate and it represented one of the most significant bottlenecks in 2DGE-based proteomics during the 1990s, when manual analysis and spot picking from gels was commonplace. However, there are now various software packages available that produce high-quality digitized gel images and incorporate methods to evaluate quantitative differences between spots on different gels (Chapter 4). These can be integrated with **spot excision robots** that use plastic or steel picking tips to transfer gel slices to microtiter plates for automated digestion, clean-up, concentration, and transfer to the mass spectrometer. Several commercially available systems can fully automate the analysis and processing of two-dimensional gels, and can handle 200–300 protein spots per hour. Sections of a silver-stained gel before and after processing with a FSI Flexys spot excision robot are shown, as an example, in **Figure 2.8**.

2.5 PRINCIPLES OF MULTIDIMENSIONAL LIQUID CHROMATOGRAPHY

Protein and peptide separation by chromatography relies on differing affinity for stationary and mobile phases

Any separation technique that distributes the components of a mixture between two phases, a fixed **stationary phase** and a free-moving **mobile phase**, is known as **chromatography**. There are many chromatography formats, including paper chromatography, thin-layer chromatography, liquid chromatography, and gas chromatography, but all depend on the same underlying principle. A mixture of molecules is dissolved in a solvent and fed into the chromatography process. As the mobile phase moves over the stationary phase, the components of the mixture can interact with the molecules of both the solvent and the stationary matrix. Different components in the mixture move at different rates because of their differing affinities for each phase. Molecules with the lowest affinity for the stationary phase will move the most quickly because they tend to remain in the solvent, whereas molecules with the highest affinity move slowly because they tend to stay associated with the stationary phase and are left behind. This results in the mixture being partitioned into a series of fractions that can be eluted and collected individually.

In proteomics, **liquid chromatography** (**LC**) is used more often than other chromatography formats because of its versatility and compatibility with mass spectrometry (Chapter 3). Unlike gel electrophoresis, liquid chromatography is suitable for the separation of both proteins and peptides, and can therefore be applied upstream of 2DGE to pre-fractionate the sample, downstream of 2DGE to separate the peptide mixtures from single excised spots, or instead of 2DGE as the major protein/peptide separation technology (**Figure 2.9**). Mass spectrometers have an upper limit on mass detection, so proteins are usually digested into peptides prior to separation so that the fractions can be injected directly into the mass spectrometer, although this generates huge numbers of fragments. Therefore, like 2DGE, multidimensional chromatographic separations based on different principles are used to resolve the peptide mixture. Alternative LC methods can exploit different separation principles, such as size, charge, hydrophobicity, and affinity for particular ligands.

In the liquid chromatography methods used in proteomics, the stationary phase is a porous matrix, usually in the form of packed beads that are supported on some form of column. The mobile phase, a solvent containing dissolved proteins or peptides, flows through the column under gravity or is forced through under high pressure. The sensitivity of separation depends to a certain extent on the internal diameter of the column because analytes can be introduced at a higher concentration in a smaller volume. MDLC separations in proteomics are therefore often accomplished by nanoflow-LC with nanoscale capillaries (column internal diameter < 0.1 mm, flow rate < 1 μl/min). Although this defines the mass flow rate, the rate at which any particular protein or peptide flows through the column depends on its

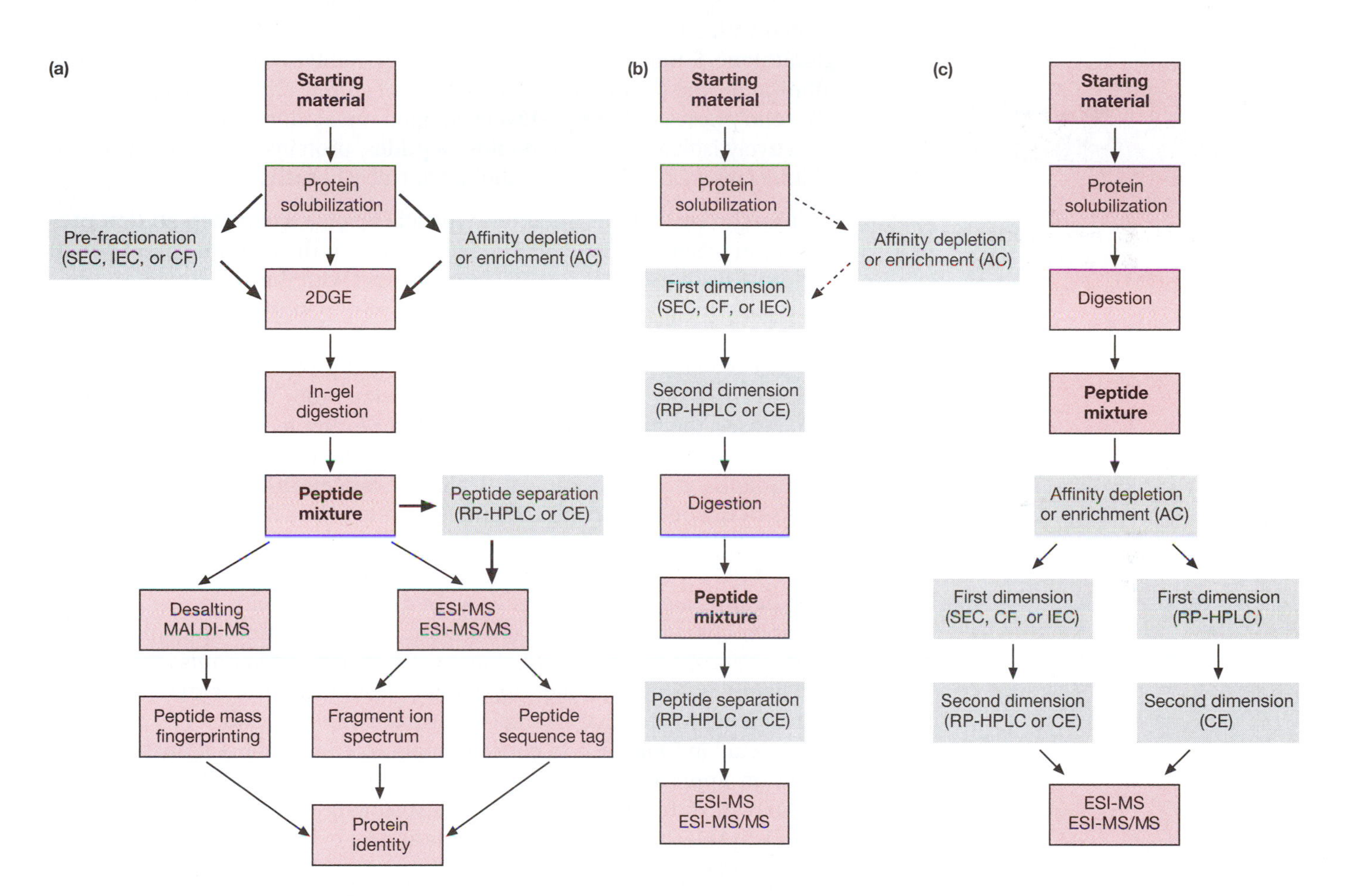

FIGURE 2.9 Protein/peptide separation strategies. (a) Liquid chromatography used in combination with 2DGE in standard proteomic analysis. Prior to 2DGE, the protein sample may be subject to affinity chromatography (AC) to deplete abundant proteins or enrich for certain types of protein. Pre-fractionation may then be carried out by size exclusion chromatography (SEC), ion exchange chromatography (IEC), or chromatofocusing (CF) to select proteins covering a particular range of isoelectric points or molecular masses for separation by 2DGE. After 2DGE, spots are digested in the gel with trypsin. The resulting peptides may be desalted and transferred to a MALDI mass spectrometer, or separated by microcapillary electrophoresis (CE) or microcapillary reversed-phase HPLC (RP-HPLC) before injection into the ESI mass spectrometer. Multidimensional liquid chromatography can also be used instead of 2DGE for the separation of proteins and peptides. (b) Chromatography steps with different separative principles may be used as a direct replacement for 2DGE for the separation of proteins, with or without a prior affinity depletion or enrichment step. On-column digestion with trypsin is followed by a further round of RP-HPLC to feed individual peptide fractions into the mass spectrometer. (c) Multidimensional liquid-phase separations can also be applied directly to complex peptide mixtures. This strategy almost always involves an affinity depletion or enrichment step because of the very complex nature of the peptide mixture. Favored approaches include AC-SEC-RPHPLC-MS, AC-IEC-RPHPLC-MS, and AC-RPHPLC-CEMS. The analysis of proteins and peptides by mass spectrometry is discussed in Chapter 3.

affinity for the matrix, and matrices with different chemical and physical properties can be used to separate proteins or peptides according to different selective principles. These principles, and how they are applied, are discussed in the following sections.

Affinity chromatography exploits the specific binding characteristics of proteins and/or peptides

Affinity chromatography partitions proteins or peptides on the basis of their specific, ligand-binding affinities. The matrix on an affinity column contains ligands that are highly selective for particular proteins or classes of proteins. Beads conjugated with antibodies, for example, can be used to isolate a single protein or peptide from a complex mixture, whereas beads coated with glutathione can be used to capture fusion proteins containing glutathione-*S*-transferase (GST) affinity tags. Similarly, immobilized metal-affinity chromatography (IMAC) is a form of affinity chromatography where the solid phase contains positively charged metal ions. This can be used to selectively isolate phosphoproteins/peptides, proteins with oligo-histidine affinity tags such as His_6, and other negatively charged proteins.

Affinity chromatography methods typically involve a **two-step elution** procedure in which the first fraction emerging from the column comprises all the proteins or peptides that failed to interact with the affinity matrix, and the second fraction comprises all the proteins or peptides that were retained on the column. This is achieved by sequential washing with two solutions, the first of which flushes out all the unbound proteins and the second of which causes the bound proteins to dissociate from the affinity matrix. In some cases, the first fraction is required (for example, the aim is to remove an abundant protein from a sample to simplify the analysis of the remaining proteins, a process known as **affinity depletion**). In other cases, the aim is to isolate the second fraction, which contains the proteins that bind selectively to the affinity matrix (**affinity purification**). The objective may be to isolate a specific protein or class of proteins, to isolate fusion proteins bearing a particular affinity tag, or to isolate proteins or peptides with common characteristics. The application of affinity chromatography to the study of phosphoproteins and other post-translational variants is discussed in Chapter 8. Another major application of affinity chromatography is the isolation of proteins that interact to form a complex, a subject discussed in more detail in Chapter 7. In conventional proteomics separations, the most common application of affinity chromatography is to simplify extremely complex mixtures of peptides while ensuring most proteins are still represented, by selecting those with histidine and/or cysteine residues (see below and Chapter 4).

Size exclusion chromatography sieves molecules on the basis of their size

Size exclusion chromatography (also known as **gel filtration chromatography**) is a profiling technique used to separate proteins according to their size. The column is packed with inert beads made of a porous compound such as agarose. Small proteins can enter the pores in the beads and so they take longer to find their way through the column than larger proteins, which do not fit in the pores and find a quicker path by moving through the gaps between beads. This separative principle is known as **molecular exclusion** and does not require any chemical interaction between the solutes and the stationary phase. Commercial preparations of size exclusion chromatography beads, for example Sepharose, have different sized pores suitable for the optimal separation of protein or peptide mixtures over different size ranges.

Ion exchange chromatography exploits differences in net charge

Unlike affinity chromatography, the other forms of chromatography used in proteomics are nonspecific, that is, they are used to profile the sample and separate proteins according to general physicochemical properties. **Ion exchange (IEX) chromatography** separates proteins or peptides according to their charge. It is based on the reversible adsorption of solute molecules to a solid phase that contains charged chemical groups. Cationic or anionic resins may be used (**Table 2.1**) and these attract molecules of opposite charge in the solvent. Variants of IEX chromatography therefore include **cation exchange (CAX)**, **anion exchange (AEX)**, **strong cation exchange (SCX)**, and **strong anion exchange (SAX)**, the latter with more highly charged resins. Instead of a two-step elution procedure, multi-step or **gradient elution** is achieved by washing the column with buffers of gradually increasing ionic strength or pH (**Figure 2.10**). The output of such a procedure is a **chromatogram**, where the *x* axis displays retention time and the *y* axis shows **absorption peaks** that correspond to individual components of the sample (**Figure 2.11**). The resolution of a chromatographic separation

TABLE 2.1 FUNCTIONAL GROUPS USED ON ION EXCHANGERS

Anion exchangers	Functional group
Diethylaminoethyl (DEAE)	$-O-CH_2-CH_2-N^+H(CH_2CH_3)_2$
Quaternary aminoethyl (QAE)	$-O-CH_2-CH_2-N^+(C_2H_5)_2-CH_2CHOH-CH_3$
Quaternary ammonium (Q)	$-O-CH_2-CHOH-CH_2-O-CH_2-CHOH-CH_2-N^+(CH_3)_3$
Cation exchangers	**Functional group**
Carboxymethyl (CM)	$-O-CH_2-COO^-$
Sulfopropyl (SP)	$-O-CH_2-CHOH-CH_2-O-CH_2-CH_2-CH_2SO_3^-$
Methylsulfonate (S)	$-O-CH_2-CHOH-CH_2-O-CH_2-CHOH-CH_2SO_3^-$

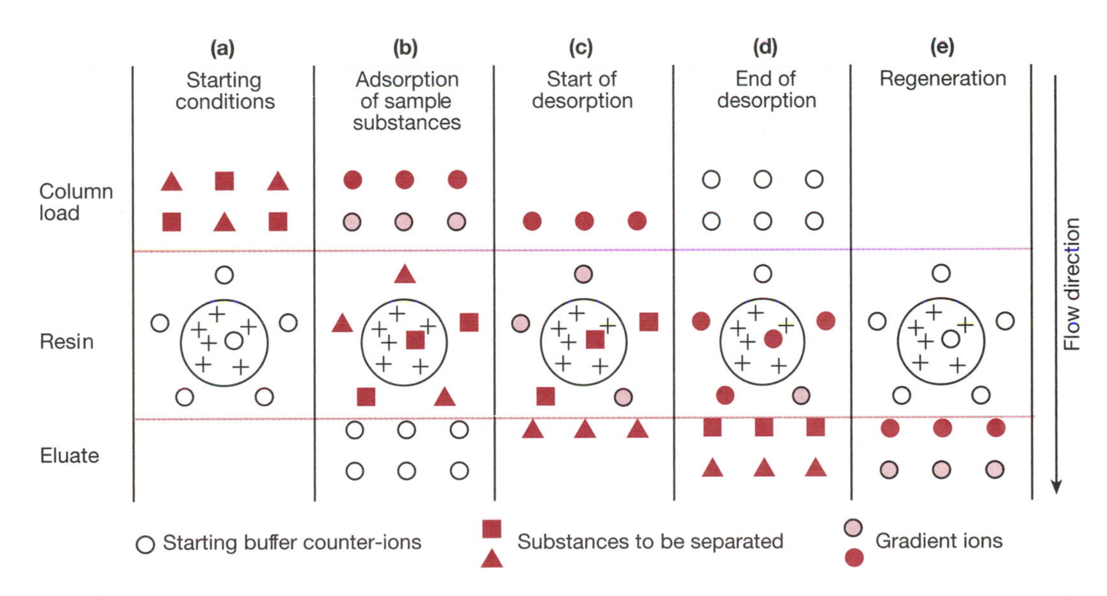

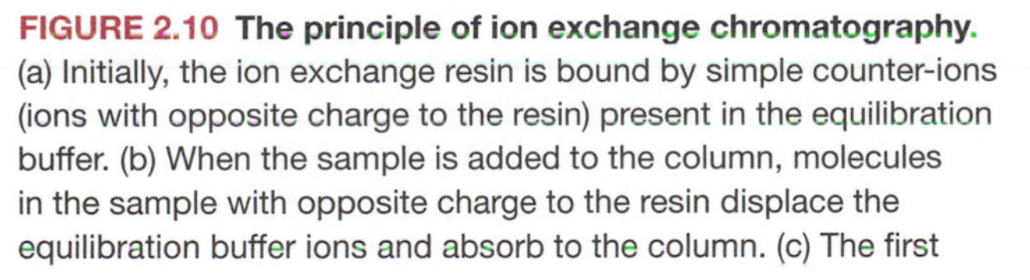

FIGURE 2.10 The principle of ion exchange chromatography. (a) Initially, the ion exchange resin is bound by simple counter-ions (ions with opposite charge to the resin) present in the equilibration buffer. (b) When the sample is added to the column, molecules in the sample with opposite charge to the resin displace the equilibration buffer ions and absorb to the column. (c) The first elution buffer displaces those components of the sample that are bound most weakly. (d) As the ionic strength of the elution buffer increases (or as the pH changes), more strongly associated solute ions are displaced. (e) After all solute ions have been displaced, the column is regenerated with equilibration buffer.

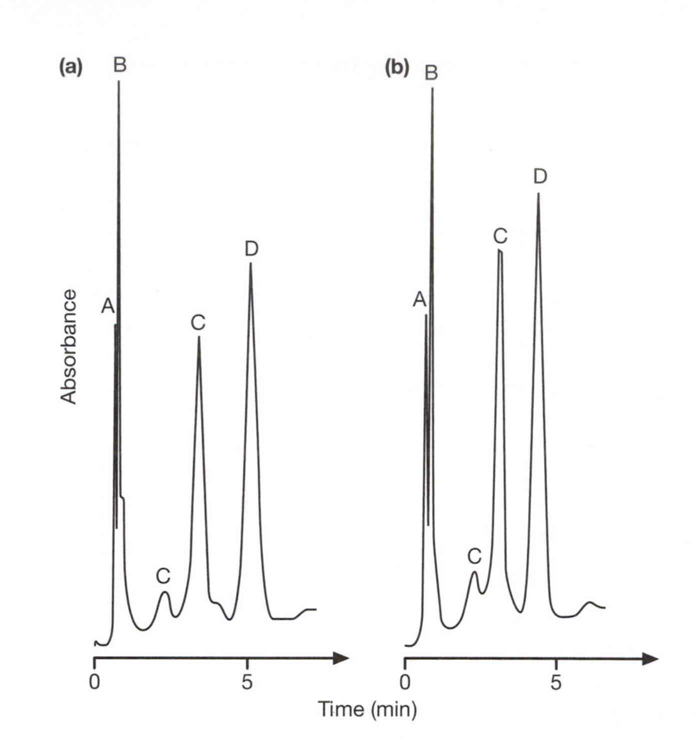

FIGURE 2.11 Chromatograms showing the results of separations of protein mixtures by ion exchange chromatography. The lettered peaks correspond to different proteins (A = ovalbumin, B = conalbumin, C = cytochrome *c*, D = lysozyme). The first separation (a) was performed at pH 5.85, whereas the second (b) was performed at pH 6.5. Conditions such as pH and temperature have a significant effect on the output. (Courtesy of Rebecca Carrier and Julie Bordonaro, Rensselaer Polytechnic Institute.)

is expressed as the **peak capacity**, that is, the number of peaks that can be resolved from the baseline over the full elution spectrum. The number of peaks on the chromatogram reveals the complexity of the sample, whereas quantitative data may be obtained by comparing peak areas.

A similar technique, **chromatofocusing** (**CF**), involves the use of an ion exchange column adjusted to one pH with a buffer adjusted to a second pH. This generates a pH gradient along the column, which can be used to elute proteins in order of their isoelectric points. Focusing effects taking place during the procedure produce sharp peaks and help to concentrate individual fractions.

Reversed-phase chromatography and hydrophobic interaction chromatography exploit the affinity between peptides and hydrophobic resins

Like ion exchange chromatography, **reversed-phase** (**RP**) **chromatography** involves the reversible adsorption of proteins or peptides to the stationary phase matrix, and multiple fractions are produced by gradient elution. In this case, however, the proteins and peptides are separated according to their hydrophobicity, and the reversed-phase resin consists of hydrophobic ligands, such as C_4 to C_{18} alkyl groups (**Figure 2.12**). In proteomics, reversed-phase separations are usually carried out using **high-performance liquid chromatography** (**RP-HPLC**) in which the mobile phase is forced through the column under high pressure. Although the separative principle is hydrophobicity, RP-HPLC results in a quasi-mass-dependent separation because retention tends to increase with molecular mass. Gradient elution is achieved by gradually increasing the amount of an organic modifier in the elution buffer, which disrupts the weakest hydrophobic interactions first (**Figure 2.13**). Of all the chromatography techniques used in proteomics, RP-HPLC is the most powerful method and has the highest resolution (a peak capacity of up to 100 components in practice) and because different resins show distinct retention profiles with different buffers, it is possible to use RP-HPLC with different resins in the same buffer, or the same resins and different buffers, to perform multidimensional separations. RP-HPLC is widely used for the separation of peptides following tryptic digestion and

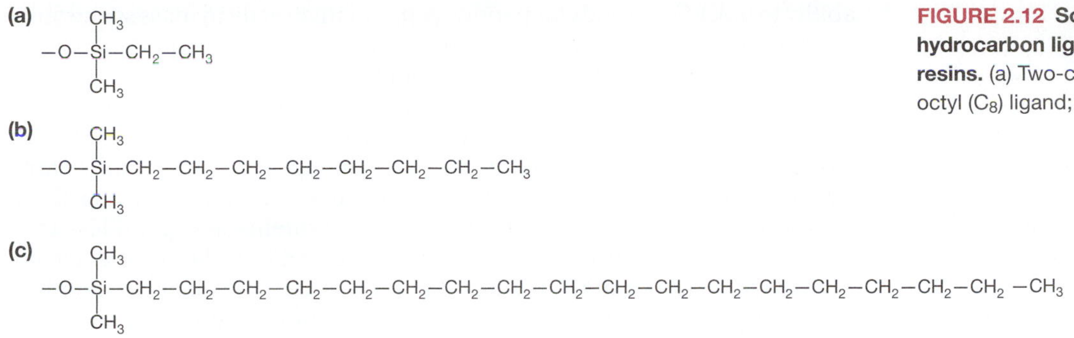

(a)

$$CH_3$$
$$-O-Si-CH_2-CH_3$$
$$CH_3$$

(b)

$$CH_3$$
$$-O-Si-CH_2-CH_2-CH_2-CH_2-CH_2-CH_2-CH_2-CH_3$$
$$CH_3$$

(c)

$$CH_3$$
$$-O-Si-CH_2-CH_2-CH_2-CH_2-CH_2-CH_2-CH_2-CH_2-CH_2-CH_2-CH_2-CH_2-CH_2-CH_2-CH_2-CH_2-CH_2-CH_3$$
$$CH_3$$

FIGURE 2.12 Some commonly used *n*-alkyl hydrocarbon ligands on reversed-phase resins. (a) Two-carbon capping group; (b) octyl (C_8) ligand; (c) octadecyl (C_{18}) ligand.

HPLC columns are often linked directly to electrospray ionization mass spectrometers to facilitate fully automatic peptide separation and analysis by LC-MS or LC-MS/MS (liquid chromatography–mass spectrometry or liquid chromatography–tandem mass spectrometry; Chapter 3). **Hydrophobic interaction chromatography** is a similar technique, which also separates proteins on the basis of their hydrophobic properties, although using different resin compositions (C_2–C_8 alkyl groups, or aryl ligands) and more polar elution buffers. Similarly, **hydrophilic interaction chromatography** uses a polar solid phase to separate proteins on the basis of their hydrophilic properties.

2.6 MULTIDIMENSIONAL LIQUID CHROMATOGRAPHY STRATEGIES IN PROTEOMICS

Multidimensional liquid chromatography is more versatile and more easily automated than 2DGE but lacks a visual dimension

As discussed above, liquid chromatography is often used either upstream or downstream of 2DGE to pre-fractionate samples and to separate tryptic peptides prepared from individual gel spots. However, the flexibility of LC methods in terms of combining different separative principles, and the

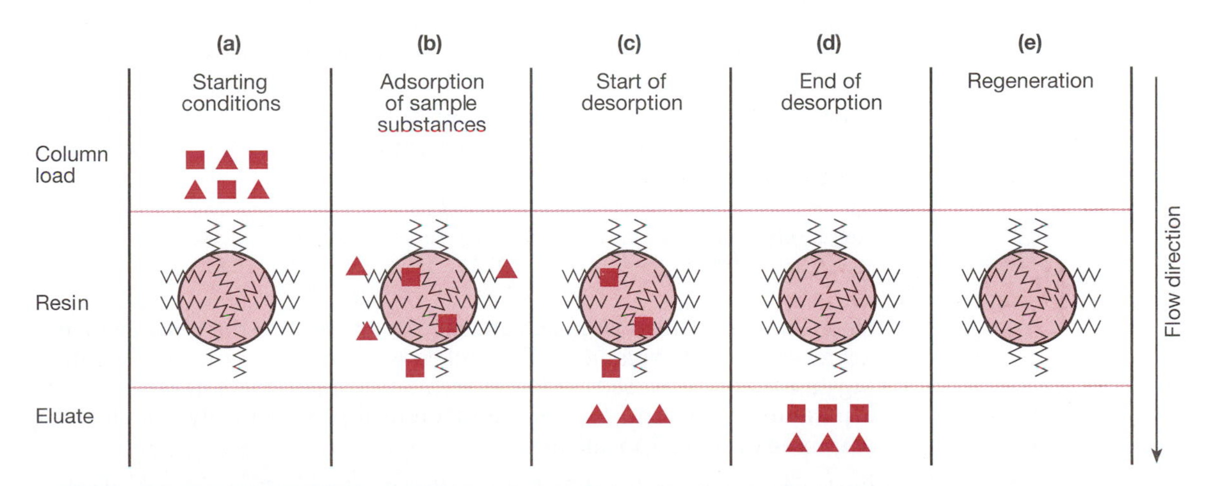

FIGURE 2.13 The principle of reversed-phase chromatography. (a) Initially, the reversed-phase resin is unoccupied. (b) When the sample is added to the column, most of the molecules in the sample bind to the resin because of the nature of the sample loading buffer. (c) The first elution buffer displaces those components of the sample that are bound most weakly to the hydrophobic resin. (d) As the level of organic modifier (for example, acetonitrile) in the elution buffer increases, more strongly associated solute molecules are displaced. (e) After all the sample molecules have been displaced, the column is regenerated.

ability to link LC methods for peptide separation directly to mass spectrometry without the need for hands-on sample transfer, makes multidimensional liquid chromatography an attractive solution to many of the drawbacks of 2DGE discussed above. HPLC columns allow large sample volumes to be loaded and concentrated on the column, making low-abundance proteins easier to detect. Peptides from many of the proteins that are difficult to analyze by 2DGE (for example, membrane proteins and very basic proteins) can be separated easily using appropriate resins. Proteins and peptides separated in the liquid phase do not need to be stained in order to be detected. Perhaps most importantly, the fact that LC methods can separate peptides as well as proteins, and the ability to couple LC columns directly to the mass spectrometer, means that the entire analytical process from sample preparation to peptide mass profiling can be automated. The disadvantages of LC methods are that the visual aspects of protein separation by 2DGE are lost, including the pI and molecular mass data that can be determined from the positions of spots on the gel, and that LC is a serial analysis technique so it is difficult to run parallel experiments without access to parallel sets of identical apparatus (including the mass spectrometer). However, the precision of contemporary LC-MS equipment means that, like 2DGE, protein/peptide separation is very well controlled and that most errors arise upstream, in sample preparation, or downstream, in data analysis and interpretation.

The most useful MDLC systems achieve optimal peak capacity by exploiting orthogonal separations that have internally compatible buffers

The resolving power of a chromatographic process was defined by Eli Grushka in 1970 in terms of **peak capacity** using the following equation:

$$p = 1 + \frac{\sqrt{N}}{4} \ln \frac{t_n}{t_A}$$

Here, p is the peak capacity, N is the number of plates, and t_n and t_A are the final and void peak times respectively.

In principle, MDLC should be able to multiply the peak capacity in consecutive separations, but this is only possible if the properties affecting peptide separation in one dimension do not affect separation in the other dimension, a situation that could be described as **complete orthogonality**. In practice, there is always some correlation between the solute retention characteristics of proteins/peptides on different resins, and MDLC methods fall short of the theoretical resolving power. To reflect this, Gilar and colleagues derived a function known as the **practical peak capacity** that takes correlated fractions into account:

$$N_p = P_1 P_2 \frac{\sum \text{bins}}{P_{\text{max}}}$$

Here, N_p is the practical peak capacity, P_1 and P_2 are the peak capacities achieved in each separation, $\sum_{\text{bins}}$ refers to the number of "bins" containing data that can be used to identify estimate orthogonality in a nonuniform separation space, and P_{max} is the theoretical peak capacity calculated by summing the data from all bins.

Such studies have identified three forms of two-dimensional chromatography that provide the best resolving power for proteomics: namely, orthogonal reversed-phase HPLC using different pH buffers in each separation (RP-RPLC), ion exchange followed by reversed-phase chromatography (particularly SCX-RPLC), and hydrophobic interaction followed by reversed-phase chromatography (HILIC-RPLC). However, these studies do not take into account the time taken to achieve maximum separation, that is, a method that achieves high orthogonal resolution but takes several days to complete would not be as useful as a method with a slightly lower resolution that is complete in one hour. A compromise can be achieved by plotting peak capacity and the speed of separation in what is known as a **Poppe plot**.

Many different combinations of separation methods have been reported, some involving two orthogonal chromatography steps and others that combine chromatography with a different separation method such as capillary electrophoresis (Box 2.2). The main requirements for pairing different separation techniques are that the first separation should have the highest loading capacity, that the two techniques should be configurable (that is, it must be possible to connect them to allow fractions from the first separation to be fractionated further in the second dimension), and that the buffers should be compatible. For example, the sequential use of ion exchange and reversed-phase chromatography in a technique such as SCX-RPLC is popular because the elution buffer from the cation exchange step is suitable as a reversed-phase chromatography loading buffer; otherwise each fraction would have to be individually re-buffered prior to the second separation. RP-HPLC is preferred as the second separation in part because the elution buffer is compatible with the solvents used in both MALDI-MS and ESI-MS (matrix-assisted laser desorption/ionization mass spectrometry and electrospray ionization mass spectrometry; Chapter 3).

MudPIT shows how MDLC has evolved from a laborious technique to virtually hands-free operation

Initially, MDLC was achieved by a discontinuous process in which fractions were collected from the ion exchange or gel filtration column and then manually injected into the HPLC column (**Figure 2.14**). Although they are labor-intensive, **discontinuous systems** are not affected by time constraints. The fractions eluting from the first column can be stored off-line indefinitely, and fed one-by-one into the HPLC column, which is directly coupled to the mass spectrometer. A further advantage is that large sample volumes can be applied to the first column in order to obtain sufficient amounts of low-abundance proteins for analysis in the second dimension.

However, the need for manual sample injection can be circumvented by equipping the first column with an automatic fraction collection system and a column-switching valve. Fractions are then collected from the first column across the elution range, and the switching valve can bring the RP-HPLC column in-line to receive the fractions sequentially. Alternatively, some researchers have developed apparatus comprising a single ion exchange column coupled, via an appropriate set of switching valves, to multiple HPLC columns arranged in parallel (**Figure 2.15**). In this scheme, fractions emerging from the first column are directed sequentially to the multiple HPLC columns, and the cycle is repeated when the first column has been regenerated.

The emphasis on higher throughput through increased automation and reduced handing led to the development of **multidimensional protein identification technology** (**MudPIT**) using a **biphasic column**, in which the top of the column contains SCX resin and the distal part contains reversed-phase resin. Because the solvents are compatible (see above), this allows the stepped elution of fractions from the first resin and the gradient elution of second-dimension fractions from the second. This technique was pioneered as **direct analysis of large protein complexes** (**DALPC**) and was modified later into the MudPIT platform (see Washburn et al., Further Reading). As shown in **Figure 2.16**, peptide mixtures loaded

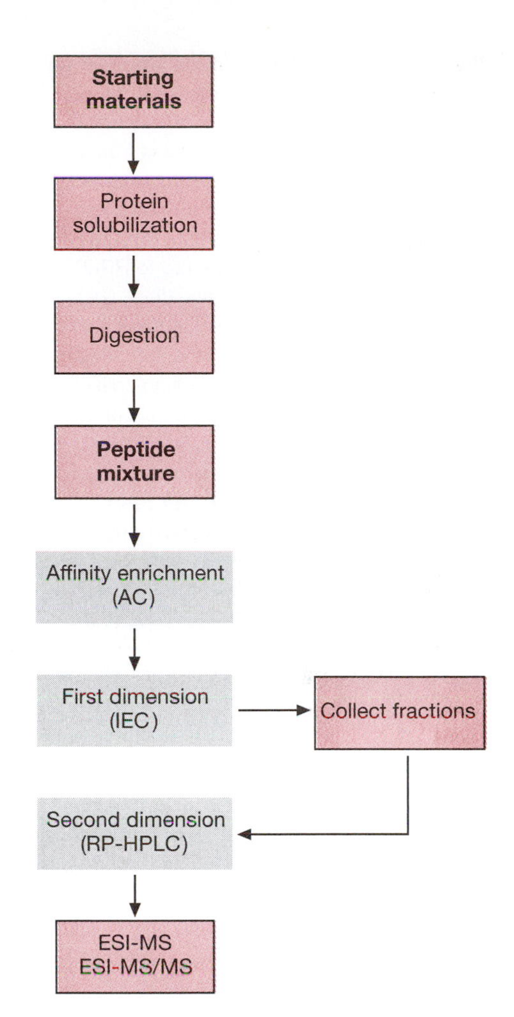

FIGURE 2.14 Discontinuous multidimensional chromatography for peptide separation. In this experiment, following the example of Gygi et al. (see Further Reading), proteins from two different yeast samples were labeled with mass-coded biotin affinity tags that reacted specifically with cysteine residues (this labeling strategy is not shown in detail; it is used for comparative protein quantitation between samples and further details can be found in Chapter 4, p. 79). The proteins from each sample were digested with trypsin and mixed together to produce a single peptide pool. Affinity chromatography was then used to select peptides carrying the affinity tag, reducing the complexity of the mixture about tenfold. However, since most proteins contain at least one cysteine residue, the remaining population of peptides still provided coverage of about 90% of the yeast proteome. The recovered peptides were separated by ion exchange chromatography using a strong cation exchange resin. Thirty individual samples were collected off-line, and four of these were subjected to second-dimension separation by RP-HPLC.

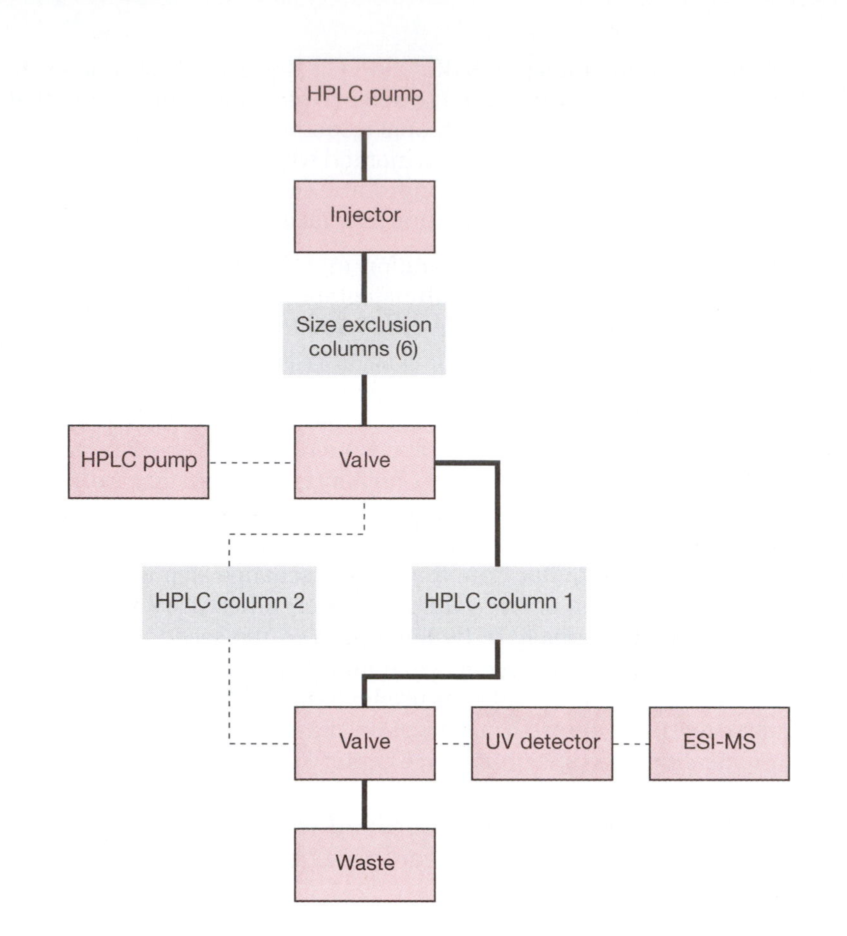

FIGURE 2.15 Continuous multidimensional chromatography with column-switching. In this example, two HPLC columns working in parallel receive alternating eluates from a bank of six size exclusion columns in series. After sample injection and separation by size exclusion chromatography, eluate from the size exclusion columns is directed to HPLC column 1 using a four-port valve (*thick line*). While the peptides are trapped in this column, HPLC column 2 is eluted and the sample is directed to the detector and fraction collector (*broken line*). After flushing and equilibrating column 2, the valves are reversed allowing column 2 to be loaded with the next fraction from the size exclusion separation, while column 1 is eluted. This cycle continues until the fractions from the size exclusion separation are exhausted. (Adapted from Opiteck GJ, Ramirez SM, Jorgenson JW & Moseley MA 3rd (1998) *Anal. Biochem.* 258, 349. With permission from Elsevier.)

onto the SCX resin are eluted using a stepped gradient of salt, resulting in the release of first-dimension fractions into the reversed-phase resin. Second-dimension fractions are then eluted from the reversed-phase resin into the mass spectrometer using a gradient of acetonitrile. This process, and the subsequent regeneration step, does not interfere with the SCX chromatography step, and after regeneration another fraction can be released from the SCX resin by increasing the salt concentration. When this method was developed in 2001, it achieved the highest-resolution analysis of the yeast proteome at the time, resolving and identifying 1484 proteins, including those normally underrepresented in proteomics experiments (**Box 2.4**). The orthogonality of the original MudPIT method has been improved through various adjustments such as the use of a mixed bed AEX/CAX chromatography step for the first separation step, the use of a semi-continuous salt gradient for elution, and the use of combined salt/

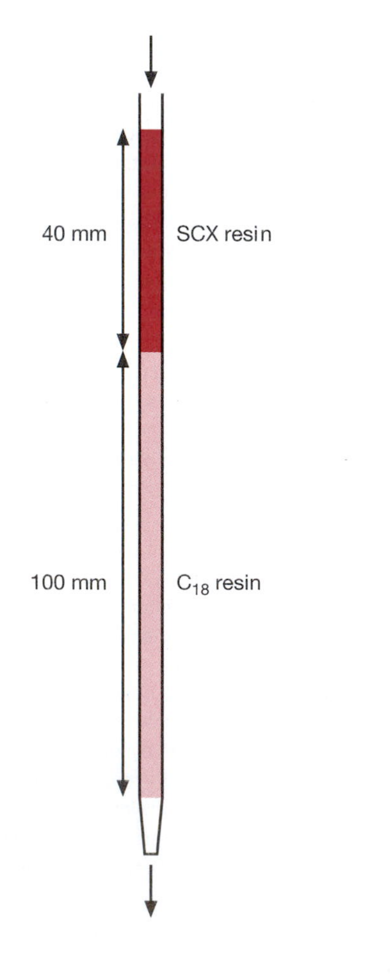

FIGURE 2.16 Continuous multidimensional chromatography using a biphasic column. In this example, simplified from the MudPIT method developed by Yates and colleagues (see Washburn et al., Further Reading) a 140 mm × 0.1 mm fused silica capillary is packed at the distal end with 5-μm C_{18} (reversed-phase) particles and at the proximal end with 5-μm strong cation exchange (SCX) particles. After introduction of the sample (*top arrow*), fractions are eluted from the SCX resin with a stepped salt gradient. After each salt step elution, the ion exchange fraction flows into the reversed-phase material and is eluted using a gradient of acetonitrile (*bottom arrow*). Reversed-phase elution and re-equilibration does not affect the SCX resin. This cycle is repeated until the SCX resin is exhausted.

BOX 2.4 CASE STUDY.
Analysis of the yeast proteome by MudPIT.

The first MudPIT experiment was described by Michael Washburn, Dirk Wolters, and John Yates in 2001. At the time, the most successful multidimensional liquid chromatography experiments could resolve and identify approximately 200 yeast proteins, whereas the greatest number of yeast proteins resolved by 2DGE was 279 (with the typical underrepresentation of scarce proteins and hydrophobic proteins). MudPIT was the first fully automated high-throughput method for protein separation and identification, that is, all operations were carried out continually and in-line with no human intervention required after sample loading. Proteins from mid-log-phase yeast cells were digested and the peptides separated in a biphasic SCX/RP column feeding directly into the mass spectrometer for MS/MS analysis followed by interpretation using the SEQUEST algorithm (see Chapter 3). This analysis resulted in the assignment of 5540 peptides to mass spectra, leading to the identification of 1484 proteins. Most remarkably, this included 131 proteins with three or more transmembrane domains, as well as other typically underrepresented proteins such as low abundance transcription factors and kinases, proteins with extreme pI values, or molecular weights greater than 180 kDa (**Figure 1**). At the time, the largely unbiased nature of this proteome sampling method was a true breakthrough.

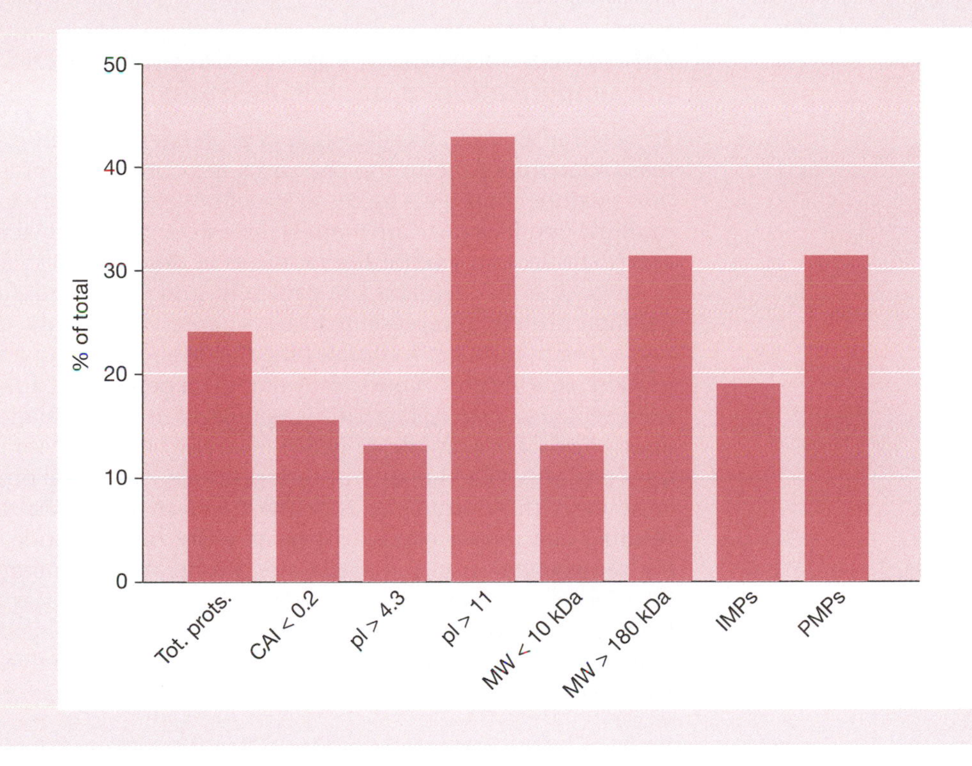

FIGURE 1 Sensitivity of the original MudPIT method to diverse classes of proteins. The number of proteins in each class identified in the experiment are compared with the total number in the yeast proteome predicted sequence databases. CAI = codon adaptation index, MW = molecular weight, IMP = integral membrane proteins with three or more transmembrane domains, PMP = peripheral membrane proteins.

pH gradients for elution. The latter recently allowed the separation and identification of 14,105 unique proteins from a mouse liver sample.

Although it is arguable whether multidimensional chromatography will ever displace 2DGE as a technology platform in proteomics, it is clear that it is a useful technique both alone and in combination with electrophoresis for protein and peptide separation. The combined advantages of sensitivity, representation, resolution, and, perhaps most importantly, the potential to automate LC-MS and LC-MS/MS procedures may offset the disadvantages of losing the visual data provided by 2DGE. The use of mass spectrometry to identify proteins separated by 2DGE and/or LC methods is discussed in Chapter 3.

RP-RPLC and HILIC-RP systems offer advantages for the separation of certain types of peptide mixtures

RP-RPLC systems have become popular because they achieve high peak capacities in the first separation, ensuring minimal overlap between fractions, and the orthogonality can be tailored by adjusting buffer conditions while using the same resin in both columns (for example, high-resolution separations with C_{18} resins have been achieved by using a pH 10 buffer in one column and a pH 2.6 buffer in the other). The overall orthogonality of RP-RPLC systems is somewhat lower than that of SCX-RPLC, but the separation is more efficient as well as being easier to handle.

HILIC-RPLC is particularly suitable for the separation of complex samples that are rich in polar compounds. The pairing of hydrophobic interaction and reversed-phase chromatography is highly compatible in terms of buffer requirements and also provides a very large practical peak capacity, as seen with the zwitterionic HILIC-RPLC procedure, which resembles SCX-RPLC at low pH but achieves better separation of charged peptides at pH 7–8. A HILIC-RPLC system has also been developed based on carbamoyl silica gel for the selective separation and purification of phosphoproteins (see Chapter 8).

Affinity chromatography is combined with MDLC to achieve the simplification of peptide mixtures

As noted in Box 2.1, the digestion of proteins into peptides is required for mass spectrometry but, this increases the complexity of an already complex mixture of proteins by 20- to 50-fold. We will discuss in Chapter 3 how multiple peptides per protein need to be detected to provide confident protein identification, but for the purposes of this chapter it is only necessary to know that 3–5 peptides are generally sufficient to confirm that a given parental protein is represented in a sample. This means that only a fraction of the peptides in a complex mixture are required to achieve confident protein identification. One way to simplify a peptide mixture without losing representation is to isolate those peptides containing relatively uncommon amino acids. For example, only 10% of tryptic peptides contain cysteine residues and only 17% contain histidine residues. Almost all proteins contain at least one of these amino acids, so the selection of peptides containing these residues can reduce complexity by an order of magnitude without biasing the representation of the proteome. Affinity chromatography, specifically IMAC, can be used to isolate histidine-containing peptides, and specialized tagging methods have been developed to label cysteine residues that simultaneously allow affinity enrichment and also quantitative determination. We discuss these methods in more detail in Chapter 4.

FURTHER READING

Di Palma S, Hennrich ML, Heck AJ & Mohammed S (2012) Recent advances in peptide separation by multidimensional liquid chromatography for proteome analysis. *J. Proteomics* 75, 3791–3813.

Gao M, Qi D, Zhang P et al. (2010) Development of multidimensional liquid chromatography and application in proteomic analysis. *Expert Rev. Proteomics* 7, 665–678.

Gygi SP, Rist B, Gerber SA et al. (1999) Quantitative analysis of complex protein mixtures using isotope-coded affinity tags. *Nat. Biotechnol.* 17, 994–999.

Hames BD & Rickwood D (1990) Gel Electrophoresis of Proteins: A Practical Approach. IRC Press.

Krenkova J & Foret F (2012) On-line CE/ESI/MS interfacing: recent developments and applications in proteomics. *Proteomics* 12, 2978–2990.

O'Farrell PH (1975) High-resolution two-dimensional electrophoresis of proteins. *J. Biol. Chem.* 250, 4007–4021.

Opiteck GJ, Ramirez SM, Jorgenson JW & Moseley MA (1998) Comprehensive two-dimensional high-performance liquid chromatography for the isolation of overexpressed proteins and proteome mapping. *Anal. Biochem.* 258, 349–361.

Rabilloud T (2002) Two-dimensional gel electrophoresis in proteomics: old, old fashioned, but still it climbs up the mountains. *Proteomics* 2, 3–10.

Rabilloud T (2009) Membrane proteins and proteomics: love is possible, but so difficult. *Electrophoresis* 30 (Suppl. 1), S174–S180.

Rabilloud T (2010) Variations on a theme: changes to electrophoretic separations that can make a difference. *J. Proteomics* 73, 1562–1572.

Rabilloud T, Chevallet M, Luche S & Lelong C (2010) Two-dimensional gel electrophoresis in proteomics: past, present and future. *J. Proteomics* 73, 2064–2077.

Rabilloud T, Vaezzadeh AR, Potier N et al. (2009) Power and limitations of electrophoretic separations in proteomics strategies. *Mass Spectrom. Rev.* 28, 816–843.

Ramautar R, Heemskerk AA, Hensbergen PJ et al. (2012) CE-MS for proteomics: advances in interface development and application. *J. Proteomics* 75, 3814–3828.

Wang H & Hanash S (2003) Multi-dimensional liquid phase-based separations in proteomics. *J. Chromatogr. B* 787, 11–18.

Washburn M, Wolters D & Yates J (2001) Large-scale analysis of the yeast proteome by multidimensional protein identification technology. *Nat. Biotechnol.* 19, 242–247.

Wu Q, Yuan H, Zhang L & Zhang Y (2012) Recent advances on multidimensional liquid chromatography–mass spectrometry for proteomics: from qualitative to quantitative analysis—a review. *Anal. Chim. Acta* 731, 1–10.

Zhang X, Fang A, Riley CP et al. (2010) Multi-dimensional liquid chromatography in proteomics—a review. *Anal. Chim. Acta* 664, 101–113.

Zhao SS, Zhong X, Tie C & Chen DD (2012) Capillary electrophoresis–mass spectrometry for analysis of complex samples. *Proteomics* 12, 2991–3012.

CHAPTER 3

Strategies for protein identification

3

3.1 INTRODUCTION

The techniques described in Chapter 2 allow complex protein mixtures to be separated into their components but do not allow those components to be identified. Indeed, the individual fractions produced by such methods are almost always anonymous. Each spot on a two-dimensional gel, and each fraction emerging from a liquid chromatography column, appears very much like any other. In the case of 2DGE, even differences in spot distribution provide only vague clues about protein identity, for example, apparent molecular mass and pI. Proteomic analysis must therefore incorporate a method to characterize the fractions and determine which proteins are actually present.

In the early days of proteomics, when 2DGE was the only available separation method, the typical approach was to select a small number of proteins for careful analysis and identification, especially if those proteins were abundant in one sample but not in another. Initially, this analysis was achieved by low-throughput means such as peptide sequencing by Edman degradation and later by higher-throughput mass spectrometry methods. A more recent approach is **shotgun proteomics** (also known as **discovery proteomics**), which in its ideal form means the unbiased separation and identification of all proteins in a sample by digestion into peptides then systematic fractionation followed by mass spectrometry. Shotgun proteomics has become possible because complex mixtures of peptides can be fractionated by multidimensional liquid chromatography (MDLC), sometimes in concert with other methods such as capillary electrophoresis, and the fractions can be fed automatically into a mass spectrometer for identification. Although several technologies have been developed for protein identification on a small scale, there is no doubt that contemporary proteomics would be impossible without advances in mass spectrometry that allow thousands of samples to be processed and characterized in a single day.

3.2 PROTEIN IDENTIFICATION WITH ANTIBODIES

In the early days of protein analysis, the spots on two-dimensional gels could be identified using only two methods. One was to run parallel gels, one experimental and one with purified protein standards, and identify proteins on the main gel that migrated in the same manner as the known standards. The other was to transfer the proteins from two-dimensional gels onto a suitable membrane or support and identify the proteins *in situ* with probes, typically antibodies (an approach known as western blotting or immunoblotting). In the absence of practical methods to characterize more than a few well-known proteins at a time, the patterns of protein spots on two-dimensional gels became important in their own right and were used as diagnostic fingerprints. Statistical analysis methods that are today commonly associated with microarrays were developed in the 1980s so that protein spot maps from two-dimensional gels could be compared, and bespoke analysis platforms such as TYCHO devised by Anderson and colleagues were in use as early as 1981.

The use of antibodies for the detection of proteins has become more sophisticated since the 1980s, reflecting the availability of more convenient methods to select and refine antibodies and produce them as recombinant proteins. Antibody-based detection methods are used in a number of proteomic technologies, including affinity enrichment/depletion (Chapter 2), quantitative assays (Chapter 4), the purification of protein complexes (Chapter 7), the identification of post-translational modifications (Chapter 8), and the construction of antibody microarrays (Chapter 9). However, although antibodies are powerful tools for the isolation and identification of individual proteins and can be applied as multiplex antibody arrays for the identification of hundreds of proteins simultaneously, there is no current antibody-based platform that can be used to identify and quantitate all the proteins in the proteome of a complex organism. Many antibodies recognize only particular conformational variants of proteins and it is difficult to envisage a procedure that would ensure that all proteins folded in the same manner in the same buffer. Antibodies can also cross-react with other proteins nonspecifically, albeit with lower affinity. For these reasons, the high-throughput identification of proteins cannot solely be based on the parallel use of specific detection reagents such as antibodies and must instead be achieved by the determination of **protein sequences** using reagents that can be applied to all proteins regardless of their structure, origin, or physical and chemical properties.

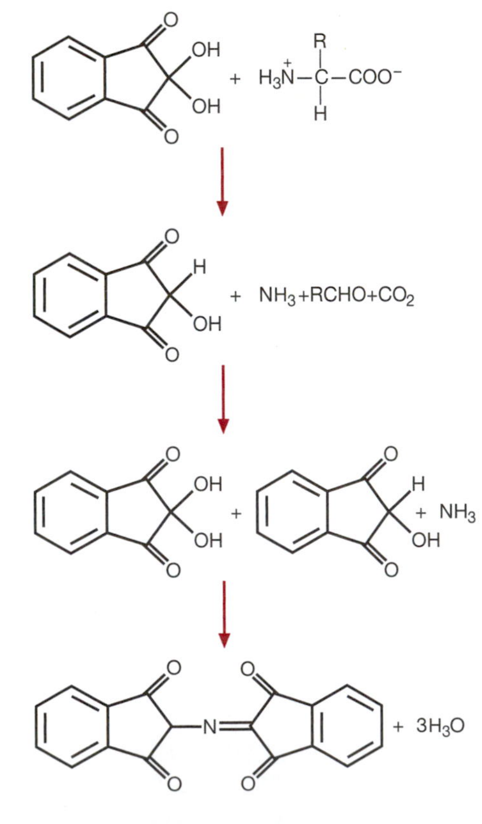

FIGURE 3.1 Chemical structure of ninhydrin, which reacts with the primary amine groups of amino acids as shown.

3.3 DETERMINATION OF PROTEIN SEQUENCES BY CHEMICAL DEGRADATION

Complete hydrolysis allows protein sequences to be inferred from the content of the resulting amino acid pool

Proteins can be completely broken down into their constituent amino acids by boiling in highly concentrated hydrochloric acid for 24–72 hours. The amino acids can then be labeled with an agent such as ninhydrin (**Figure 3.1**) or fluorescamine (**Figure 3.2**), separated by HPLC (Chapter 2), and detected as they elute from the column using a panel of standard amino acids as a reference. The sensitivity of fluorescamine labeling is such that as little as 1 ng of an amino acid can be detected, allowing the analysis of very small quantities of purified protein. The acidic and polar amino acids are eluted first (that is, Asp, Thr, Ser, and Glu) and the basic amino acids are eluted last (that is, Lys, His, and Arg). In each case, the height of the absorption peak is proportional to the relative abundance of the amino acid in the sample.

Although this method reveals the **amino acid composition** of a protein, it does not indicate the sequence, because all the peptide bonds in the protein are broken so consecutive residues cannot be identified directly. However, algorithms such as AACompIdent can predict protein sequences on the basis of amino acid compositions by searching protein sequence databases for entries that would give a similar composition profile. Such correlative search methods are only useful where there are significant existing sequence data and where a pure protein is available, so they have been most widely used in the microbial proteome projects. As we shall see below, protein

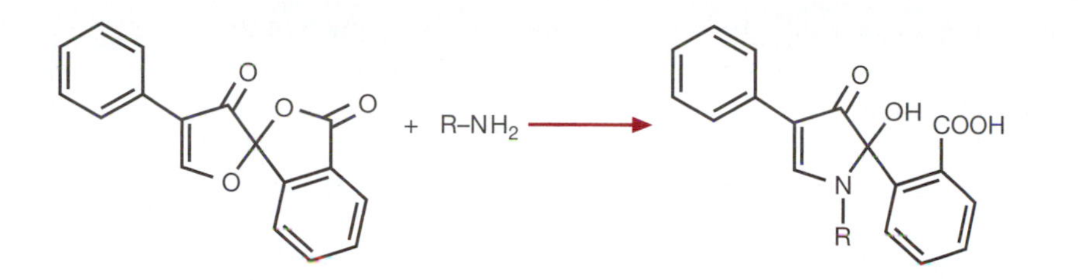

FIGURE 3.2 Chemical structure of fluorescamine, which reacts with the primary amine groups of amino acids as shown.

identification by mass spectrometry is another approach that involves correlative searching, but it can also determine protein sequences *de novo* (without reference to known sequences). Complete hydrolysis experiments do not allow *de novo* sequencing, but further evidence for protein identification can be obtained by taking into account properties such as apparent molecular mass and pI, or by derivatizing the N- and C-terminal amino acids, allowing the first and last residues to be identified positively. For example, the N-terminal amino acid can be modified by dansyl chloride or 9-fluorenylmethyl chloroformate, resulting in a predictable shift in the position of the corresponding fraction. Even with terminal residue identification, however, it may be difficult to identify the protein with confidence. Annotations are much more reliable if at least some sections of contiguous sequence can be determined directly, and this is a prerequisite for the *de novo* sequencing of proteins.

Edman degradation was the first general method for the *de novo* sequencing of proteins

Any method for the direct chemical sequencing of a protein must remove amino acids selectively and progressively from one end of the molecule by breaking only the terminal peptide bond. Each amino acid can then be identified by HPLC as discussed above. There are several methods for sequencing proteins from either the N- or C-terminus using broad-specificity exopeptidases, but the most reliable and therefore the most widely used method is **Edman degradation**, which was developed by Pehr Edman in 1960. Edman degradation involves labeling the N-terminal amino acid of a protein or peptide with **phenyl isothiocyanate** (**Figure 3.3**). Mild acid hydrolysis then results in the cleavage of the peptide bond immediately adjacent to this modified residue, but leaves the rest of the protein intact. The terminal amino acid (or rather its phenylthiohydantoin derivative) can then be identified by chromatography, and the procedure is repeated on the next residue and the next, thus building up a longer sequence (**Figure 3.4**).

With automation, Edman degradation can sequence a small peptide (10 residues) in about 24 hours and a larger peptide (30–40 residues) in 3 days. It is not suitable for sequencing proteins larger than 50 residues in a single run, because each cycle of degradation is less than 100% efficient. This means that after a large number of cycles there is a mixed population of molecules in the analyte rather than a pure sample, so single rounds of Edman degradation produce multiple peaks. The problem is addressed by cleaving large proteins into peptides, using either chemical reagents or specific endoproteases. For example, **trypsin** is an endoprotease that cleaves specifically at the C-terminal side of the basic amino acid residues lysine and arginine as long as the next residue is not proline. Since both lysine and arginine are common amino acids, a protein of about 500 residues might contain 20–50 sites and would be broken into an equivalent number of peptides each less than 25 residues in length. The determination of these individual sequences would allow the entire sequence of the protein to be built up as a series of fragments. However, without further information, there is no way to assemble the fragments in the correct order. This information can be obtained in two ways:

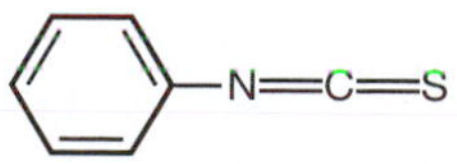

FIGURE 3.3 Chemical structure of phenyl isothiocyanate.

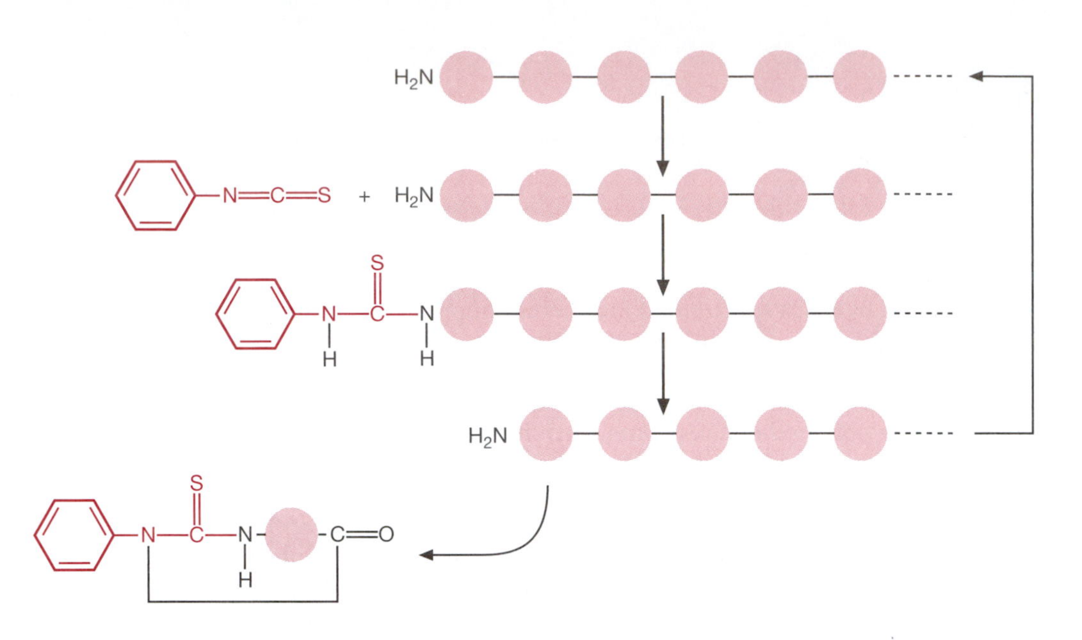

FIGURE 3.4 Edman degradation. The N-terminal amino acid of a peptide is derivatized with phenyl isothiocyanate (shown in *red*). Then, under acidic conditions, the adjacent peptide bond is cleaved and the terminal amino acid is released as a cyclic derivative that can be identified by HPLC. The peptide is now one residue shorter and the cycle begins again.

- Overlapping fragments can be sequenced after digesting the protein using a reagent with different specificity to trypsin (**Figure 3.5**). It is important to choose a reagent that cuts frequently, because a reagent with a very long and specific recognition site would not generate enough informative peptides (**Table 3.1**).
- The solved amino acid sequences of the peptides can be used to design degenerate PCR primers, which can be used to isolate a corresponding genomic or cDNA sequence. This can then be translated to predict the full-length protein sequence. In the early 1980s, when directly solved protein sequences outnumbered DNA sequences, it was common to use protein sequences to design degenerate primers. Now that the number of DNA sequences vastly outnumbers the number of protein sequences, this approach is usually unnecessary.

Edman degradation was the first protein identification method to be applied in proteomics, but it is difficult to apply on a large scale

In the mid-1980s, a technique was developed for the transfer of proteins separated by 2DGE onto polyvinylidene difluoride (PVDF) membranes, followed by *in situ* digestion with trypsin and sequencing by Edman degradation (**Figure 3.6**). However, even with improved membranes and efficient digestion and processing methods, the technique was still laborious because

Digestion with trypsin

Leu-Asp-Glu-Trp-Gly-Val-Ile-Lys

Ala-Val-Ile-Leu-Ser-Glu-Ile-Lys

His-Thr-Val-Glu-Val-Arg

Digestion with Glu-C (alkaline)

Trp-Gly-Val-Ile-Lys-Ala-Val-Ile-Leu-Ser-Glu

Ile-Lys-His-Thr-Val-Glu

Sequence deduced

Leu-Asp-Glu-Trp-Gly-Val-Ile-Lys | Ala-Val-Ile-Leu-Ser-Glu-Ile-Lys | His-Thr-Val-Glu-Val-Arg

Trp-Gly-Val-Ile-Lys-Ala-Val-Ile-Leu-Ser-Glu | Ile-Lys-His-Thr-Val-Glu

FIGURE 3.5 Protein sequences can be obtained by Edman sequencing of overlapping peptides generated with proteases of differing specificities.

TABLE 3.1 CLEAVAGE OF PROTEINS INTO PEPTIDES USING CHEMICAL AND ENZYMATIC REAGENTS

Reagent	Cleavage properties
Chemical agents	
70% formic acid	Asp-↓-Pro
Cyanogen bromide in 70% formic acid	Met-↓
2-Nitro-5-thiocyanobenzoate, pH 9	↓-Cys
Hydroxylamine, pH 9	Asn-↓-Gly
Iodobenzoic acid in 50% acetic acid	Trp-↓
Endoprotease	
Trypsin	Arg/Lys-↓
Lys-C	Lys-↓
Arg-C	Arg-↓
Glu-C (bicarbonate)	Glu-↓
Glu-C (phosphate)	Asp /Glu-↓
Asp-N	↓-Asp
Chymotrypsin	Phe/Tyr/Trp/Leu/Met-↓ (also Ile/Val-↓)

The cleavage properties of all the endoproteases except Asp-N are dependent on the residue after the cleavage site not being proline.

of the length of time taken to complete a single Edman chemistry cycle. Furthermore, many proteins are not amenable to Edman chemistry because the α-amino group of the N-terminal amino acid is often modified (for example, by acetylation) and therefore does not react with phenyl isothiocyanate. Because of these intrinsic limitations, it was soon realized that proteins could never be sequenced on the same scale as DNA without a completely new approach. Toward the end of the 1980s, developments in mass spectrometry provided the necessary technological breakthrough (see below). Despite the impact of mass spectrometry, Edman degradation remains the most convenient method for determining the N-terminal sequence of a protein. It is also extremely sensitive, in that robust sequences can routinely be obtained from as little as 0.5–1 pmol of pure protein, with some groups achieving sensitivities in the several hundred femtomole range.

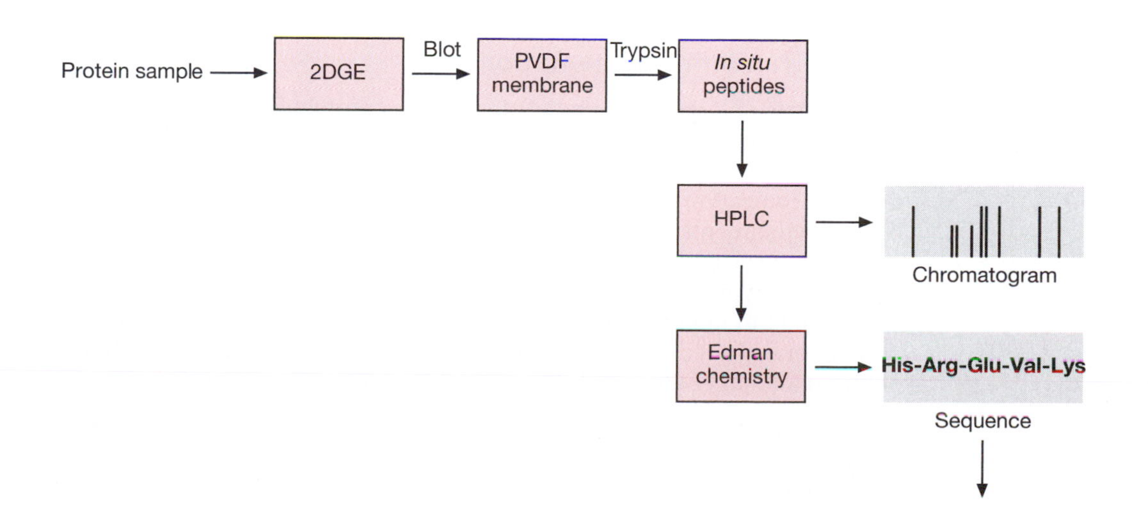

FIGURE 3.6 Early pipeline for protein identification using Edman degradation.

3.4 MASS SPECTROMETRY—BASIC PRINCIPLES AND INSTRUMENTATION

Mass spectrometry is based on the separation of molecules according to their mass/charge ratio

A **mass spectrometer** is an instrument that can measure the **mass/charge ratio** *m*/*z* of ions in a vacuum. From these data, molecular masses can be determined with a high degree of accuracy, allowing the molecular composition of a given sample or analyte to be determined. In proteomics, the analyte is usually a collection of peptides derived from a protein sample by digestion with trypsin or a similar reagent. Three types of analysis can be carried out:

- The analysis of intact peptide ions. This allows the masses of intact peptides to be calculated, and these masses can be used to identify proteins in a sample by correlative database searching.
- The analysis of fragmented peptide ions. This allows the masses of peptide fragments to be determined, and these can be used in correlative database searching, or to derive *de novo* sequences, or in hybrid approaches.
- The analysis of fragmented whole proteins. This is known as **top-down proteomics** because one begins with an intact protein. The approach becomes more challenging with proteins whose mass is greater than 50 kDa (approximately 500 amino acids). This is the upper limit for reliable mass detection using standard mass spectrometers due to the inefficient ionization and poor protein stability. Proteins whose mass is less than 50 kDa can be ionized and fragmented in a procedure analogous to the analysis of fragmented peptide ions, although producing larger fragments. The analysis of peptides or their fragments is therefore known as **bottom-up proteomics** because one starts with parts of proteins rather than whole proteins, although newer techniques that allow the analysis of larger peptides (**middle-down proteomics**) are now attracting more attention.

Mass spectrometers have three principal components: a source of ions, a mass analyzer, and an ion detector. The function of the ionization source is to convert the analyte into gas-phase ions in a vacuum. The ions are then accelerated in an electric field toward the analyzer, which separates them according to their *m*/*z* ratios on their way to the detector. The function of the detector is to record the impact of individual ions.

The integration of mass spectrometry into proteomics required the development of soft ionization methods to prevent random fragmentation

The analysis of large macromolecules such as proteins by mass spectrometry was initially difficult because there was no reliable way to produce intact gas-phase ions. Generally, larger molecules were broken up by the volatization and ionization process, producing a collection of random fragments. Although fragments derived from single proteins and peptides can be informative, nonselective fragmentation of the 50 or so tryptic peptides that constitute a typical protein yields a mass spectrum that is far too complex and difficult to interpret. This began to change in the 1990s with the development of so-called **soft-ionization methods** that achieve the ionization of peptides and other large molecules without significant fragmentation.

The first soft-ionization methods to be embraced in proteomics were **MALDI** and **ESI**, which are compared in **Figure 3.7**. MALDI means **matrix-assisted laser desorption/ionization**, a process in which the analyte is initially mixed

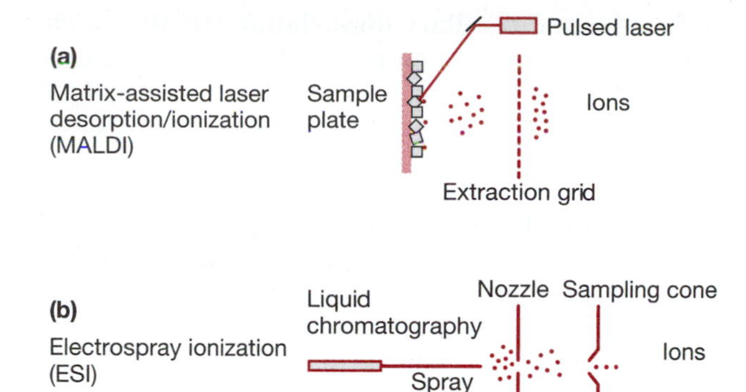

FIGURE 3.7 Soft ionization methods in proteomics. (a) MALDI involves heating crystals of analyte on a sample plate using laser pulses. (b) ESI involves forcing the sample through a narrow spray needle, resulting in a fine spray of ions.

with a large excess of an aromatic "matrix compound" that can absorb laser energy. For example, the matrix compound α-cyano-4-hydroxycinnamic acid can absorb the energy from a nitrogen UV laser (337 nm). The analyte and matrix are then dissolved in an organic solvent and placed on a metallic probe. The solvent evaporates, leaving the analyte embedded in matrix crystals, which are placed in the vacuum chamber of the mass spectrometer and exposed to a high voltage. At the same time, the crystals are targeted with a short laser pulse. The laser energy is absorbed by the crystals and emitted (desorbed) as heat, resulting in rapid sublimation that converts the analyte into gas-phase ions. These accelerate away from the target through the analyzer towards the detector. MALDI is used predominantly for the analysis of simple peptide mixtures, such as the peptides derived from a single spot on a two-dimensional gel, and is particularly useful for the analysis of intact peptides.

ESI refers to **electrospray ionization**, in which the analyte is dissolved and forced through a narrow needle held at a high voltage. A fine spray of charged droplets emerges from the needle and is directed into the vacuum chamber of the mass spectrometer through a small orifice. As the droplets enter the mass spectrometer, they are dried using a stream of inert gas, resulting in gas-phase ions that are accelerated through the analyzer toward the detector. Because ESI produces gas-phase ions from solution, it is readily integrated with upstream protein separation by liquid-phase methods, particularly capillary electrophoresis and liquid chromatography. Therefore, whereas MALDI-MS is primarily used to analyze simple peptide mixtures, ESI-MS is more suited to the analysis of complex peptide mixtures as in shotgun proteomics.

Controlled fragmentation is used to break peptide bonds and generate fragment ions

Although the random fragmentation of proteins and peptides is uninformative, selective fragmentation by breaking peptide bonds generates fragments whose mass has a predictive value that can be used for protein identification. Fragmentation is often achieved using a process known as collision-induced dissociation (CID) in which the fragmentation of intact peptide or protein ions is induced by repeated collision and thus heating by neutral gas atoms. This usually breaks the C–N bond, resulting in the generation of so called ***b*-series** and ***y*-series ions**, the former with the charge retained on the N-terminal fragment, and the latter with it retained on the C-terminal fragment (**Figure 3.8**). CID often causes a proportion of *b*-ions to lose the -CO group, resulting in pairs of *b*-ions and *a*-ions differing by 28 atomic mass units, which can help to identify the *b*-ions and facilitate the interpretation of CID spectra (see below). Similar effects are caused by

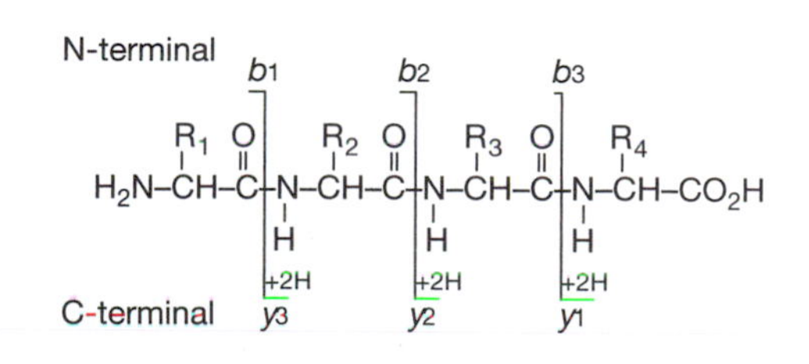

FIGURE 3.8 Fragment ion nomenclature for the most common positive N- and C-terminal ions.

alternative methods such as **heated capillary dissociation** (**HCD**), which also involves collision with neutral atoms, and **infrared multiphoton dissociation** (**IRMPD**), which involves collision with photons. The resulting ions can be used to query sequence databases to find matches by correlative searching, or they can be arranged in nested sets to determine the peptide sequence *de novo* based on the incremental mass increases compared with the mass of different amino acids. There may also be multiple breakages producing internal fragments of several contiguous amino acids, as well as immonium ions representing single amino acids, plus the neutral loss of water, ammonia, and certain post-translational modifications. Certain ion types are unique to high-energy CID, including the *d*, *v*, and *w* series, which can differentiate between isoleucine and leucine based on the formation of β-carbon substituents.

Neutral loss caused by CID (as well as the limited information available from long peptides and intact proteins reflecting the uneven distribution of fragmentation sites) has led to the development of two novel fragmentation techniques that can be more informative. Electron-capture dissociation (ECD) involves the capture of a thermal electron by a multi-protonated peptide or protein cation that causes fragmentation of the N–Cα bond to produce alternative ***c*-series** and ***z*-series** fragment ions. These are analogous to the *b*-series and *y*-series ions discussed above and can be used in the same way to generate sequence information, but ECD is advantageous because it achieves more frequent and evenly distributed fragmentation and does not tend to cause neutral losses to the same degree as CID. ECD is widely used in middle-down and top-down proteomics because it provides more information from large peptides and whole proteins. One disadvantage of ECD is that it works best with high-end FT-ICR analyzers (see below) and is generally unsuitable for use with more accessible mass spectrometers. This has been addressed by the development of electron transfer dissociation (ETD) in which the electron is transferred from a radical anion to the multi-protonated peptide or protein to generate *c*-series and *z*-series fragment ions. This method is also suitable for the analysis of post-translational modifications (Chapter 8). Another fragmentation method that can be used with FT-ICR analyzers is **SORI-CID** (**sustained off-resonance irradiation collision-induced dissociation**), which involves accelerating the ions in the cyclotron and inducing collisions by increasing the pressure.

Five principal types of mass analyzer are commonly used in proteomics

There are five main types of mass analyzer used in proteomics, each with its own strengths and weaknesses in terms of accuracy, sensitivity, and resolution. These can be combined in many different ways to generate a large variety of hybrid instruments. Conceptually, the simplest instruments are the quadrupole (Q) and time-of-flight (TOF) analyzers. The more sophisticated instruments are the ion trap (IT), and Fourier-transform ion cyclotron resonance (FT-ICR) analyzers, and the more recently developed Orbitrap, which is similar to the FT-ICR. Ion-mobility spectrometry can also be used to separate ionized molecules, as discussed in Box 3.1.

A **quadrupole** is a set of four parallel metal rods, opposite pairs of which are electrically connected so that a voltage can be applied across the space between them. A quadrupole can be operated in **RF-only mode** (RF referring to radio frequency), which allows ions of any m/z ratio to pass through, or in **scanning mode**, where a potential difference is applied and the instrument acts as a **mass filter**, that is, selecting ions with a specific m/z ratio and discarding others. When the mass filter is applied, ions of a selected m/z ratio are allowed through to the detector whereas all others are deflected from their linear flight path and are eliminated from subsequent analysis. By

BOX 3.1 ALTERNATIVE METHODS.
Ion-mobility spectrometry–mass spectrometry (IMS-MS).

Ion-mobility spectrometry involves the separation of gas-phase ions based on their mobility (drift time) in a neutral carrier gas. When coupled to mass spectrometry (IMS-MS) the technique provides higher-resolution separations and thus a better signal-to-noise ratio than standard LC-MS, and the better identification of charge states. The technique was pioneered in proteomics to characterize the human plasma proteome, using a combination of strong cation-exchange chromatography and reversed-phase liquid chromatography with IMS-MS. The increase in separation capacity associated with IMS produced one of the most extensive proteome maps to date, including the preliminary identification of 9087 proteins (2928 with high confidence) from 37,842 unique peptide assignments.

varying the voltage over time, ions with different *m/z* ratios can be sequentially allowed through to the detector and a **mass spectrum** of the analyte can be obtained.

The simplest configuration of the quadrupole in proteomics is the **triple quadrupole** where three such devices are arranged in series, usually connected to an ESI source. Triple-quadrupole instruments can be set up for the analysis of either intact peptides or their fragment ions (**Figure 3.9**). In the former case, the instrument is operated in **standard MS mode** where only one of the quadrupoles is used for scanning (designated Q), the others remaining in RF mode (designated q). In standard MS mode, the first quadrupole scans the intact peptide ion stream and sequentially directs ions of different *m/z* values through the second and third quadrupoles in RF-only mode onto the detector. This configuration is therefore designated **Qqq**. In the latter case, the instrument is operated in **tandem MS** or **MS/MS mode**. The first quadrupole scans the intact peptide ion stream and sequentially directs ions of different *m/z* values into the second quadrupole, which operates in RF mode and acts as a **collision cell** (that is, a gas stream fragments the peptide ions by CID as discussed above). The resulting fragment ions are scanned in the third quadrupole to generate a **CID spectrum**, that is, a mass spectrum of the fragments derived from each specific peptide. When the analysis of one peptide ion is complete, the first quadrupole directs a

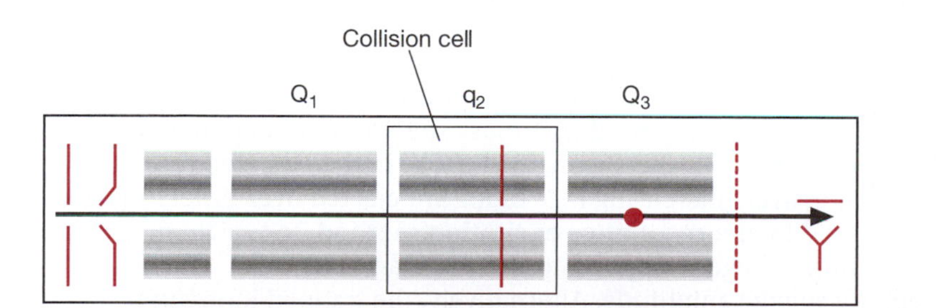

MS			
Scanning	RF	RF	Standard MS mode
Fixed *m/z*	RF	RF	Single-ion monitoring
MS/MS			
Fixed *m/z*	Collision	Scanning	Product-ion scan
Scanning	Collision	Fixed *m/z*	Parent-ion scan
Scanning	Collision	Scanning	Neutral-loss scan
Fixed *m/z*	Collision	Fixed *m/z*	Selected reaction monitoring

FIGURE 3.9 Layout of a triple-quadrupole mass spectrometer, in which ions are selected by varying the electric fields between four rods. In this diagram, the mass spectrometer is running in product-ion scan mode, where ions are selected in the first quadruple (Q_1), q_2 is used as a collision cell to induce fragmentation, and the resulting fragment ions are scanned in Q_3.The table shows the settings for each quadrupole for different MS and MS/MS nodes.

different intact peptide ion into the collision cell and the fragmentation and analysis process is repeated. This configuration is therefore designated **QqQ**. The use of two analyzers in series to separate intact ions and then their fragments is described as product-ion scanning or **daughter-ion scanning**. Several other operational modes are also available in QqQ instruments, including neutral-loss scanning and precursor-ion scanning. These are often used to distinguish between phosphorylated and nonphosphorylated versions of the same protein and are discussed in Chapter 8. If the m/z values of both quadrupoles are fixed, this enables selected reaction monitoring (SRM), which is highly sensitive and can provide quantitative data (Chapter 4).

Unlike quadrupole instruments, no electric field is required to separate ions in a **time-of-flight** (**TOF**) analyzer. Instead, this instrument exploits the fact that in any mixture of ions carrying the same charge, heavy ions will take longer to travel down a field-free flight tube than lighter ones (**Figure 3.10**). TOF analyzers have, until recently, been used almost exclusively with MALDI ion sources for the analysis of intact peptide ions (**MALDI-TOF**). This is because the MALDI process tends to produce singly-charged peptide ions, and under these conditions the time of flight of any ion is inversely proportional to the square root of its molecular mass. More recently, MALDI sources have been coupled with tandem TOF-TOF analyzers or hybrid quadrupole-TOF analyzers separated by a collision cell (**QqTOF**), allowing the analysis of CID spectra from MALDI-derived precursor ions with much higher sensitivity than possible with the QqQ and single TOF instruments. TOF analyzers typically have 10 times the resolving power of QqQ analyzers and are 10–100 times more accurate (10–20 ppm compared with 100–1000 ppm).

Although fragment analysis typically requires a collision cell, it can also be achieved in a standard MALDI-TOF instrument by exploiting a phenomenon known as **post-source decay** (**PSD**), which involves increasing the laser power to about twice the level used to obtain a normal mass spectrum, causing multiple collisions between the peptides and the matrix compound during ionization. This delays fragmentation in a large proportion of the peptides, and the fragmented ions can be separated from the intact peptides using ion gates and mirrors (**reflectrons**).

The **quadrupole ion trap** (**QIT**) is a more sophisticated analyzer, consisting of a quadruple in RF mode connected to an **ion trap** (**IT**), which is a chamber surrounded by a ring electrode and two end-cap electrodes (**Figure 3.11**). The voltage applied to the ring electrode determines which ions are trapped. Ions above the threshold m/z ratio remain in the trap while others are ejected through small holes in the distal end-cap electrode (this is known as **mass instability mode**). A mass spectrum of intact peptides can be obtained by gradually increasing the voltage in the ring electrode so that ions of progressively increasing m/z ratios are ejected over time. Ions may also be ejected by the **resonance excitation method**, in which an oscillatory excitation voltage is applied to the end-cap electrodes while varying the trapping voltage amplitude. Alternatively, the trapped ions can be fragmented by injecting a stream of helium gas, and the resulting fragments can be ejected by ramping the voltage of the ring electrode to generate a CID

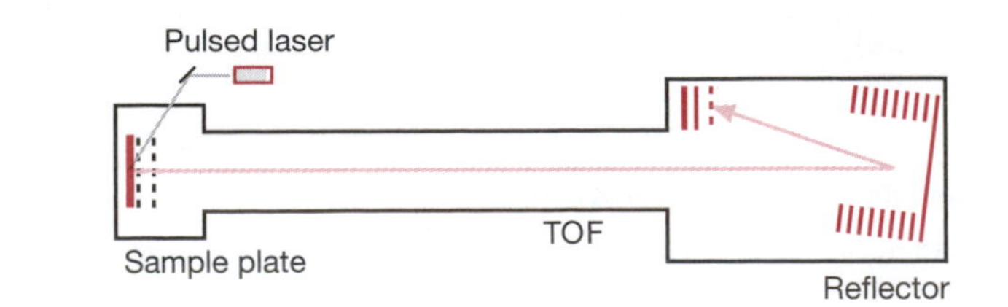

FIGURE 3.10 Layout of a MALDI-TOF mass spectrometer (reflector mode), in which ions are separated by virtue of the time taken to travel down a field-free flight tube to the detector.

spectrum. Multiple rounds of analysis can be carried out, because one of the fragment ions from the first analysis can be retained in the trap and subject to further collision. Up to three rounds of fragmentation are routinely used, particularly for the analysis of glycoproteins, but up to 12 rounds of fragmentation have been reported. This form of tandem mass spectrometry is termed **MS^n** where n is the number of rounds of fragmentation. The standard QIT instrument traps ions in a three-dimensional field but there are also linear field versions such as the Thermo Fischer Scientific **LTQ** (**linear trap quadrupole**), which offer greater storage capacity and faster scanning, and the hybrid **QqLIT** instrument, which combines quadrupole scanning with a linear ion trap to exploit the capabilities of both and thus offers a greater dynamic range. The LTQ also allows fragmentation of low-mass fragment ions by **pulsed *q* dissociation** (**PQD**), which involves activating the precursor ion for a very short time at a high q value (proportional to the RF voltage in the ion trap), and **higher-energy collision dissociation** (**HCD**, also known as **high energy C-trap dissociation** because of its location in the instrument), which is particularly suitable for iTRAQ experiments (see Chapter 4). Despite the power of ion traps, one major limitation is the ratio between the precursor ion m/z and the lowest m/z of a trapped fragment ion, which never increases above 0.3 (this is generically known as the **one-third rule**), which means that for a parent ion of 600 m/z it is not possible to detect fragment ions below 200 m/z, thus limiting the potential for *de novo* peptide sequencing.

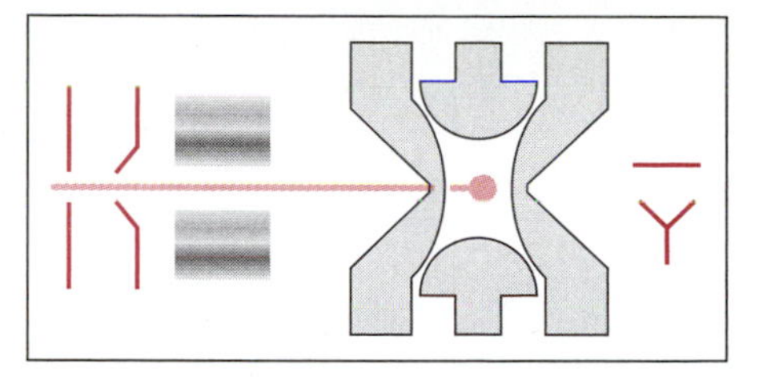

FIGURE 3.11 **Layout of an ion-trap mass spectrometer, in which ions are constrained within a chamber surrounded by ring electrodes.**

The most sophisticated apparatus is the **FT-ICR** analyzer, which is also the most complex and difficult to operate, but has by far the highest resolution, mass accuracy, dynamic range, and sensitivity, and is particularly suitable for top-down proteomics. The operating principle is that ions in a magnetic field will orbit at a frequency that is related to their mass m, their charge z, and the strength of the magnetic field B. This is called the cyclotron frequency f_c. The relationship can be described by the following equation:

$$m/z = B/2\pi f_c$$

Therefore, all ions with the same m/z value will orbit with the same cyclotron frequency in a uniform magnetic field, and this collection of ions is known as an **ion packet**.

Orbiting ions of a particular m/z value are then excited by an applied RF field, which causes the cyclotron radius to expand. If the frequency of the applied field is the same as the cyclotron frequency of the ions, the ions absorb energy, thus increasing their velocity (and the orbital radius) but keeping a constant cyclotron frequency. As the selected ions cycle between the two electrodes, electrons are attracted first to one plate and then the other, with the same frequency as the cycling ions (that is, in resonance with the cyclotron frequency). This movement of electrons is detected as an image current on a detector. The image current is then converted, by Fourier transformation, into a series of component frequencies and amplitudes of the individual ions. Finally, the cyclotron frequency values are converted into m/z values to produce the mass spectrum.

The **Orbitrap mass analyzer** is similar in principle to the FT-ICR instrument because masses are detected by sensing oscillations, but in this case the ions are injected tangentially into the electric field between the electrodes and maintained cycling around the central electrode because their electrostatic attraction to it is balanced by centrifugal forces. The ions also move back and forth along the axis of the central electrode, causing harmonic oscillations whose frequency is inversely proportional to the square root of the m/z ratio. The Orbitrap has been produced commercially as a hybrid **LTQ-Orbitrap** instrument by Thermo Fischer Scientific and carries all the advantages of the FT-ICR but has an even greater dynamic range and a faster scan rate.

3.5 PROTEIN IDENTIFICATION USING DATA FROM MASS SPECTRA

Peptide mass fingerprinting correlates experimental and theoretical intact peptide masses

The first routine method for protein identification using mass spectra was **peptide mass fingerprinting** (**PMF**), which refers to the identification of proteins using data from intact peptide masses (**Figure 3.12**). This method is compatible with 2DGE and MALDI-TOF mass spectrometry, where proteins are separated before digestion into peptides. The principle of the technique is that each protein can be uniquely identified by the masses of its constituent peptides, this unique signature being known as the peptide mass fingerprint. Algorithms allowing database searching on the basis of peptide mass data were developed simultaneously by several groups in the early 1990s and have been implemented in a number of software packages, the most commonly used of which are **Mascot**, MS-Fit, and ProFound, each of which can be accessed and queried over the Internet (**Table 3.2**). PMF involves the following steps:

- The sample of interest should comprise a single protein or a simple mixture, for example, an individual spot from a two-dimensional gel or a single LC fraction. The sample is digested with a specific cleavage reagent, usually trypsin (Table 3.1).
- The masses of the peptides are determined, for example, by MALDI-TOF mass spectrometry.
- The experimenter chooses the software and one or more protein sequence databases to be used for correlative searching (Chapter 5).
- The algorithm carries out a virtual digest of each protein in the sequence database using the same cleavage specificity as trypsin (or whichever other reagent has been used experimentally) and then calculates theoretical peptide masses for each protein.
- The algorithm attempts to correlate the theoretical peptide masses with the experimentally determined ones.
- Proteins in the database are ranked in order of best correlation, usually with a significance threshold based on a minimum number of peptides matched.

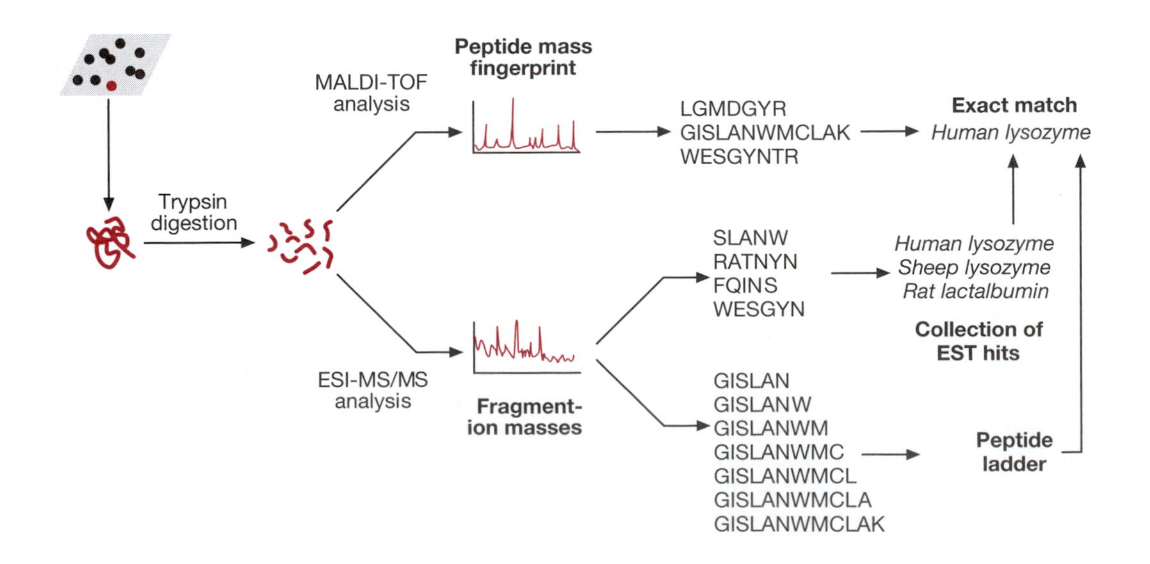

FIGURE 3.12 Protein identification by mass spectrometry. In a typical strategy, digested peptides are analyzed by MALDI-TOF mass spectrometry in order to determine the masses of intact peptides. These masses can be used in correlative database searches to identify exact matches. If this approach fails, ESI-MS/MS analysis can be used to generate peptide fragment ions. These can be used to search less robust data sources (for example, EST databases) and to produce *de novo* peptide sequences.

TABLE 3.2 KEY TOOLS FOR THE ANALYSIS OF MASS SPECTROMETRY DATA

Program	URL
Peptide mass fingerprinting	
Mascot	http://www.matrixscience.com/search_form_select.html
MS-Fit	http://prospector.ucsf.edu/prospector/cgi-bin/msform.cgi?form=msfitstandard
ProFound	http://prowl.rockefeller.edu/prowl-cgi/profound.exe
MassSearch	http://www.cbrg.ethz.ch/services/MassSearch_new
MS/MS database searches	
Sequest	http://fields.scripps.edu/sequest/
Mascot	http://www.matrixscience.com/search_form_select.html
ProteinProspector	http://prospector.ucsf.edu
ProbID	http://tools.proteomecenter.org/wiki/index.php?title=Software:ProbID
X! Tandem	http://www.thegpm.org
SpectrumMill	http://www.chem.agilent.com/
Phoenyx	http://www.genebio.com/products/phenyx/
VEMS	http://yass.sdu.dk/
ProteinPilot	http://www.absciex.com
MyriMatch	http://fenchurch.mc.vanderbilt.edu/software.php
PepSplice	http://www.ti.inf.ethz.ch/pw/software/pepsplice/
RAId_DbS	http://www.ncbi.nlm.nih.gov/CBBresearch/qmbp/raid_dbs/
Mass Matrix	http://www.massmatrix.net/mm-cgi/home.py
Sequence tag/hybrid methods	
InsPecT	http://proteomics.ucsd.edu/Software/Inspect.html
Popitam	http://code.google.com/p/popitam/
TagRecon	http://fenchurch.mc.vanderbilt.edu/software.php
ByOnic	http://proteinmetrics.com/software-products/byonic-software/
Spectral Networks	http://proteomics.ucsd.edu/Software/SpectralNetworks.html
MODi	http://www.massmatrix.net/mm-cgi/home.py
MS-Tag	http://prospector.ucsf.edu/prospector/cgi-bin/msform.cgi?form=mstagstandard
De novo *sequencing*	
Lutefisk	http://www.hairyfatguy.com/Lutefisk
PepNovo	http://proteomics.ucsd.edu/Software/PepNovo.html
PEAKS	http://www.bioinformaticssolutions.com
Sequit	http://www.sequit.org/
Spectral matching	
SpectraST	http://www.peptideatlas.org/spectrast/
X! P3	http://ppp.thegpm.org/tandem/thegpm_ppp.html
BiblioSpec	https://skyline.gs.washington.edu/labkey/project/home/software/BiblioSpec/begin.view
Post-search data analysis	
PeptideProphet and ProteinProphet	http://www.proteomecenter.org/software.php
Scaffold	http://www.proteomesoftware.com/
IDPicker	http://fenchurch.mc.vanderbilt.edu/software.php
MassSieve	http://www.ncbi.nlm.nih.gov/staff/slottad/MassSieve/
MS-GF	http://proteomics.ucsd.edu/Software/MSGeneratingFunction.html
MaxQuant	http://maxquant.org
PeptideClassifier	http://www.mop.unizh.ch/software.html

The masses of intact peptides are extremely discriminatory, making the PMF technique robust as a means of protein identification. However, because PMF relies on correlative searching, the likelihood of finding a matching protein depends on both the quality of the experimental data and the availability of sequence information for the organism from which the experimental sample was obtained. Data attributes that need to be considered for reliable protein identification include the quality and relative intensity of the peaks in the mass spectrum, the mass accuracy of the instrument, the coverage of the protein, and possible interfering factors such as post-translational modifications and mis-cleavages (**Box 3.2**). These factors influence the likelihood that a match is genuine rather than spurious, a probability often expressed as a **MOWSE score** (from one of the original algorithms developed for PMF—Molecular Weight Search, now part of **Mascot**, see Table 3.2). PMF is best suited to those organisms for which large amounts of genomic, cDNA, and protein sequence data are available, and particularly for species with completed genome sequences.

Even in species with abundant sequence data, there are potential pitfalls in PMF analysis. Many protein sequences are modified after translation, for example, by trimming or cleavage or by the removal of inteins (Chapter 8). Even if there is perfect correspondence between a cDNA sequence and a protein sequence, there remain many other reasons for the absence of a correlation between experimentally derived masses and those predicted from database entries (Box 3.2). Because protein identification by PMF depends

BOX 3.2 BACKGROUND ELEMENTS.
Possible causes of incorrect protein identification in correlative database searching.

There are many potential reasons why an experimentally derived peptide mass or fragment does not match the corresponding theoretical mass:

- There may be an error in the sequence database, causing the algorithm to generate an incorrect predicted mass for one or more peptides.
- The mass spectrometer may not be accurate. Calibration is important because small differences in mass tolerance can make a great deal of difference to the quality of the resulting matches. Many investigators use internal calibration standards in every sample, for example, autolysis products such as the peptides derived from trypsin when it is digested by other molecules of trypsin. Different instruments also vary significantly in accuracy.
- The protein might exist as two or more polymorphic variants, and the version stored in the sequence database might not be the same as the version found in the sample. For example, single nucleotide polymorphisms (SNPs) represent an abundant form of genetic variation that may contribute as many as 50,000 single amino acid differences in the human proteome.
- Differences in mass may be caused by post-translational modifications occurring *in vivo*. Many of the algorithms used to correlate predicted and determined masses can build in anticipated mass changes brought about by known modifications such as phosphorylation. This is discussed in more detail in Chapter 8.
- Differences in mass may be caused by nonspecific modifications occurring during protein extraction, separation, or processing. For example, many chemical modifications occur as a result of gel staining in 2DGE experiments (see Box 4.1). In some cases, it may be a good idea to carry out deliberate modifications. For example, the cysteine residues in denatured proteins may be modified universally with iodoacetamide to prevent sporadic modifications that could complicate the interpretation of mass spectra. The use of affinity mass tags for protein quantitation in mass spectrometry is also a form of deliberate modification and is discussed in Chapter 4.
- There may have been nonspecific cleavage of the protein. Even highly specific cleavage reagents such as trypsin occasionally cut at nonspecific sites and ignore genuine sites. The presence of multiple adjacent or clustered lysine and arginine residues, for example, can prevent trypsin cleavage reactions reaching completion.
- The presence of multiple proteins in the analyte may make the mass spectrum too complex to interpret. In some cases, this can be due to external contamination. For example, it is easy for laboratory staff to contaminate protein samples with minute amounts of keratin from shed hair and skin cells or common protein-based laboratory reagents. Common contaminating proteins in proteomics experiments are listed in a database known as the **common repository of adventitious proteins**, handily abbreviated to **cRAP**.

entirely on the accurate correlation of determined and predicted masses, even small unanticipated differences in mass can prevent the detection of matching proteins. This is why PMF is carried out using peptides rather than whole proteins, since the former provide greater scope for database correlation. The confidence attributed to PMF searching can also be increased by using so called orthogonal datasets, for example, data obtained from the digestion of the same protein with two different proteases (either separately or in combination), data obtained from the digestion of the same protein in a native state and following some form of chemical modification or substitution, or partial sequence data obtained using alternative methods.

Shotgun proteomics can be combined with database searches based on uninterpreted spectra

The limitations of PMF can be addressed in the shotgun proteomics approach by fragmenting the intact peptides and deriving richer data from the resulting fragment ion masses. The **uninterpreted MS/MS spectrum** (with sequence data unknown) can then be compared against theoretical fragmentation spectra derived from peptides in whichever database is queried. The search is constrained to peptides that match the expected enzyme digestion pattern and ion mass tolerance, and additional limitations may be imposed, including expected post-translational or chemical modifications and the anticipated type of fragment ions (for example, *b*-series and *y*-series ions or *c*-series and *z*-series ions, depending on the fragmentation method). The search score is calculated according to the degree of similarity between the experimental and theoretical spectra, and different principles are used by different programs. In **probability-based matching**, virtual CID spectra are derived from the relevant peptides of all protein sequences in the database and these are compared with the observed data to derive a list of potential matches. In **cross-correlation**, it is the degree of overlap between the observed and predicted peaks that determines the best potential match. Several algorithms, such as Sequest and Mascot (Table 3.2) use uninterpreted data for cross-correlation. Others use empirically derived rules or statistical fragmentation frequencies. Different programs give different but overlapping results and it is often valuable to run MS data through several of the available programs to increase the likelihood of matches. It is also important to tailor the sensitivity of the database search to the accuracy of the instrumentation; for example, the most accurate FT-ICR and Orbitrap instruments can be used with searches that impose a narrow mass window, whereas a greater tolerance is required for other instruments.

Another form of uninterpreted data search that can be used with fragment ion spectra is the **spectral library search**. Spectral libraries are compiled from collections of deposited MS/MS spectra and they allow direct comparisons (using tools such as SpectraST and BiblioSpec) to identify matches or overlaps. This approach is faster and more accurate than sequence library searching, but the amount of data available for comparison are still limited, despite efforts to develop proteome-wide datasets for a number of model organisms.

MS/MS spectra can be used to derive protein sequences *de novo*

As an alternative to correlative database searching with uninterpreted MS/MS spectra, protein sequences can be deduced from the spectra directly. In its most robust form (**complete interpretation**), this allows the entire sequence of a peptide to be deduced *de novo*, but a short deduced sequence is often enough for standard sequence database searches using adapted search algorithms such as **MS-BLAST** (Chapter 5). Importantly, whereas uninterpreted spectra can only be used to search full protein sequences in the major databases, partially interpreted spectra can be used to search lower-quality sources, such as expressed sequence tags (ESTs). These are short (100–300 bp) cDNA signatures (see Chapter 1) that are generally too short to contain complete tryptic peptides but may match peptide fragments generated by MS/MS.

The interpretation of MS/MS data is complex because fragmentation produces a diverse collection of ions. To illustrate the complexity of the process, a worked example is provided in **Box 3.3**. The most informative fragments are those where fragmentation has occurred at the peptide bond (for example, corresponding to the *b*-series and *y*-series ions in CID spectra or the *c*-series and *z*-series ions in ECD spectra). The differences in mass between consecutive ions in either series should correspond to the masses of individual amino acids (**Box 3.4**), and this can be used to derive a short sequence or **peptide tag**. Tags can also be derived from ragged termini generated adventitiously or by limited exopeptidase digestion, followed by MALDI-TOF mass spectrometry. Algorithms for database searching using peptide tags include **MS-Tag** (Table 3.2) and TagIdent, and some algorithms such as Mascot allow combined sequence-based and mass-based searches to refine the results. Even tags as short as three consecutive amino acid residues are informative because these can reduce the number of candidate peptides for comparison significantly, thus reducing search times. Longer sequences allow gaps in search strings (gapped peptides) and in a more organized sense can be used to generate **spectral dictionaries**, which are analogous to the spectral libraries discussed above but are designed to bridge the gap between *de novo* sequencing and spectral database searching.

The complete interpretation of fragment-ion spectra can be difficult without some form of labeling to identify specific sets of nested fragments (for example, either the *b*- or *y*-series fragments). A useful approach is to divide the sample into two aliquots, attach a specific **mass label** to either the N- or C-terminus of the intact peptide in one of the aliquots, and then compare the mass spectra to identify the modified and unmodified forms. For example, methyl esterification of the C-terminus of a peptide adds 14 mass units. The comparison of mass spectra from treated and untreated samples therefore allows the *y*-series of ions to be identified by the specific mass displacement (**Figure 3.13a**). However, this reaction also esterifies acidic side chains, so the analysis becomes more complex for peptides containing Asp and Glu residues. An alternative strategy is to carry out trypsin digestion in a buffer in which half the water contains a heavy isotope of oxygen (^{18}O). Trypsin incorporates an oxygen atom from water into the carboxyl group of the newly generated C-terminus of each peptide. Therefore, if the above buffer is used, each *y*-series fragment will be present as a mixture of two derivatives differing by 2 mass units, and will be represented on the mass spectrum as a doublet (Figure 3.13b). A disadvantage of this approach is that the intensity of the signal for each *y*-series ion is also reduced by 50%. Derivatization of the tryptic peptide fragments can also improve the efficiency of fragmentation in MALDI PSD experiments, resulting in more sequence coverage.

The alternative *de novo* sequencing approaches provide a good example of how Edman degradation and mass spectrometry can be used together to generate sequence information. Edman chemistry is used to derive a nested set of intact N-terminal peptides and these are then analyzed by MALDI-TOF mass spectrometry to identify mass differences corresponding to specific amino acids and therefore derive the sequence. Unlike MS/MS approaches, the Edman-based methods discriminate easily between the amino acids glutamine and lysine (Box 3.4). Whereas Edman chemistry is limited to the generation of N-terminal peptides, both N-terminal and C-terminal peptide ladders can be produced by the limiting use of exopeptidases, with the resulting analyte again being subjected to MALDI-TOF mass spectrometry for sequencing.

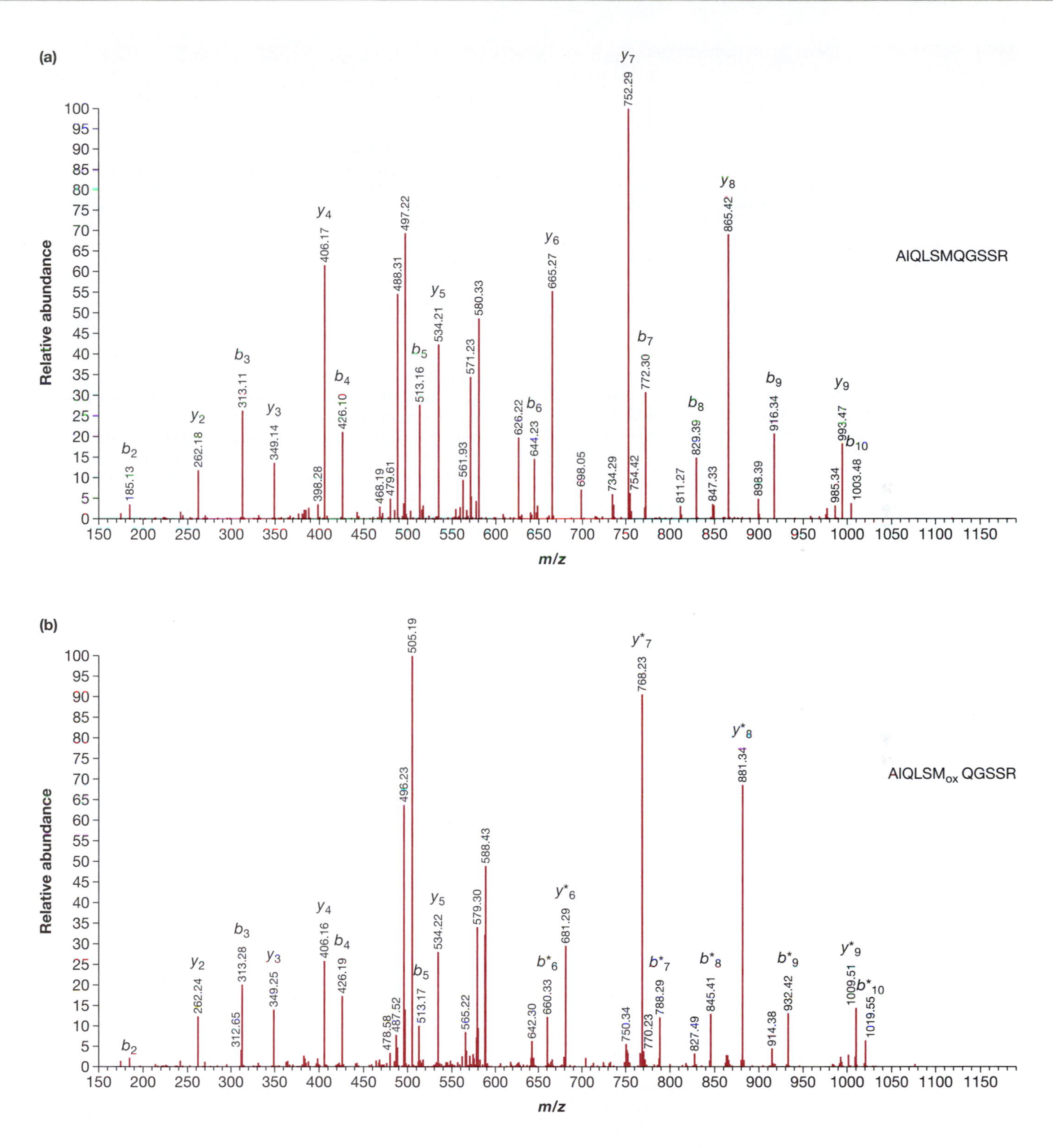

FIGURE 3.13 Peptide CID spectra. Examples of (a) unmodified and (b) modified peptide CID spectra showing the shift in mass. The precursor masses of these ions are 1177.60 and 1193.63 (Δm = 16 Da), suggesting the addition of oxygen (oxidation). Database searches identified a peptide containing a methionine. As is evident from the CID spectrum in (b), the masses of all *b*-series and *y*-series ions (denoted by *) containing methionine are shifted by 16 Da compared with the corresponding ions in (a). (Courtesy of Venkatesha Basrur, University of Michigan, USA.)

BOX 3.3 CASE STUDY.
Interpreting CID spectra.

Interpreting CID spectra is akin to assembling a jigsaw puzzle. Mass spectrometers are often used to determine the intensity and mass of parent ions in a fixed m/z window (generally 400–2000, representing intact peptides) and to select ions with a particular m/z value above a set intensity threshold that are subjected to CID, generating daughter ions whose m/z values are then recorded. This cycle can be repeated several thousand times within the chromatographic time window. The m/z values of parent ions and daughter ions constitute the puzzle pieces. Whether or not the assembled puzzle is available as a reference, the puzzle can still be assembled because the pieces contain all the necessary information.

If the assembled puzzle (that is, the protein sequence) is already available, correlative database searching can be used to identify the protein by matching experimental and theoretical peptides (see p. 58). Otherwise, CID spectra can be interpreted manually using following steps:

1. **Determine the charge state of the parent peptide.** The mass of the parent peptide must be known to identify the protein sequence but mass spectrometers measure m/z, so it is essential to determine the charge of the observed ion. If a high-resolution Orbitrap mass spectrometer is used, the spacing between the peaks in the precursor ion scan is often sufficient. For example, Figure 1 shows a precursor ion scan in the 440–700 m/z range, with two major ions at 463.24 and 679.36. The inset panels showing these regions in more detail reveal Δm values of approximately 0.33 and 0.5 daltons (Da) between the isotopic peaks of the two ions. Using the simple formula $z = 1/\Delta m$, one can assign charge states of +3 and +2, respectively, for these ions (z = charge, 1 = nominal mass of neutron). Generally, +2 charged ions are the most useful for *de novo* sequencing because more highly charged parent ions generate daughter ions that also behave as multiply charged ions, making it more difficult to interpret the CID spectrum. If a high-resolution precursor ion scan is not available, a charge state of +2 or +3 can be assumed in an attempt to work out the sequence.

2. **Determine the mass of the parent peptide.** Once the charge state is assigned or assumed, the monoisotopic mass of the parent peptide can be derived using the following equation:

   $$[\mathrm{M+H}]+1 = (\text{Observed mass} \times \text{charge state}) - (\text{charge state} - 1)$$

   Taking the +2 ion from Figure 1 as an example, the mass of the parent peptide can be calculated as 1357.72 Da.

3. **Determine the C-terminal amino acid.** Since many of the proteases commonly used for digestion cleave after a specific residue, cleavage specificity can be used to determine the C-terminal amino acid. Figure 2 is the CID spectrum derived from the +2 charged parent peptide discussed above, which was generated after trypsin digestion. Trypsin cleaves on the C-terminal side of K and R residues, therefore subtracting the mass of the K (147 Da) or R (175 Da) from the mass of the parent peptide identifies the C-terminal amino acid. In this example, the presence of a fragment ion at 1183.46 Da indicates that R is the C-terminal amino acid: (1357.72 − 175) + 1 proton = 1183.72 Da.

4. **Find the *b*-series and *y*-series ions.** CID often causes *b*-series ions to lose their –CO group (28 Da) giving rise to *a*-series ions. This doublet of peaks at the low-m/z end of the spectrum can be used to identify the *b*-series ions. Subtracting the *b*-series ion mass from the parent-ion mass should identify the corresponding *y*-series ion. The spectrum in Figure 2 shows two peaks at 235.07 and 207.1 separated by 28 Da, which suggests that peak 235.07 can be assigned as *b*-series ion. Since this peak does not correspond to a single amino acid mass (Box 3.4) it may represent a b_2-ion (dipeptide, for example, AY, HP, MC, FS, or $M_{ox}S$). MC can be ruled out if cysteine alkylation was included during the sample preparation step, and FS and $M_{ox}S$ can be distinguished by the presence or absence of an immonium ion at 120 Da, because this is unique to F. Notably, CID spectra representing peptides generated by proteases that cleave after basic residues are generally dominated by intense *y*-series ions peaks. Thus if a/b-series ion peaks are not observed, one can assume that the high-intensity peak at the high-mass end of the spectrum represents *y*-series ions.

5. **Identify the *b*-series and *y*-series ions.** Continue subtracting high-intensity, high-mass end peaks from the parent ion mass as long as possible and label them as ion pairs, for example, a,a' and b,b'. This avoids the assignment of the same m/z peak to two different ion series. In Figure 2 for example, a,a' = 1123, 235 Da and b,b' = 895, 463 Da. It is useful to omit sequential ion pairs if the peaks are not evident (for example, in the case of 1010, 348 there is a clear peak at m/z 1010 but not m/z 348.

6. **Reconstruct the sequence.** When many ion pairs have been established, the difference between sequential ions should correspond to the mass of an amino acid. In Figure 2, [*a*,*b*,*c*,*d*,*e*,*f*,*g*,*h*,C-term] corresponds to [LDELVLHR].

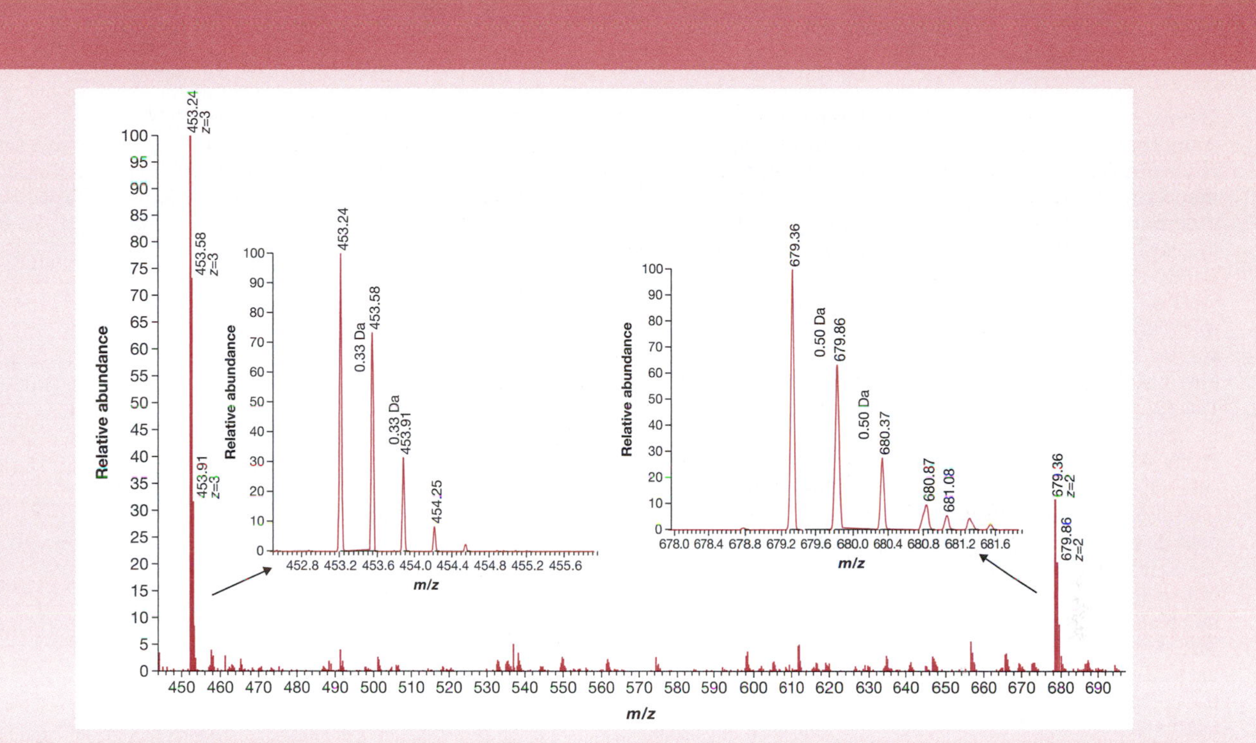

FIGURE 1 Mass spectrum of an isotopically resolved peptide ion (AYLDELVELHR). The insets show more detailed views of the same ion with two different charge states. The spacing between the isotopic peaks can help to determine the charge state. In this example, the 0.33 and 0.5 Da spacings of the left and right insets indicate +3 and +2 charge states of the peptide. (Courtesy of Venkatesha Basrur, University of Michigan, USA.)

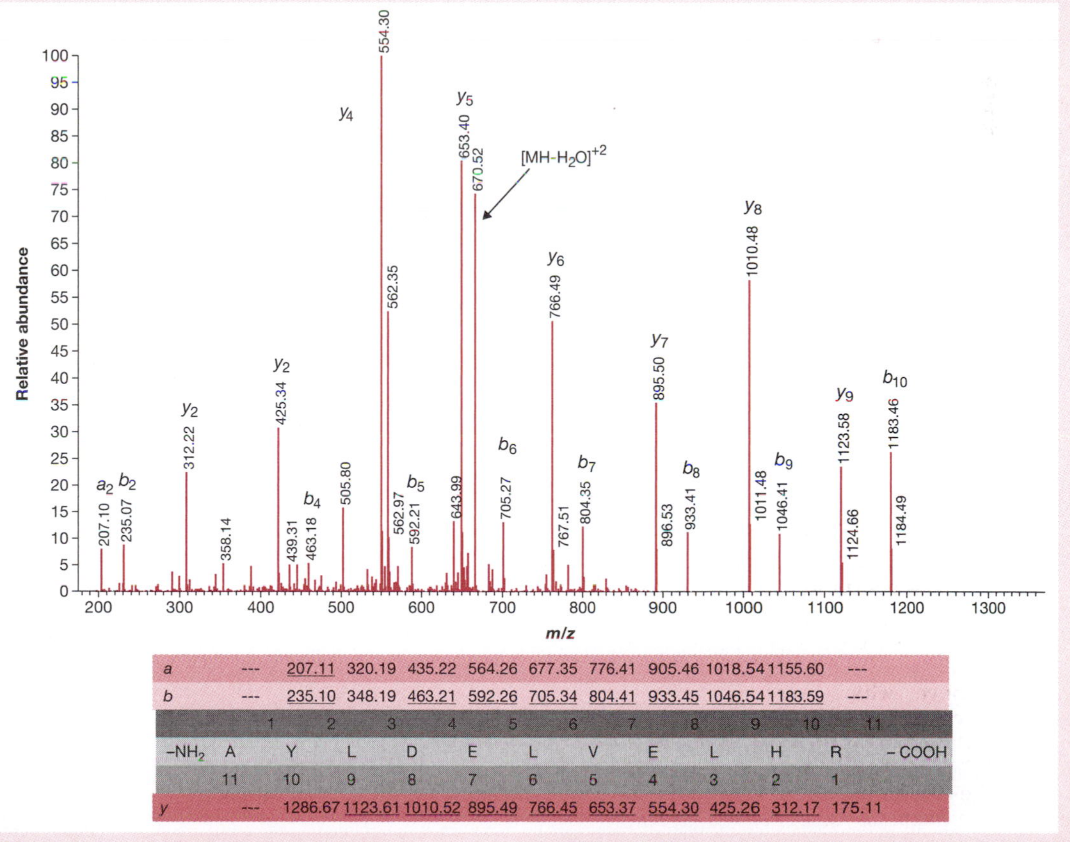

a	---	207.11	320.19	435.22	564.26	677.35	776.41	905.46	1018.54	1155.60	---	
b	---	235.10	348.19	463.21	592.26	705.34	804.41	933.45	1046.54	1183.59	---	
	1	2	3	4	5	6	7	8	9	10	11	
$-NH_2$	A	Y	L	D	E	L	V	E	L	H	R	–COOH
	11	10	9	8	7	6	5	4	3	2	1	
y	---	1286.67	1123.61	1010.52	895.49	766.45	653.37	554.30	425.26	312.17	175.11	

FIGURE 2 MS/MS spectrum of +2 charged peptide ion (AYLDELVELHR). The bottom panel shows the theoretical *a*/*b*-series and *y*-series ion masses expected for the peptide. There is a substantial difference in intensity (relative abundance) between the *b*-series and *y*-series ions. (Courtesy of Venkatesha Basrur, University of Michigan, USA.)

BOX 3.4 BACKGROUND ELEMENTS.
Ladder sequencing by mass spectrometry.

The *de novo* sequencing of peptides by mass spectrometry relies on the assembly of nested sets of N-terminal or C-terminal peptides (or peptide fragments) differing in length by a single residue. The differences in mass between consecutive members of a series can be compared with a standard table of amino acid residue masses to work out the sequence (**Table 1**). Note that two pairs of residues—glutamine and lysine, leucine and isoleucine—have very similar or identical masses. In the case of glutamine and lysine, there is a slight difference (128.13 and 128.17, respectively), which is very difficult to detect in fragment ion spectra but can be established quite easily when Edman chemistry is used, because the side chain of lysine is modified. Other difficulties with MS/MS spectra include the fact that two adjacent glycine residues (mass = 57) could be mistaken for a single asparagine residue (mass = 114) if the ion series is incomplete, whereas the complete ladder produced by conventional sequencing removes this ambiguity. Leucine and isoleucine have identical masses and can be distinguished by inspection of the corresponding cDNA sequence or by high-energy CID (see page 54).

TABLE 1 AMINO ACID RESIDUE MASSES

Residue	Chemical formula	1-letter code	3-letter code	Monoisotopic mass
Alanine	C_3H_5ON	A	Ala	71.03711
Arginine	$C_6H_{12}ON_4$	R	Arg	156.10111
Asparagine	$C_4H_6O_2N_2$	N	Asn	114.04293
Aspartic acid	$C_4H_5O_3N$	D	Asp	115.02694
Cysteine	C_3H_5ONS	C	Cys	103.00919
Glutamic acid	$C_5H_7O_3N$	E	Glu	129.04259
Glutamine	$C_5H_8O_2N_2$	Q	Gln	128.05858
Glycine	C_2H_3ON	G	Gly	57.02146
Histidine	$C_6H_7ON_3$	H	His	137.05891
Isoleucine	$C_6H_{11}ON$	I	Ile	113.08406
Leucine	$C_6H_{11}ON$	L	Leu	113.08406
Lysine	$C_6H_{12}ON_2$	K	Lys	128.09496
Methionine	C_5H_9ONS	M	Met	131.04049
Phenylalanine	C_9H_9ON	F	Phe	147.06841
Proline	C_5H_7ON	P	Pro	97.05276
Serine	$C_3H_5O_2N$	S	Ser	87.03203
Threonine	$C_4H_7O_2N$	T	Thr	101.04768
Tryptophan	$C_{11}H_{10}ON_2$	W	Trp	186.07931
Tyrosine	$C_9H_9O_2N$	Y	Tyr	163.06333
Valine	C_5H_9ON	V	Val	99.06841

FURTHER READING

Aebersold R & Mann M (2003) Mass spectrometry-based proteomics. *Nature* 422, 198–207.

Ahrens CH, Brunner E, Qeli E et al. (2010) Generating and navigating proteome maps using mass spectrometry. *Nat. Rev. Mol. Cell Biol.* 11, 789–801.

Bensimon A, Heck AJ & Aebersold R (2012) Mass spectrometry-based proteomics and network biology. *Annu. Rev. Biochem.* 81, 379–405.

Domon B & Aebersold R (2006) Mass spectrometry and protein analysis. *Science* 312, 212–217.

Griffen TJ, Goodlet DR & Aebersold R (2001) Advances in proteome analysis by mass spectrometry. *Curr. Opin. Biotechnol.* 12, 607–612.

Gstaiger M & Aebersold R (2009) Applying mass spectrometry-based proteomics to genetics, genomics and network biology. *Nat. Rev. Genet.* 10, 617–627.

Han X, Aslanian A & Yates JR 3rd (2008) Mass spectrometry for proteomics. *Curr. Opin. Chem. Biol.* 12, 483–490.

Holčapek M, Jirásko R & Lísa M (2012) Recent developments in liquid chromatography–mass spectrometry and related techniques. *J. Chromatogr. A* 1259, 3–15.

Maiolica A, Jünger MA, Ezkurdia I & Aebersold R (2012) Targeted proteome investigation via selected reaction monitoring mass spectrometry. *J. Proteomics* 75, 3495–3513.

Mann M, Hendrickson RC & Pandey A (2001) Analysis of proteins and proteomes by mass spectrometry. *Annu. Rev. Biochem.* 70, 437–473.

Mann M & Pandey A (2001) Use of mass spectrometry-derived data to annotate nucleotide and protein sequence databases. *Trends Biochem. Sci.* 26, 54–61.

Mueller LN, Brusniak MY, Mani DR & Aebersold R (2008) An assessment of software solutions for the analysis of mass spectrometry based quantitative proteomics data. *J. Proteome Res.* 7, 51–61.

Nesvizhskii AI (2007) Protein identification by tandem mass spectrometry and sequence database searching. *Methods Mol. Biol.* 367, 87–119.

Nesvizhskii AI (2010) A survey of computational methods and error rate estimation procedures for peptide and protein identification in shotgun proteomics. *J. Proteomics* 73, 2092–2123.

Uetrecht C, Rose RJ, van Duijn E et al. (2010) Ion mobility mass spectrometry of proteins and protein assemblies. *Chem. Soc. Rev.* 39, 1633–1655.

CHAPTER 4

Strategies for protein quantitation

4

4.1 INTRODUCTION

The objective of some proteomics experiments is simply to catalog the proteins that are found in a given sample, and this form of analysis can be described as predominantly descriptive. The comparison of related samples may reveal proteins that are present in one sample but not in another, due to the samples representing different cell types, developmental stages, and cell states (for example, stage of the cell cycle or in response to changes in the environment) so there may be numerous proteins uniquely expressed under particular circumstances. There are also proteomic changes associated with disease; for example, a comparison of normal skin and squamous cell carcinoma might reveal a set of protein spots unique to the disease. Once identified, these proteins could be useful as disease markers and might even represent potential new therapeutic targets (Chapter 10). However, there are very few proteins that show such unambiguous on/off changes. More often, the difference between samples is one of degree. Therefore, the accurate quantitation of proteins is now a vital aspect of proteomics.

There are several well-established methods for the quantitation of individual proteins, either in solution or using a solid-phase assay, which are based on the use of labeled antibodies (**Box 4.1**). The adaptation of such assays for proteomic analysis is difficult because even if antibodies could be found to bind to every protein in the proteome, the signal intensity for each antigen–antibody interaction would depend not only on the abundance of the target protein but also on the strength of the antigen–antibody binding (that is, the affinity of the antibody). Despite these technical hurdles, some analytical protein microarrays have been manufactured that are arrayed with thousands of antibodies, and these are described in Chapter 9. Generally, the most successful microarrays contain a small number of well-characterized antibodies. The more complex the device, the greater the problems with sensitivity and specificity, and the resulting quantitative data are less reliable.

Large-scale protein quantitation in proteomics relies primarily on the use of general labeling or staining, or on the selective labeling or staining of particular classes of proteins. There are various methods for measuring the total amount of protein in a solution (**Box 4.2**). However, it is necessary to compare the abundances of thousands of proteins in parallel across multiple samples in typical proteomics experiments. The chosen strategy depends largely on how the protein samples are prepared and fractionated, and can be divided into two broad categories: those based on the analysis of two-dimensional gel images and those based on comparing the abundance of ions across samples by mass spectrometry.

BOX 4.1 RELATED TECHNOLOGIES. Quantitation of individual proteins.

The quantity of an individual protein can be detected using **immunoassays** in which specific antibodies are used as labeled probes. The **western blot** or **immunoblot** is a convenient way to compare the abundances of gel-separated proteins. In this technique, proteins separated by gel electrophoresis, with different samples represented in different lanes, are blotted onto a sheet of polymeric material (usually nitrocellulose, nylon, or polyvinylidene difluoride) where they are immobilized. A general stain can be applied to reveal all the protein bands, for example, Ponceau S, silver nitrate, India ink, colloidal gold, or Amido black, but particular proteins are detected by flooding the sheet with a solution containing a specific antibody. The antibody may be conjugated to its own radioactive, fluorescent, or enzymatic label (**direct detection**) or a **secondary antibody** may be used that recognizes the primary antibody and therefore amplifies the signal (**indirect detection**, also known as a **sandwich assay**). This allows very small amounts of protein to be detected.

The accurate quantitation of individual proteins in solution can be achieved using a solid-phase immunoassay in which a capture antibody specific for the target protein is immobilized on a polymeric sheet or plastic dish. A drop of solution (for example, serum or cell lysate) containing the target protein is added to the sheet and antigen–antibody complexes are allowed to form. The solution is washed away and the target protein is detected with a second antibody that recognizes a different epitope to the capture antibody. As with the western blot, this detection antibody may be labeled or it may be recognized in a further reaction with a secondary antibody. The most popular version of this technique is the **enzyme-linked immunosorbent assay** (**ELISA**) in which the detection antibody carries an enzyme that converts a colorless substrate into a colored compound, or a nonfluorescent substrate into a fluorescent compound.

BOX 4.2 RELATED TECHNOLOGIES. Measuring the total protein concentration of a solution.

There are several methods for determining the total concentration of proteins in solution, each of which exploits properties that are general to all proteins. One widely used method is the measurement of UV (ultraviolet) absorbance. This is a nondestructive method, allowing the proteins to be recovered for further analysis. Therefore, it is used not only for quantitation but also to detect protein and peptide fractions eluting from HPLC columns. The UV light is absorbed by aromatic amino acid residues (tyrosine and tryptophan) as well as by peptide bonds. It cannot be used in two-dimensional gels because polyacrylamide absorbs UV light over the same range of wavelengths.

Other protein assay methods are colorimetric or fluorometric and are based on covalent or noncovalent dye binding or chemical reactions. The **Bradford assay** measures the degree of binding to Coomassie Brilliant Blue dye, which changes color from brown to blue in the presence of proteins. Other protein-binding agents are fluorescent and more sensitive, for example, OPA (*o*-phthaldialdehyde), fluorescamine, and NanoOrange. The **Lowry assay** and the related **BCA** (**bicinchoninic acid**) **assay** are based on the reduction of copper ions in the presence of proteins, resulting in the chelation of a colorless substrate and the production of a colored complex that can be detected using a spectrophotometer.

4.2 QUANTITATIVE PROTEOMICS BASED ON 2DGE

The quantitation of proteins in two-dimensional gels involves the creation of digital data from analog images

The abundance of different proteins on a two-dimensional gel is reflected by the shape, size, and intensity of the corresponding spots. Assuming that spots are well resolved, protein quantitation requires the conversion of an analog gel image into digital data, resulting in a catalog of individual spots listed as *x*, *y* positions, shape parameters, and quantitative values (**integrated spot intensities**). It is then possible to carry out objective comparisons of equivalent spots on different gels and thus to determine whether a particular protein is more or less abundant in one sample compared with another. It can be difficult to reproduce the exact conditions for protein separation in 2DGE, so the identification of corresponding spots even on two-dimensional gels containing the same original sample can be a challenge. Robust methods are therefore required for the analysis of gels representing different samples if many of the spots differ in abundance and some spots are present on one gel and absent on another.

The first stage in protein quantitation is **image acquisition**, and the method used depends on how the proteins were labeled or stained. Radioactively labeled proteins can be detected on an X-ray film or by phosphorimaging. The X-ray film may then be scanned by a CCD camera or a densitometer, whereas phosphorimagers come with their own scanning devices. A **charge-coupled device** (**CCD**) is simply a solid-state electrical component that is divided into a series of light-sensitive areas or photosites composed of a material that emits electrons when struck by a photon of light. The image from a CCD camera is generated by a microprocessor that counts the electrons at each photosite. A **densitometer** is a scanning device that works on a similar principle, that is, light reflected from or transmitted through the surface of a film is detected by a photodiode, which thus records the density of the light and dark areas on the image. Coomassie-stained and silver-stained gels may also be scanned with a CCD camera or densitometer, whereas gels stained with the fluorescent reagents or gels containing fluorescently labeled proteins may be scanned using a CCD camera or a fluorescence imager.

The quality of the digital data depends critically on the resolution of the scanned image, which can be considered in terms of both spatial resolution (expressed as pixels per unit length or area) and **densitometric resolution** (that is, the range of gray values that can be interpreted). However, the densitometric resolution also depends on the labeling or staining method. **Silver staining** was the major non-radioactive detection method used for separated proteins because it is 10–100 times more sensitive than Coomassie Brilliant Blue. However, silver stains do not detect glycoproteins very efficiently and the most sensitive detection methods lead to chemical modification of cysteine residues, thereby interfering with downstream analysis by mass spectrometry (this reflects the use of formaldehyde for stain development, and its replacement has helped to increase the compatibility between silver staining and MS). In terms of comparative protein quantitation, the major disadvantage of silver staining is its narrow linear range (about one order of magnitude). This means that it is possible to accurately determine whether one protein is twice as abundant as another (or more importantly, if one protein is twice as abundant in one sample compared with another), but it is not possible to accurately compare protein abundance if there is a tenfold or greater difference. Fluorescent stains such as SYPRO Ruby, Deep Purple, and Flamingo are now strongly preferred as these are at least as sensitive as silver staining but share none of its disadvantages. That is, they detect glycoproteins efficiently, they do not cause any covalent protein modifications, and they have an extensive linear range (over three orders of magnitude), which means they can be used to compare protein abundances very effectively. The different ways for detecting proteins in two-dimensional gels are summarized in **Box 4.3**.

Spot detection, quantitation, and comparison can be challenging without human intervention

Spots on protein gels are not uniform in shape, size, or density. Some spots appear as discrete entities while others overlap to a greater or lesser degree. The edges of some spots are clearly defined while those of others may be blurred. Small spots may appear as shoulders on larger ones, or several spots may be joined together in a line. The densitometric landscape within different spots (that is, the distribution of gray values) is not always consistent. These variations may be compounded by nonspecific changes in the gel background.

BOX 4.3 BACKGROUND ELEMENTS.
Detecting proteins *in situ* in gels.

The *in situ* detection of proteins within gels can be achieved by labeling the proteins prior to electrophoresis or staining them after electrophoresis. In both cases, it is important to make the procedure as sensitive as possible without interfering with downstream analysis by mass spectrometry (Chapter 3).

Pre-labeling with organic fluorophores

A number of different organic molecules can be covalently attached to proteins prior to electrophoretic separation, allowing the direct detection and quantitation of labeled proteins within two-dimensional gels. Methods utilizing well-characterized fluorophores such as fluorescamine and fluorescein isothiocyanate have been available since the 1970s, but in the context of proteomics these methods have a number of drawbacks, including the altered solubility and/or mobility of labeled proteins and the variable sensitivity of labeling depending on the number of functional groups available for modification. However, the use of two or more different fluorophores, for example propyl-Cy3 and methyl-Cy5, to label different protein samples allows the abundance of proteins in the samples to be compared on the same gel (difference in-gel electrophoresis, see p. 75.

Coomassie Brilliant Blue

Coomassie Brilliant Blue is an organic dye that is commonly used to stain proteins in polyacrylamide gels. There are many variations on the staining protocol but staining is generally carried out using the dye in a mixture of concentrated acid with ethanol or methanol. This produces a colloidal suspension that stains proteins strongly with low background. Although widely used for general protein analysis, Coomassie Brilliant Blue and related organic dyes lack the sensitivity for proteomic analysis, having a detection limit of 10–30 ng. Depending on the exact make-up of the stain, the dye can also modify glutamic acid side chains, which can complicate the interpretation of mass spectrometry data (although adjustments to the search criteria can accommodate this).

Silver staining

Silver staining is one of the most popular techniques for staining proteins in polyacrylamide gels and many different protocols have been used. The best methods are about 10 times more sensitive than Coomassie staining and although silver stains can modify cysteine residues and alkylate exposed amino groups in formaldehyde buffers, the use of different buffers addresses this problem. Other disadvantages of silver staining include the poor linear dynamic range, which makes quantitative analysis problematical, and the fact that certain types of protein, including many glycoproteins, stain rather poorly.

Reverse stains

Both Coomassie and silver staining methods involve a protein fixing step, which reduces the recovery of protein from the gel for subsequent analytical steps. Reverse stains, which stain the gel rather than the proteins and generate a negative image, were developed to enhance the recovery of proteins from such gels. Many different formulations have been used, the most popular based on metal salts such as copper chloride or zinc chloride.

Fluorescent stains

A number of fluorophores are known to bind noncovalently to proteins, which makes them particularly compatible with downstream mass spectrometry analysis. These stains generally demonstrate little fluorescence in aqueous solution but fluoresce strongly when associated with SDS-protein complexes, and therefore produce a very low background in stained gels. The most versatile of these molecules include SYPRO Ruby (Invitrogen), Lucy (Sigma-Aldrich), Deep Purple (GE Healthcare), Krypton (Pierce), and Flamingo (Bio-Rad). These agents are very sensitive and show a broad linear dynamic range. SYPRO Ruby is one of the most widely used, and matches the sensitivity of the best silver staining techniques but has a superior linear dynamic range (extending over three orders of magnitude) and stains proteins that do not show up well with silver stains, for example many glycoproteins. The staining protocol is also simple and rapid, unlike Coomassie and silver staining techniques, which each require a lengthy de-staining step. The proprietary fluorescent dyes can be expensive but compounds with similar properties can be synthesized in the laboratory.

The human eye can generally tell the difference between a spot and background artifacts on a two-dimensional gel but humans are too subjective in their judgment to define spots rigorously. Machines can apply a fixed set of rules and parameters to the definition of individual spots and therefore interpret spot patterns more objectively. However, getting machines to see the spots in the same way that humans do can be challenging. Normally, the first stage in automated spot detection is digital image enhancement, which helps to clear the background and improve the contrast of the image to make the spot boundaries easier to delineate. Smoothing is used to eliminate variable background noise and the background is then subtracted from the rest of the image. The contrast in the subtracted image is enhanced by reassigning gray values from the mid-range to make the pixels either darker or lighter. In many cases, **edge detection filters** are used that aim to identify regions of the image in which there is a sharp change in pixel intensity.

Once a processed image is available, a number of different algorithms can be applied to detect and quantitate individual spots. These must take all the possible variations in spot morphology into account and calculate the integrated spot intensities, which are essentially absolute values that represent protein abundances. The algorithms generally use either **Gaussian fitting** (which assumes the gray values in the spot have a normal distribution along both the x and y axes) or **Laplacian of Gaussian (LOG) spot detection** methods. Other algorithms are based on the **watershed transformation method** in which a grayscale image is converted into a topographic surface with darker sections representing peaks and lighter sections representing troughs. The idea is then to "flood" the image from the minima, which divides the image into catchment basins representing individual spots and watershed lines representing divisions (**Figure 4.1a** and color plates). In practice, the indiscriminate flooding of gel images in this manner leads to over-segmentation due to background variation in pixel intensity (Figure 4.1a and color plates). To avoid this outcome, flooding can be initiated from a previously defined set of markers, which avoids any over-segmentation (Figure 4.1b and color plates). Another useful method is **line analysis** in which the computer focuses on individual vertical scan lines to identify density peaks. The density peaks in adjacent scan lines can be assembled into chains and these represent the centers of spots.

Once the two-dimensional gel has been reduced to a series of digital data representing spot intensities, the comparison of different gels is a simple process of comparing data values and determining whether the abundance of a given protein differs significantly, according to some predefined threshold, among two or more samples. A prerequisite for this type of analysis is the identification of equivalent spots on different gels, which may be challenging because gel-running conditions cannot be reproduced exactly. This may be due to several factors:

- Differences in sample preparation.
- Differences in gel composition. This can be minimized by preparing several gels from the same mixture at the same time, or by using commercially available pre-cast gels.
- Variations in running conditions. As discussed in Chapter 2, this is a significant problem for isoelectric focusing (IEF) gels with carrier ampholytes, particularly non-equilibrium gels, but it also applies to a lesser degree to immobilized pH gradient gels. The problems can be addressed to some extent by running several gels in parallel, but this is not always possible.

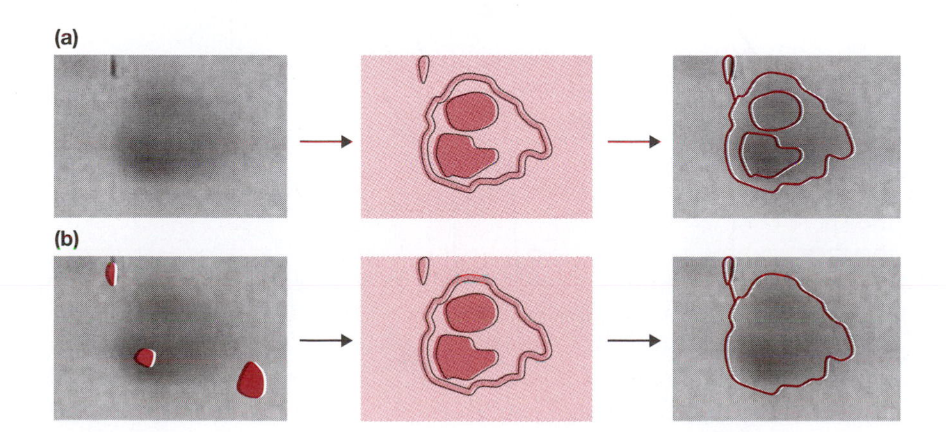

FIGURE 4.1 The watershed method for contour finding on two-dimensional gel images. (a) Any grayscale image can be considered as a topographic surface. If flooded from its minima without allowing water from different sources to merge, the image is partitioned into catchment basins and watershed lines, but in practice this leads to over-segmentation. (b) Therefore, markers (*bright red shapes*) are used to initiate flooding, and this reduces over-segmentation considerably. See also color plates. (Adapted from images by Serge Beucher, CMM\École Nationale Supérieure des Mines de Paris.)

- Minor variations within each gel that lead to regional differences in protein mobility. Again this is a major problem with carrier ampholyte IEF gels, but also applies to others.

In the absence of gels, or images thereof, showing perfect spot-to-spot correspondence, it becomes necessary to force equivalent gels into register, a process known as **gel matching**. This process makes use of **landmarks**, that is, spots that are present on all gels in the comparison and can be used as a common frame of reference. Gel matching algorithms then apply image transformation procedures such as stretching, skewing, and rotating, at both local and global levels, to bring multiple gel images into register and make them comparable. This can be thought of as a procedure in which several equivalent gels are stacked above each other and a pin is used to pierce the center of the first landmark spot through all the stacked gels. Further pins are inserted through other landmarks. When the gels are held in position by a number of pins, flexible wires can be inserted to link equivalent spots that are not perfectly in register (Figure 4.2). In some gels, a given spot may be absent, but with a number of matched landmarks surrounding the space, the algorithm can assign a zero value to the spot with reasonable confidence. As an alternative to matching gels at the spot level, other algorithms perform essentially the same task at the pixel level. An extension to the use of landmarks is a gel matching method known as **propagation**. In this approach, the algorithm begins at a known landmark and then maps the nearby spots and returns a list of x, y displacement values. Other gels are scrutinized for spots at the same displacements relative to the landmark and matches are identified. These matches can then be used as new landmarks for recursive searching.

The end result of spot detection, quantitation, and gel matching should be a table of spot values (x, y coordinates, shape parameters, and integrated spot intensities) arranged as an $N \times M$ matrix where N represents all the different spots that have been identified and M represents all the gels (Figure 4.3). M should be divided into groups based on the experimental conditions. For example, M_1, ..., M_{15} might represent five control gels, five from

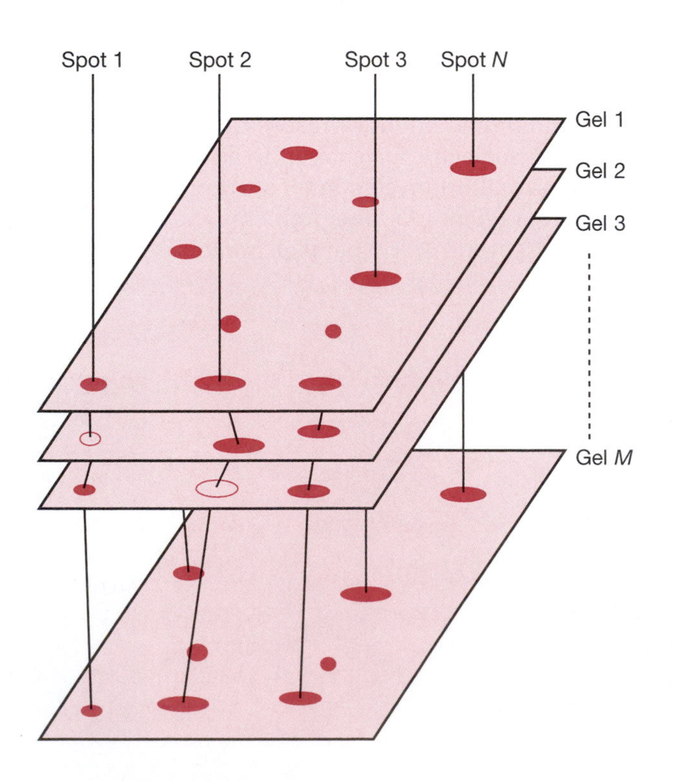

FIGURE 4.2 **Principle of spot matching to identify corresponding spots on multiple gels.** Empty circles represent absent spots.

	Control group			Experimental group		
Gel / Spot	Gel 1	Gel 2	■ ■ ■	■ ■ ■	Gel $M-1$	Gel M
Spot 1	$I_{1,1}$	$I_{1,2}$			$I_{1,M-1}$	$I_{1,M}$
Spot 2	$I_{2,1}$	$I_{2,2}$			$I_{2,M-1}$	$I_{2,M}$
■ ■ ■ ■						
Spot N	$I_{N,1}$	$I_{N,2}$			$I_{N,M-1}$	$I_{N,M}$

FIGURE 4.3 A generic data analysis matrix containing integrated spot densities (I). N is the number of spots (*rows*) and M is the number of gels (*columns*).

experimental condition 1, and five from experimental condition 2 (perhaps different stages of a disease or different time points after drug administration). The quantitative values must be normalized for any differences in the overall signal intensities on the gels (for example, due to different exposure times) and then various statistical methods can be used to identify protein spots whose abundance varies over the experimental conditions. Recent developments in proteomic gel imaging technology allow matched gels to be overlain in false color so that protein spots with differential abundance over two or more gels can be visually identified. This is essentially an artificial method for generating difference gel electrophoresis data (see below) from samples separated on different gels.

4.3 MULTIPLEXED IN-GEL PROTEOMICS

Multiplexed proteomics in the context of 2DGE is the use of fluorescent stains or probes with different excitation and emission spectra to detect different groups of proteins simultaneously on the same gel. This helps to reduce the number of duplicate gels that are required to compare different proteins and, at least in theory, obviates the need for gel matching to identify corresponding proteins. Gel matching is necessary because the staining methods discussed above are intrinsically limited to a single-color display.

Difference in-gel electrophoresis involves the simultaneous separation of comparative protein samples labeled with different fluorophores

In Chapter 1, we discussed the comparative analysis of mRNA levels in different samples by labeling each population with a different fluorophore and hybridizing both populations simultaneously to the same DNA microarray. By scanning the microarray twice, at the emission wavelengths of each fluorophore, it is possible to determine the relative abundance of different mRNAs within each sample and the relative abundance of the same mRNA between samples. The signals can be rendered in false color and combined to provide a composite image that immediately identifies differentially expressed genes.

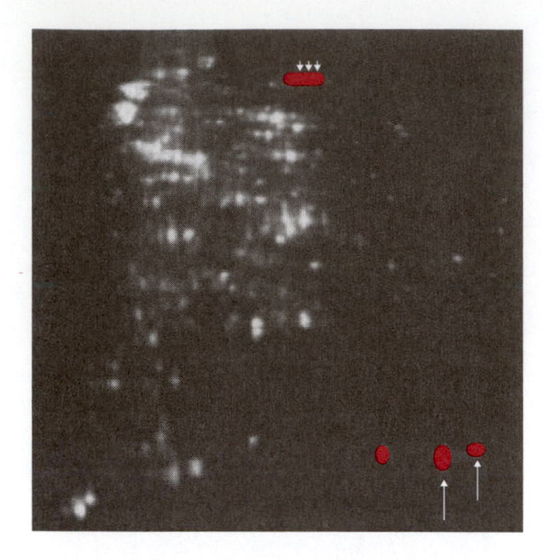

FIGURE 4.4 Two-dimensional DIGE. Overlay image of Cy3- (*green*) and Cy5- (*red*) labeled test-spiked *Erwinia carotovora* proteins. The protein test spikes were three conalbumin isoforms (*arrowheads*) and two myoglobin isoforms (*arrows*). Spots that are of equal intensity between the two channels appear *white* in the overlay image. As spike proteins were eight times more abundant in the Cy5 channel, they appear as *red* spots in the overlay. The gel is oriented with the acidic end to the left. See also color plates. (From Lilley KS, Razzaq A & Dupree P (2002) *Curr. Opin. Chem. Biol.* 6, 46. With permission from Elsevier.)

Difference in-gel electrophoresis (**DIGE**) is an analogous method in proteomics based on 2DGE (see Chapter 2). The protein extracts from related samples (for example, healthy versus diseased tissue or stimulated versus unstimulated cells) are labeled on lysine side chains with succinimidyl esters of propyl-Cy3 and methyl-Cy5, two cyanine family fluorophores with the same mass and charge but different emission wavelengths. The protein samples are mixed prior to separation and loaded onto the IEF gel for separation in the first dimension, then transferred to an SDS-polyacrylamide gel for orthogonal separation. After electrophoresis, the gel is scanned using a CCD camera or fluorescence reader fitted with two different filters and two sets of data are obtained. The images from each filter can be pseudocolored and combined, immediately revealing the spots representing proteins whose abundance differs across the sample (**Figure 4.4** and color plates). The use of further labels, for example Cy2, can allow even more samples to be run concurrently. Because the samples run together, all potential differences in gel preparation, running conditions, and local gel structure are eliminated.

DIGE has many advantages in terms of simplified data analysis, but the technique also has several drawbacks. The fluorescent labels are less sensitive than both SYPRO dyes and silver staining. This primarily reflects the fact that only a small proportion of the proteins in each sample can be labeled before solubility is compromised such that the proteins precipitate during electrophoresis. A further consequence of partial labeling is that the bulky fluorescent conjugate reduces the mobility of the proteins during SDS-PAGE so the gels must be post-stained, for example with Coomassie Brilliant Blue, to identify the "true" protein spot to be excised for downstream analysis by mass spectrometry. Such registration errors between the labeled and unlabeled protein populations are minimized during isoelectric focusing because the dyes carry a single positive charge that replaces the positive charge on the lysine side chain to which they bind and thus the labeled and unlabeled proteins have the same pI. Alternative cyanine reagents with a maleimide reactive group are designed to bind covalently to the thiol group of cysteine residues via a thioether linkage.

Accurate protein quantitation may be difficult because proteins differ in their labeling efficiency, solubility when conjugated to the label, and the extent to which they might exhibit **quenching** (a phenomenon in which there is energy transfer between two fluorophores that are close together on the same molecule, thus preventing the emission of light). Therefore, bright spots and dim spots may represent abundant and scarce proteins, or may represent proteins that are present at approximately the same level but show differential labeling efficiency or quenching effects.

Parallel analysis with multiple dyes can also be used to identify particular structural or functional groups of proteins

The sensitivity of standard gels can be combined with the convenience of multiplex fluorescence by using fluorescent reagents such as SYPRO Ruby to stain and compare protein spots on different gels plus more selective reagents that identify specific classes of proteins. These proteins can be used as landmarks for gel matching or to identify subsets of proteins that share specific structural or functional attributes. A number of stains have been developed that recognize various structurally or functionally related proteins: for example, glycoproteins and phosphoproteins (these are discussed in more detail in Chapter 8), oligo-histidine tagged proteins, calcium-binding proteins, and even proteins that have the capability to bind or metabolize particular drugs (see Chapter 9). For example, penicillin analogs have been produced carrying BODIPY dyes, which are relatively nonpolar and have a neutral chromophore and therefore do not interfere with the structure or chemical behavior of the antibiotic. These so-called BOCILLIN reagents can efficiently identify penicillin-binding proteins on a two-dimensional gel with SYPRO Ruby used as a general counterstain. Similarly, BODIPY dyes have been used to generate analogs of the cysteine protease inhibitor *trans*-epoxysuccinyl-L-leucylamido(4-guanidino)butane, thus allowing cysteine proteases to be identified and changes in their expression levels following different types of cell treatment to be investigated.

4.4 QUANTITATIVE MASS SPECTROMETRY

Label-free quantitation may be based on spectral counting or the comparison of signal intensities across samples in a narrow *m*/*z* range

Because sample processing, separation, and transfer to the mass spectrometer are generally automated, quantitative data can only be obtained from LC-MS and LC-MS/MS experiments by determining the abundance of different proteins from their mass spectra. **Quantitative mass spectrometry** is often achieved by comparing samples that have been labeled with alternative discriminatory **mass tags**, but it is also possible to achieve an accurate quantitative comparison between unlabeled samples. The principles of the available strategies are summarized in **Table 4.1** and their relative merits and drawbacks are compared in **Table 4.2**. However, it is important to emphasize that the vast majority of quantitative proteomics experiments provide relative rather than absolute quantitative data, which makes proteomics as a research approach distinct from the use of similar methods to measure protein levels in body fluids, for example (see Chapter 10).

The first label-free approach is **spectral counting**, which is conceptually similar to the census sequencing approach discussed in Chapter 1. The basis of this approach is that the number of recorded spectra corresponding to a particular peptide correlates with the abundance of that peptide in the original sample. This is a sensitive method for detecting differentially expressed proteins, although precise quantitation is affected by peptide ionization and fragmentation characteristics and becomes less accurate in the case of scarce peptides. The other major label-free strategy is based on the measurement of precursor ion signal intensity, which can be achieved in standard MS experiments by isolating *m*/*z* values representing one or more analytes of interest from a standard chromatogram. This is known as the **extracted ion chromatogram (XIC) method**. More sensitive quantitation is possible by preselecting the ions for analysis. This can be achieved by **selected ion monitoring (SIM)** in MS instruments, in which the *m*/*z* values for analysis are selected and only this restricted *m*/*z* range appears in the dataset. The preferred method is **selected reaction monitoring (SRM)** in triple quadrupole instruments using **transition pairs** (precursor and product ions) because the latter does not require full mass spectra to be recorded. By carrying out multiple SRM experiments (**multiple reaction monitoring, MRM**) and spiking the sample with isotopically labeled peptides as concentration standards, MRM can be used to construct a calibration curve that achieves absolute rather than relative quantitation.

Label-based quantitation involves the incorporation of labels that allow corresponding peptides in different samples to be identified by a specific change in mass

Quantitative proteomics is often based on the incorporation of **stable isotopes** or **mass tags** into different samples, allowing equivalent peptides (or peptide fragments) to be identified by a specific increase in mass. The general approach is to label alternative protein or peptide samples with equivalent

TABLE 4.1 PRINCIPLES OF QUANTITATIVE MASS SPECTROMETRY METHODS

Method	Principle	Comments
Label-free methods		
Spectral counting	Counting the frequency of particular mass spectra	Unreliable for rare peptides
Precursor ion peak intensity (preferably SRM)	Direct comparison of peaks between spectra	Sensitive to instrument accuracy. More reliable in FT-ICR and Orbitrap mass analyzers
Selective labeling of proteins and peptides in vitro		
ICAT	Cysteine residues labeled with isotopic/normal mass tags containing reactive iodoacetamide (or acrylamide) groups. Carries biotin tag for affinity purification	Purification of cysteine-peptides simplifies analysis but presence of biotin complicates it (addressed by the development of a cleavable derivative). Restricts analysis to cysteine-containing proteins
Nonselective labeling of proteins and peptides in vitro		
ICPL	Derivatization of amines with isotopic/normal chemical groups	Labeling can be inefficient
MCAT	Derivatization of amines with *O*-methylisourea in one peptide population only	Inexpensive because stable isotopes are not used, but lacks accuracy
Proteolysis	Proteolysis in presence of ^{18}O incorporates isotope into peptide carboxyl groups	Theoretically labels all peptides except C-terminal one, but incorporation often incomplete. Post-digestion labeling with immobilized trypsin is more efficient
TMT, iTRAQ	Derivatization of amines with isobaric mass tags	Eliminates quantitation problems caused by peak overlaps in first mass spectrum
Nonselective labeling of proteins in vivo		
SILAC	Incorporation of isotopically labeled amino acids during metabolic activity	Corrects for preparation artifacts but only applicable to microbes and cultured cells

TABLE 4.2 COMPARISON OF QUANTITATIVE MASS SPECTROMETRY METHODS

Method	Application	Accuracy	Quantitative coverage	Linear dynamic range[a]
Metabolic protein labeling	Complex biochemical workflows Comparison of 2–3 states Cell culture systems only	+++	++	1–2 logs
Chemical protein labeling (MS)	Medium to complex biochemical workflows Comparison of 2–3 states	+++	++	1–2 logs
Chemical peptide labeling (MS)	Medium-complexity biochemical workflows Comparison of 2–3 states	++	++	2 logs
Chemical peptide labeling (MS/MS)	Medium-complexity biochemical workflows Comparison of 2–8 states	++	++	2 logs
Enzymatic labeling (MS)	Medium-complexity biochemical workflows Comparison of 2 states	++	++	1–2 logs
Spiked peptides	Medium-complexity biochemical workflows Targeted analysis of few proteins	++	+	2 logs
Label-free (ion intensity)	Simple biochemical workflows Whole-proteome analysis Comparison of multiple states	+	+++	2–3 logs
Label-free (spectrum counting)	Simple biochemical workflows Whole-proteome analysis Comparison of multiple states	+	+++	2–3 logs

[a]In MRM mode, dynamic range may be extended to 4–5 logs.

From Bantscheff M, Schirle M, Sweetman G et al. (2007) *Anal. Bioanal. Chem.* 389, 1017-1031. With permission from Springer Science and Business Media.

reagents, one of which contains a heavy isotope and one of which contains a light isotope, or one of which contains a heavy mass tag and the other a light mass tag (or no tag at all). The samples are mixed, fractionated, and analyzed by mass spectrometry. The ratio of the two isotopic or mass tag variants can be determined from the peaks in the mass spectra and used to identify proteins that differ in abundance. Several variants of the approach can be used, which are discussed below and summarized in **Figure 4.5**.

ICAT reagents are used for the selective labeling of proteins or peptides

One of the first developments in quantitative mass spectrometry was a class of reagents known as **isotope-coded affinity tags (ICATs)**. These are biotinylated derivatives of iodoacetamide (and later acrylamide) both of which react with the cysteine side chains of denatured proteins. Originally, the reactive group and biotin were joined by a linker that was available in two versions, one normal or light form and one heavy or **deuterated** form in which hydrogen atoms were replaced by deuterium. The heavy and light forms were used to label different protein samples and then the proteins were combined and digested with trypsin. The biotin allowed cysteine-containing peptides to be isolated from the complex peptide mixture through affinity to streptavidin, thereby considerably simplifying the peptide mixture and reducing the number of different peptides introduced into the mass spectrometer (**Figure 4.6**).

The original deuterated ICAT reagents were prone to partial peak separation during chromatography and the presence of the biotin group interfered with database searching. Therefore, a new cleavable ICAT reagent was introduced in which the biotin could be removed by acid treatment before mass spectrometry, and the heavy version incorporated ^{13}C rather than deuterium. A solid-phase cleavable ICAT reagent has also been developed containing a photolabile linker arm so that cysteine-containing peptides from a complex mixture can be captured onto small plastic beads and then released by exposure to light. However, the main drawback of ICAT reagents that bind cysteine residues is that approximately 10% of proteins do not contain cysteine and are excluded from subsequent analysis.

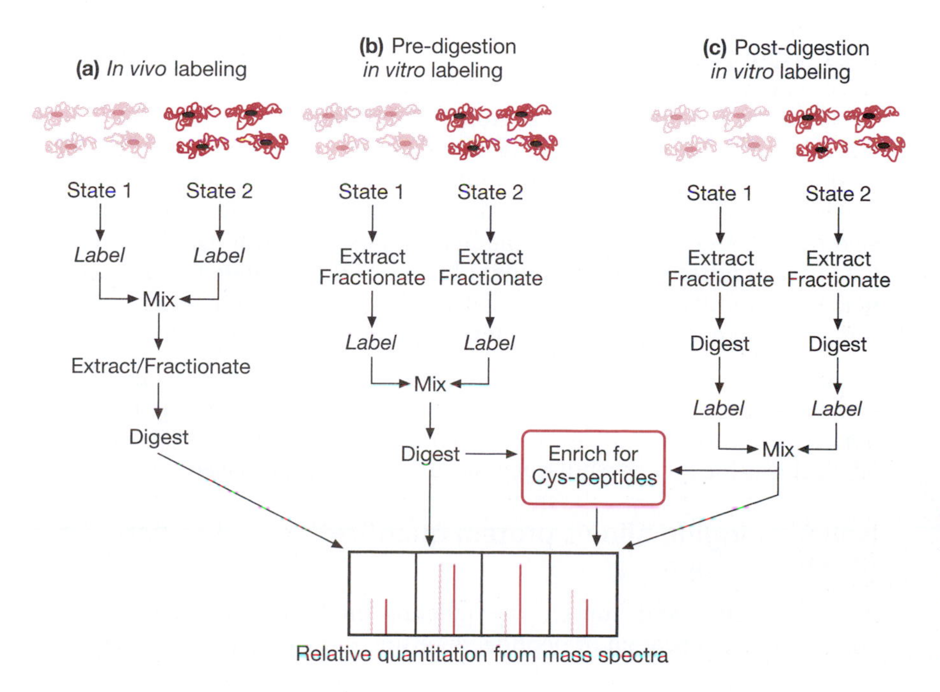

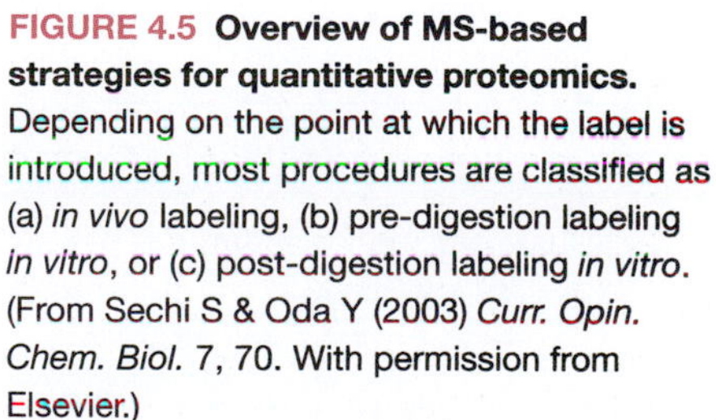

FIGURE 4.5 Overview of MS-based strategies for quantitative proteomics. Depending on the point at which the label is introduced, most procedures are classified as (a) *in vivo* labeling, (b) pre-digestion labeling *in vitro*, or (c) post-digestion labeling *in vitro*. (From Sechi S & Oda Y (2003) *Curr. Opin. Chem. Biol.* 7, 70. With permission from Elsevier.)

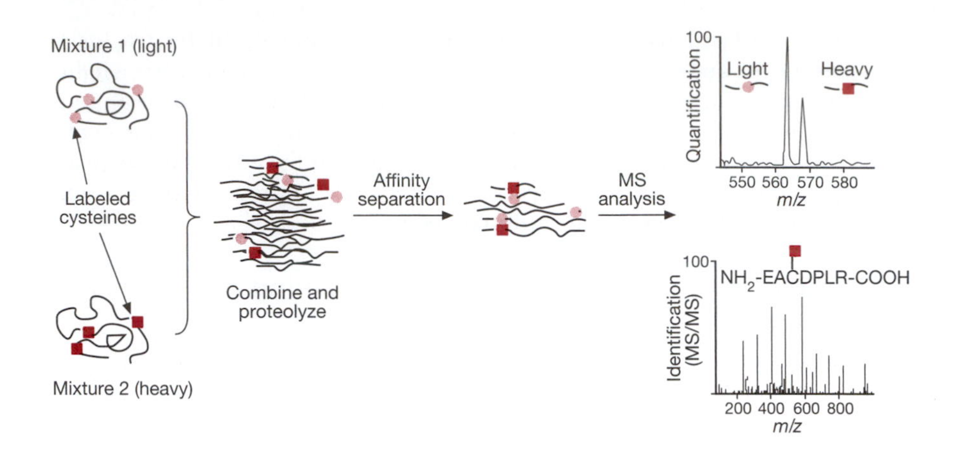

FIGURE 4.6 The ICAT reagent strategy for protein quantitation. Two protein mixtures representing two different cell states are treated with the isotopically light (*pink*) or heavy (*red*) ICAT reagents, respectively. The labeled protein mixtures are then combined and proteolyzed; tagged peptides are selectively isolated and analyzed by MS. The relative abundance is determined by the ratio of signal intensities of the tagged peptide pairs. The CID spectra are recorded and searched against large protein sequence databases to identify the protein. Therefore, in a single operation, the relative abundance and sequence of a peptide are determined. (From Tao WA & Aebersold R (2003) *Curr. Opin. Biotechnol.* 14, 110. With permission from Elsevier.)

Proteins and peptides can also be labeled nonselectively

More versatile systems have been introduced that allow nonselective protein and peptide labeling. For example, **isotope-coded protein labeling** (**ICPL**) is similar in principle to ICAT labeling but in this case the reagent labels lysine side chains by taking advantage of the ability of *N*-hydroxysuccinimide (NHS) ester derivatives to fully derivatize primary amino groups in intact proteins. ICPL reagents also have variants with different numbers of deuterium atoms to allow multiplex quantitative analysis. Similar approaches include the specific labeling of the exposed N-termini of peptides with alternative light and heavy versions of chemicals such as *N*-acetoxysuccinimide and 1-nicotinoyloxysuccinimide. There are also methods that enrich for the N-termini of proteins, allowing global analysis of the **N-terminome** (**Box 4.4**).

Nonselective labeling can also be achieved by taking advantage of the catalytic properties of proteases such as trypsin. As discussed in Chapter 3, when trypsin cleaves a protein and generates peptides, it uses oxygen atoms derived from water to create the new carboxyl group of each peptide C-terminus. This reaction can be exploited to identify *y*-series ions in fragment ion spectra (see p. 62), but it can also be used to differentially label peptides derived from alternative protein samples if normal water is used in one buffer and water substituted with heavy oxygen (^{18}O) is used in the other (**Figure 4.7**). The abundance of the peptides can then be compared, since they will appear as doublets separated by four mass units (although the C-terminal peptide of each protein is not labeled and no discrimination is possible). Whereas labeling is concurrent with digestion in the above method, it is also possible to uncouple the reactions and label the peptides after digestion by incubating the already digested peptides with immobilized trypsin and $H_2{}^{18}O$. The advantages of post-digestion labeling include the lower requirement for isotopic substrate and the ability to optimize the reaction to reduce incomplete oxygen exchange (that is, where only one ^{18}O is incorporated instead of two, generating a more complex spectrum).

The use of isotopes is avoided with the **mass-coded abundance tag** (**MCAT**) system in which the primary amine groups of one population of peptides are derivatized with *O*-methylisourea and the other population is left without a label. This is inexpensive but not as accurate as isotope-based methods.

Isobaric tagging allows protein quantitation by the detection of reporter ions

All the selective and nonselective labeling methods above generate two versions of each protein or peptide, differing in mass by a specific amount.

This produces two peaks on the first mass spectrum, and, depending on the resolution of the mass analyzer, it may be difficult to achieve accurate quantitation because the peaks may overlap to a greater or lesser degree. **Isobaric tagging** means the labeling of proteins or peptides with chemical groups that are the same in mass, so that proteins from both samples behave in the same manner during fractionation and mass spectrometry, generating a single peak in the first mass spectrum. However, the reagents

BOX 4.4 ALTERNATIVE APPLICATIONS.
Terminal amine isotopic labeling of substrates (TAILS)

TAILS is a high-throughput proteomics approach that is useful for the quantitative analysis of N-terminal peptides. There are several variants of the method, but all involve the uniform labeling of exposed amines (N-terminal amines and lysine side chains) followed by the negative selection of blocked N-terminal peptides. The use of differential labeling allows quantitative comparison between samples, and is highly useful for the analysis of protease targets by comparing the N-termini before digestion and the neo-N-termini afterwards. This method is superior to those relying on chemical modification and/or biotinylation (which do not provide reliable quantitative data) and to **combined factional diagonal chromatography** (**COFRADIC**), which requires multiple chemical processing steps before separation and analysis. The dimethylation-TAILS method only allows pairwise comparisons, but the method can be combined with SILAC or iTRAQ labeling (see p. 82) to increase the number of different tags that can be used simultaneously (Figure 1).

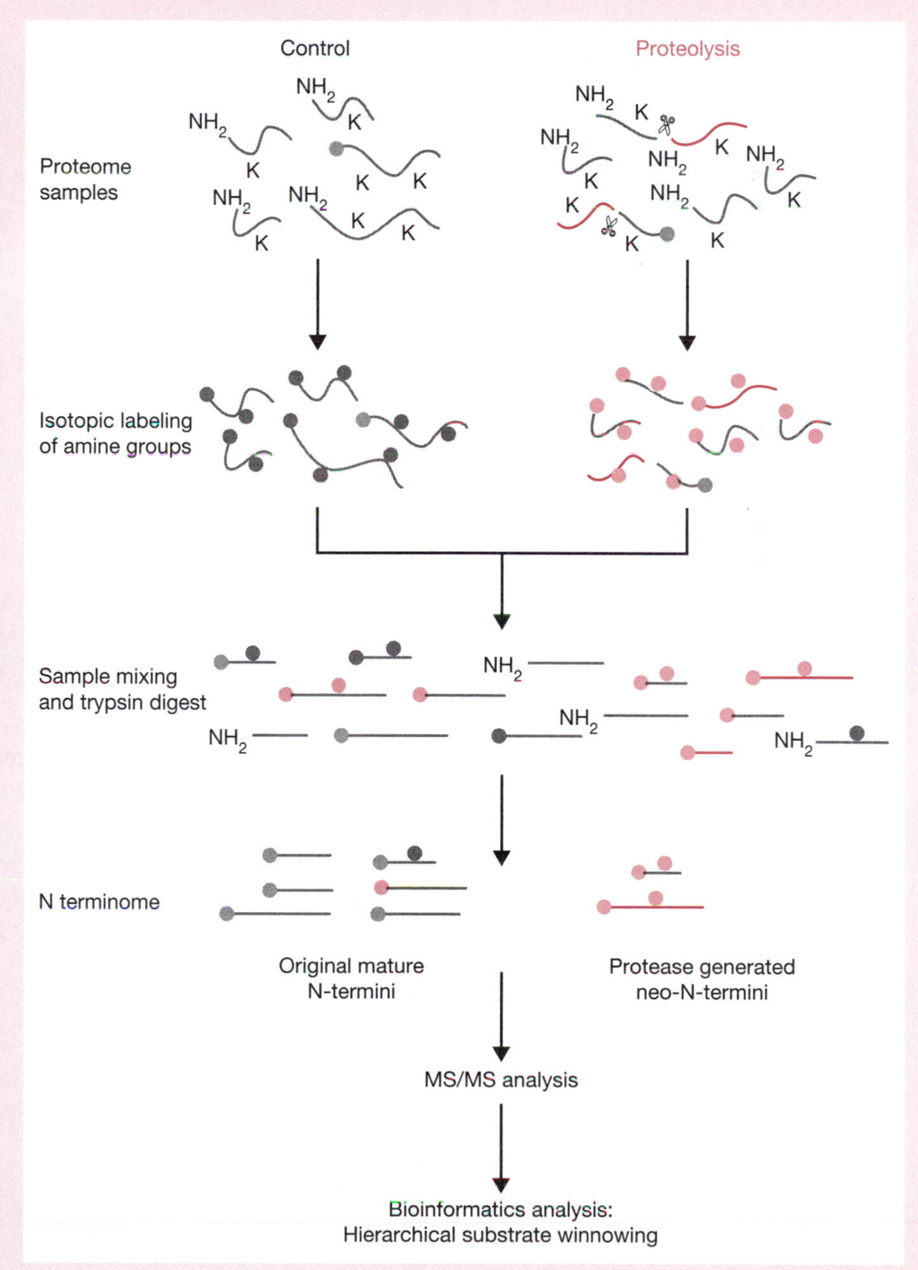

FIGURE 1 Dimethylation-TAILS. Exposed amine groups in two protein samples are isotopically labeled *in vitro* with $^{2}CH_2$-formaldehyde in one case and $^{13}CH_2$-formaldehyde in the other, simultaneously blocking the groups. The samples are then pooled, digested with trypsin, and mixed with a hyperbranched polymer that captures the exposed (unblocked) terminal amine groups of the resulting peptides, leaving a peptide mixture highly enriched for the original (blocked) N-terminal peptides.

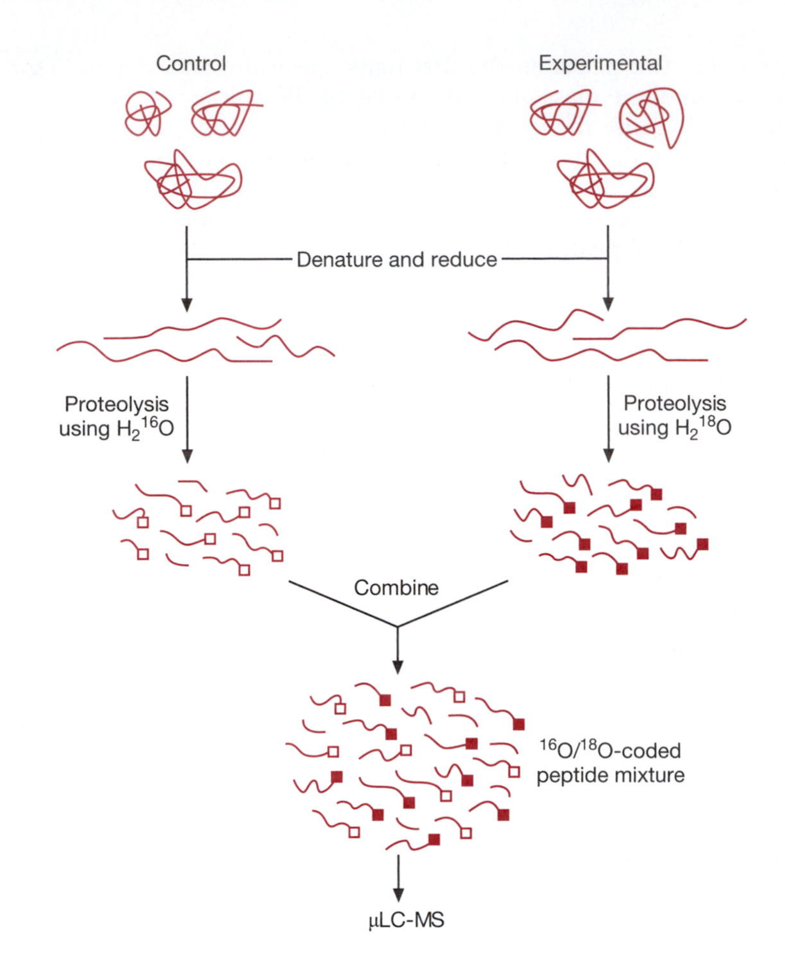

FIGURE 4.7 Enzymatic stable isotope coding of proteomes. For enzymatic labeling, proteins from two distinct proteomes are proteolytically digested in aqueous buffer containing either normal water ($H_2{}^{16}O$; *white squares*) or isotopically labeled water ($H_2{}^{18}O$; *red squares*). This encoding strategy effectively labels every C terminus produced during digestion. The samples are combined at the peptide level and then analyzed by microcapillary LC-MS. (From Goshe MB & Smith RD (2003) *Curr. Opin. Biotechnol.* 14, 101. With permission from Elsevier.)

are designed so that fragmentation during MS/MS releases **reporter ions** with different masses, allowing the abundance of the corresponding peptide to be determined. This is achieved by using mass tags comprising three regions: a reporter region, a mass balancing region, and a linker region connected to the reactive group. The mass of the reporter region plus the mass balancing region is the same in all forms of the reagent but the individual masses differ so that when the reporter ion is released it can be resolved to a particular source (**Figure 4.8**).

Two major isobaric tag platforms are available. The first is known as the **tandem mass tag (TMT) system** and comprises a mass reporter region separated from a mass normalization region via a linker that is vulnerable to fragmentation. The different forms of the label are generated by differential isotopic substitutions in the mass reporter and normalization regions, and currently there are duplex and 6-plex versions available. The other system is known as **iTRAQ** (**isobaric tags for absolute and relative quantification**) and it works on similar principles, with 4-plex and 8-plex versions available. In the 4-plex version, the four reporting groups have masses of 114, 115, 116, and 117 Da with balancing groups of 31, 30, 29, and 28 Da ensuring that all four tags have a mass of 145 Da.

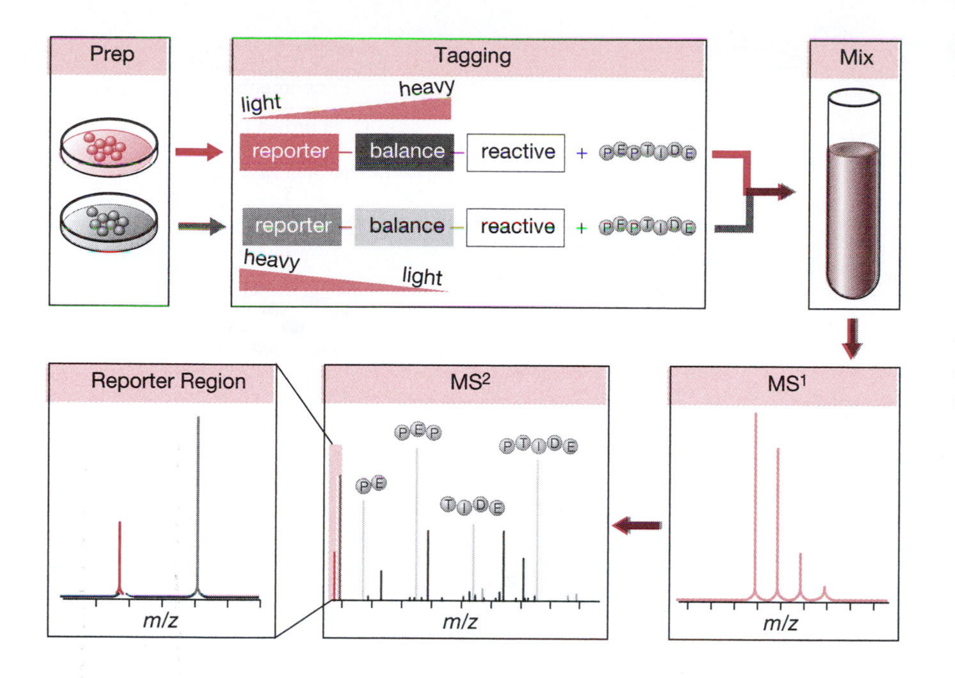

FIGURE 4.8 Workflow for the use of isobaric tagging in quantitative mass spectrometry. Protein samples are digested into peptides and labeled with isobaric tags comprising the reporter, balance, and reactive regions described in the main text. The combined mass of the reporter and balance regions is the same, producing the same mass shift in MS^1 and generating a single precursor ion. When fragmented during MS^2, the reporter regions dissociate to produce quantitative and specific ion signals for each sample. Image created by Bureta AJ.

Metabolic labeling introduces the label before sample preparation but is limited to simple organisms and cultured cells

Metabolic labeling involves the incorporation of an isotopic label into proteins while the sample is still metabolically active, for example, by growing cells in a medium containing heavy isotopes and comparing them with controls growing in normal medium. For example, Washburn and colleagues (see Further Reading) grew yeast cells in ^{14}N-minimal medium or ^{15}N-enriched medium, then pooled the cells for protein extraction, digestion, fractionation, and analysis by MS/MS. They identified more than 800 differentially expressed proteins as doublets differing in mass by one unit. The advantage of this approach is that the label is introduced early in the experiment, thereby eliminating variation arising from sample preparation and purification losses (**Figure 4.9**). One widely used variant of this approach is **stable-isotope labeling with amino acids in cell culture** (**SILAC**), which involves the inclusion of isotopically labeled amino acids (for example, [^{15}N]lysine) in the medium for one population of cells, which are then compared with controls fed with normal lysine. In more ambitious strategies, it has been possible to compare cultures fed with up to five different isotopic forms of arginine. The drawback of SILAC and other metabolic labeling methods is that they are restricted to the analysis of simple biological systems that can be maintained in a controlled environment. It is not possible to use this method with tissue explants, biopsies, body fluids, or cells that are difficult to maintain in culture.

The relative merits of different quantitative mass spectrometry methods based on SILAC have been explored in yeast, as discussed in **Box 4.5**.

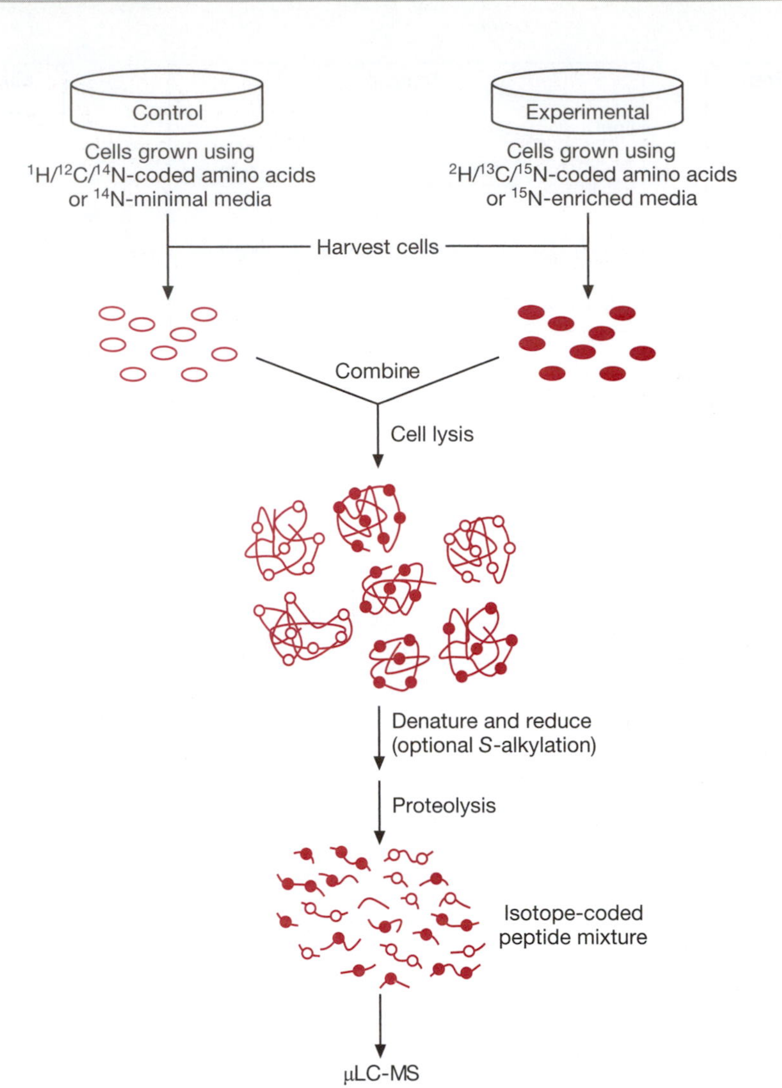

FIGURE 4.9 Metabolic stable isotope labeling. Cells from two distinct cultures are grown on media supplemented with normal amino acids ($^{1}H/^{12}C/^{14}N$) or ^{14}N-minimal media (*white spheres*) or stable-isotope amino acids ($^{2}H/^{13}C^{15}N$) or ^{15}N-enriched media (*red spheres*). These mass tags are incorporated into proteins during translation, thus providing complete proteome coverage. An equivalent number of cells for each sample is combined and processed for microcapillary LC-MS analysis. (From Goshe MB & Smith RD (2003) *Curr. Opin. Biotechnol.* 14, 101. With permission from Elsevier.)

FURTHER READING

Bantscheff M, Lemeer S, Savitski MM & Kuster B (2012) Quantitative mass spectrometry in proteomics: critical review update from 2007 to the present. *Anal. Bioanal. Chem.* 404, 939–965.

Bantscheff M, Schirle M, Sweetman G et al. (2007) Quantitative mass spectrometry in proteomics: a critical review. *Anal. Bioanal. Chem.* 389, 1017–1031.

de Godoy LMF, Olsen JV, Cox J et al. (2008) Comprehensive mass-spectrometry-based proteome quantification of haploid versus diploid yeast. *Nature* 455, 1251.

Domon B & Aebersold R (2010) Options and considerations when selecting a quantitative proteomics strategy. Options and considerations when selecting a quantitative proteomics strategy. *Nat. Biotechnol.* 28, 710–721.

Goshe MB & Smith RD (2003) Stable isotope-coded proteomic mass spectrometry. *Curr. Opin. Biotechnol.* 14, 101–109.

Gygi SP, Rist B, Gerber SA et al. (1999) Quantitative analysis of complex protein mixtures using isotope coded affinity tags. *Nat. Biotechnol.* 17, 994–999.

Holman SW, Sims PF & Eyers CE (2012) The use of selected reaction monitoring in quantitative proteomics. *Bioanalysis* 4, 1763–1786.

Leitner A & Lindner W (2006) Chemistry meets proteomics: the use of chemical tagging reactions for MS-based proteomics. *Proteomics* 6, 5418–5434.

Mallick P & Kuster B (2010) Proteomics: a pragmatic perspective. *Nat. Biotechnol.* 28, 695–709.

Marcotte EM (2007) How do shotgun proteomics algorithms identify proteins? *Nat. Biotechnol.* 25, 755–757.

Pan S, Aebersold R, Chen R et al. (2009) Mass spectrometry based targeted protein quantification: methods and applications. *J. Proteome Res.* 8, 787–797.

BOX 4.5 CASE STUDY.
Comparative quantitative proteomics of haploid and diploid yeast cells.

In 2006, Matthias Mann and colleagues showed that the SILAC method could be used to detect more than half of the proteins in the proteome of log-phase yeast cells (based on previous experiments showing that approximately 4500 proteins were expressed in such cells). In 2008, they published another ground-breaking study in which they used SILAC to differentially label the proteins in haploid and diploid yeast cells, and then carried out a comprehensive quantitative analysis to compare protein abundance in the different cell states.

Three different strategies were used as shown in **Figure 1**, resulting in more than 32% protein coverage by peptides and hence the unambiguous identification of 4399 proteins. The second strategy, which involved the digestion of proteins in solution followed by separation by IEF, was both the simplest and the most successful, yielding 3987 proteins. There was an 89% overlap between the proteome dataset produced in this experiment and previous large-scale studies based on protein tagging by homologous recombination (Chapter 7). The data did not appear to select against low-abundance proteins (indeed, several of the identified proteins are thought to be present at fewer than 50 molecules per cell) and the representation of membrane proteins was higher within the experimental dataset than within the yeast genome.

The quantitative data were based on the analysis of 1,788,451 SILAC peptide pairs, which represents more than 30 peptides per protein, and this analysis revealed 196 proteins whose abundance differed significantly between haploid and diploid cells (**Figure 2** and color plates). The affected proteins included key members of the pheromone signaling pathway that is responsible for mating in yeast as well as transposon-associated proteins and proteins associated with the cell wall. There was little agreement between the proteomic data and previous transcriptomic studies, although once low-confidence microarray results were filtered out there was better correlation, at least among the genes involved in the pheromone response pathway.

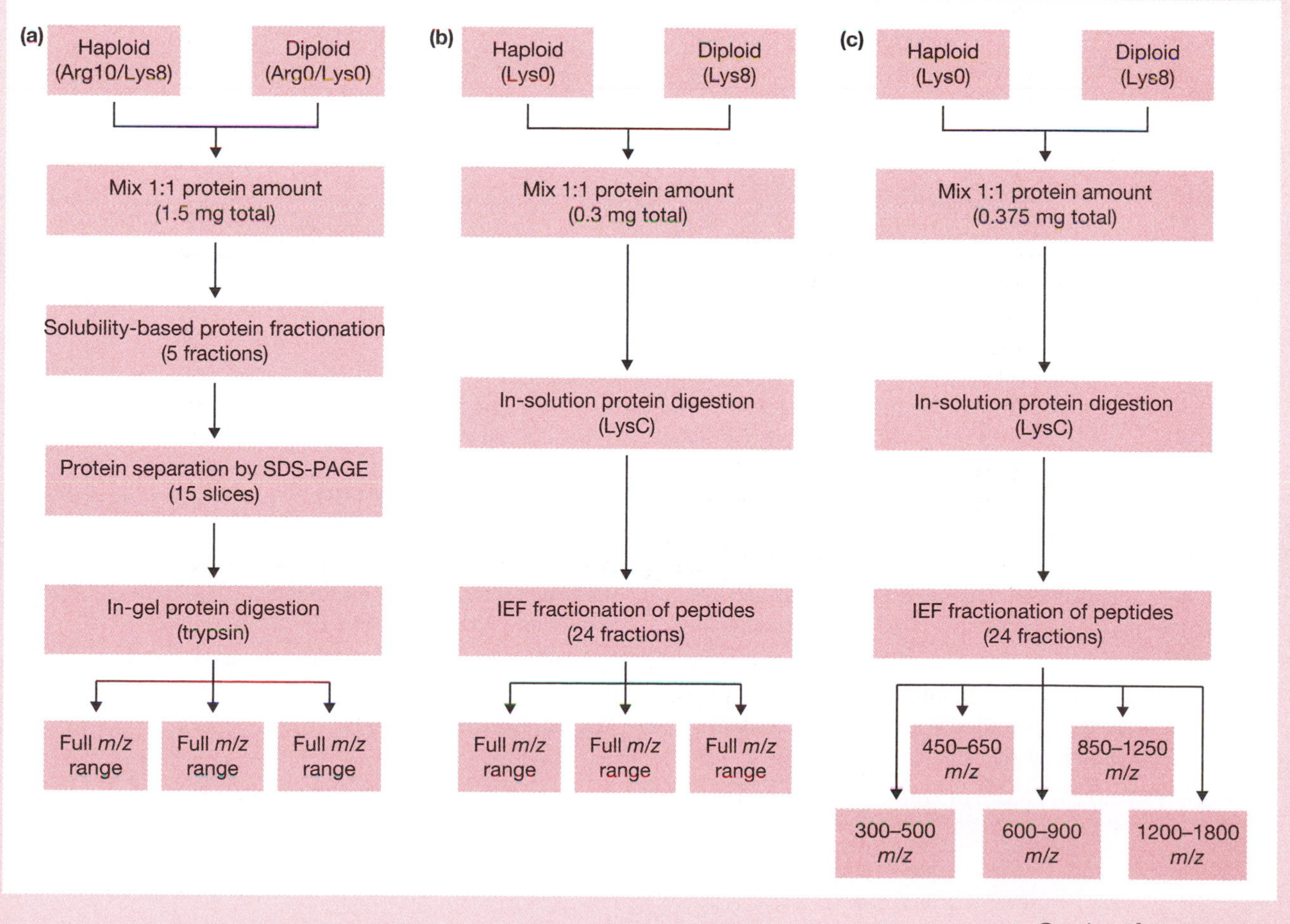

FIGURE 1 Three different strategies based on the differential incorporation of labeled amino acids to compare the proteomes of haploid and diploid yeast cells. (From de Godoy LMF, Olsen JV, Cox J et al. (2008) *Nature* 455, 1251–1254. With permission from Macmillan Publishers Ltd.)

Continued on next page

BOX 4.5 CASE STUDY.
Comparative quantitative proteomics of haploid and diploid yeast cells.

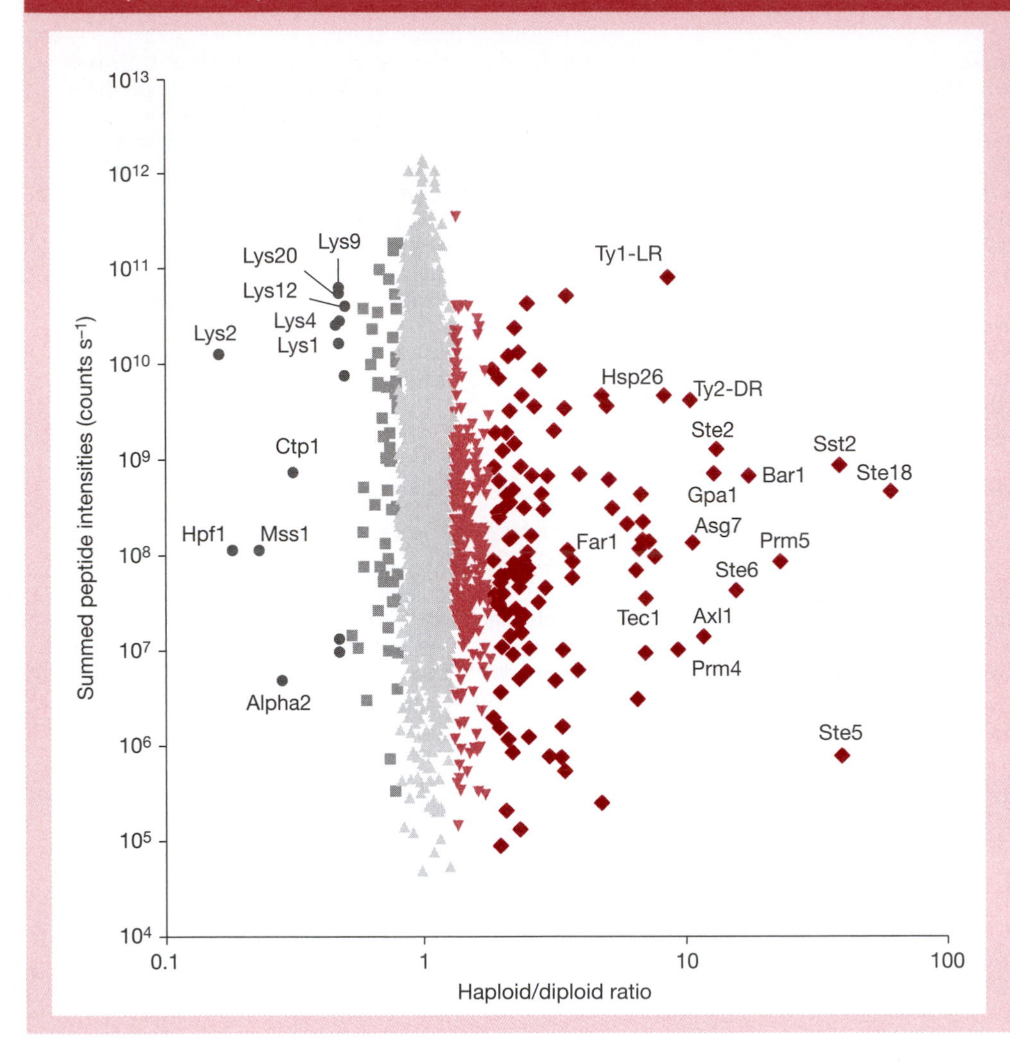

FIGURE 2: Quantitative difference between the haploid and diploid yeast proteome (overall fold change). Proteins to the left (becoming deeper gray) are more strongly represented in haploid cells. Proteins to the right (becoming deeper red) are more strongly represented in diploid cells. (From de Godoy LMF, Olsen JV, Cox J et al. (2008) *Nature* 455, 1251–1254. With permission from Macmillan Publishers Ltd.) See also color plates.

Patton W (2000) A thousand points of light; the application of fluorescence detection technologies to two-dimensional gel electrophoresis and proteomics. *Electrophoresis* 21, 1123–1144.

Patton W (2000) Making blind robots see: the synergy between fluorescent dyes and imaging devices in automated proteomics. *BioTechniques* 28, 944–957.

Patton WF & Beecham JM (2001) Rainbow's end: the quest for multiplexed fluorescence quantitative analysis in proteomics. *Curr. Opin. Chem. Biol.* 6, 63–69.

Steen H & Mann M (2004) The ABC's (and XYZ's) of peptide sequencing. *Nat. Rev. Mol. Cell Biol.* 5, 699–711.

Washburn M, Wolters D & Yates J (2001) Large-scale analysis of the yeast proteome by multidimensional protein identification technology. *Nat. Biotechnol.* 19, 242–247.

Yates JR, Ruse CI & Nakorchevsky A (2009) Proteomics by mass spectrometry: approaches, advances, and applications. *Annu. Rev. Biomed. Eng.* 11, 49–79.

CHAPTER 5

The analysis of protein sequences

5

5.1 INTRODUCTION

One of the most widely used bioinformatics techniques in proteomics is **correlative database searching**. As discussed in Chapter 3, this involves searching sequence databases for proteins containing peptides that match experimental data obtained by mass spectrometry. When successful, this process results in the definitive identification of a protein that was formerly characterized only by its position as a spot on a two-dimensional gel or as a peak on a chromatogram. However, sequence analysis can provide a great deal more information than a positive identification. By comparing the sequence of a protein with all the other sequences stored in the sequence databases, a researcher can find information about protein domain structure, physicochemical properties, interactions with other molecules, the presence of modifications, the biochemical activity/molecular function of the protein, its overall role in the cell or organism, the existence of polymorphisms and other variants in the same organism and related proteins in other organisms, the evolution of the protein family, and even a potential role in disease and/or interactions with drugs.

Many thousands of protein sequences have been entered into databases, either following *de novo* sequence determination or by the translation of nucleotide sequences. The three primary nucleotide sequence databases that comprise the **International Nucleotide Sequence Database Collaboration (INSDC)** each contain nucleotide sequence translations where appropriate, but these have been separately archived in dedicated protein sequence databases, which differ from the nucleotide sequence databases in terms of the degree of manual curation. The most comprehensive database is **UniProt**, which was launched in 2003 to combine the overlapping resources of **Swiss-Prot**, **TrEMBL**, and the Protein Information Resource protein database (PIR-PDB). Prior to the merge, these databases coexisted but differed in their protein sequence coverage and priorities/strategies for functional annotation. UniProt actually comprises four separate but interlinked databases offering different resources, as summarized in **Box 5.1**. Other useful protein sequence databases are discussed briefly in **Box 5.2**.

The number of sequences in the databases is still growing at an exponential rate, reflecting the impact of automated ultra-high-throughput next-generation DNA sequencing technologies (Chapter 1) and also the use of mass spectrometry for *de novo* protein sequencing (Chapter 3). Therefore, as well as identifying known proteins in samples, database searching based

BOX 5.1 BACKGROUND ELEMENTS.
UniProt.

UniProt (**Universal Protein Resource**) is a consortium comprising the European Bioinformatics Institute (EBI, Wellcome Trust Genome Campus, Hinxton, UK), the Swiss Institute of Bioinformatics (SIB, Geneva, Switzerland), and the Protein Information Resource (PIR, National Biomedical Research Foundation, Georgetown University Medical Center, Washington DC, USA). The consortium formed in 2002, pooling the resources of the consortium partners. These included the SIB Swiss-Prot database, the EBI TrEMBL database, and the PIR Protein Sequence Database, which had overlapping coverage and differences in annotation, and additional resources from the SIB ExPASy (Expert Protein Analysis System) bioinformatics resource portal.

Before the collaboration, Swiss-Prot had been developed as a reliable source of well-annotated protein sequence information whereas TrEMBL was simply a collection of automatically translated nucleotide sequences (Translated EMBL Nucleotide Sequence Data Library). The latter was created because the curators of Swiss-Prot were unable to match the surge in sequence data emerging from large-scale sequencing projects. The PIR Protein Sequence Database contained automatic translations and curated sequences.

UniProt was launched in December 2003 and comprises four core databases:

UniProtKB (**UniProt Knowledgebase**) is a curated database containing high-quality entries manually annotated by experts (the successor of Swiss-Prot) as well as unreviewed entries that have been annotated automatically (the successor of TrEMBL). The UniProtKB/Swiss-Prot database is therefore much smaller than UniProtKB/TrEMBL but the data are far higher in quality, and are regularly updated based on new literature with the aim of providing all known information about each protein. UniProtKB/TrEMBL contains records that are analyzed by computer and enriched with automatic annotation, fed directly from the three INSDC nucleotide sequence databases. In 2012, UniProtKB/Swiss-Prot contained more than 500,000 sequences whereas UniProtKB/TrEMBL contained more than 21 million.

UniParc (**UniProt Archive**) is a nonredundant archive of sequences from UniProtKB and a number of other databases (for example, organism-specific genome databases, patent databases), which combines identical protein sequences under a single entry defined by a unique protein identifier (UPI). UniParc contains only raw protein sequences with no annotations, but links to annotation resources in the other databases.

UniRef (**UniProt Reference Clusters**) is a set of three databases that store sequences from UniProtKB and UniParc as clusters, regardless of the organism. Each cluster is stored as a single entry and comprises sequences that have 100% identity (UniRef 100) or at least 90% or 50% identity (UniRef 90, UniRef 50) to the longest sequence in the cluster. This approach reduces the database size and accelerates searching.

UniMes (**UniProt Metagenomic and Environmental Sequences**) is specifically for metagenomic and environmental data, where the species of origin may be unclear. The predicted proteins from this dataset are automatically annotated and are also parsed with InterPro (p. 103).

on protein sequence similarity can often improve tentative genome annotations (**Box 5.3**) and provide the first leads to elucidate the function of a novel protein by linking it to related sequences whose functions are already known. This reflects the fact that the function of a protein is dependent on its three-dimensional structure (that is, the overall shape, the charge distribution, and the juxtaposition of key amino acid residues) and therefore dictates how the protein interacts with other molecules. The three-dimensional structure depends in turn on how the polypeptide folds in space, and this reflects the length of the sequence, the nature and order of the amino acids, and whether those amino acids are modified.

Proteins with similar sequences therefore often have similar structures and functions, a topic we discuss in more detail from the structural perspective in Chapter 6. In this chapter, we focus on the use of protein sequences to infer more detailed characteristics of proteins. This sequence/structure/function paradigm is a key pillar of bioinformatics methods and allows us to make confident structural and functional predictions based on a protein sequence.

BOX 5.2 BACKGROUND ELEMENTS.
Specialized protein sequence databases.

In addition to the comprehensive resource UniProt (see Box 5.1), there are many other protein sequence databases that fulfill a variety of specialized purposes, such as those focusing on particular organisms, protein functions, evolutionary relationships, annotation techniques, or intellectual property rights. Some examples are discussed below. There are also many databases of individual protein families, some of which can be accessed at: http://www.oxfordjournals.org/nar/database/subcat/3/10. Much of the information available in UniProt is also available in the NCBI resources **GenPept** (translations of GenBank nucleotide sequences, accessed directly through GenBank) and **RefSeq** (nonredundant sequences that are manually curated; http://www.ncbi.nlm.nih.gov/RefSeq/).

Direct annotation databases. Several protein sequence databases contain only proteins whose function has been established directly by experiments, and not by homology searching or other indirect methods. CharProtDB (http://www.jcvi.org/charprotdb/) is one such example. This is a curated database of biochemically characterized proteins that was established by collecting information about protein functions from the literature and then expanded by including data from other publicly available protein collections characterized by direct experimental annotation. Another example is EXProt (http://www.cmbi.kun.nl/EXProt/), a nonredundant database with sequences drawn from genome annotation projects and public databases where functions have been verified experimentally.

Indirect annotation databases. In contrast to the above, other protein sequence databases have been established to improve the rate of functional annotation using bioinformatics approaches. For example, COMBREX (http://combrex.bu.edu) aims to accelerate the functional annotation of new bacterial and archaeal genomes by processing the sequences using computational methods and then establishing procedures to validate those predictions using biochemical methods.

Organism-focused databases. Many databases exist to serve particular scientific communities, for example, Flybase for researchers working on *Drosophila*, TAIR for those working on *Arabidopsis* and many others. These databases offer diverse information (nucleotide sequences, protein sequences and structures, metabolic pathways, protein functions, mutants, transgenic lines and so forth) and are not discussed here because they cannot be considered as protein sequence databases. However, the Munich Information Center for Protein Sequences (MIPS) is exceptional, because although it is a protein database, it focuses on the annotation of proteins from selected model organisms (particularly fungi and plants) and attempts to resolve differences in coverage and annotation quality by using consistent automatic annotation methods (http://mips.gsf.de). MIPS also hosts a manually curated database of protein interactions (Chapter 7).

Specialized protein functions and characteristics. Some protein databases focus on particular characteristics of protein behavior, and a good example of this approach is PA-GOSUB (Proteome Analyst: Gene Ontology Molecular Function and Subcellular Localization). This contains sequences, predicted functions, and predicted subcellular localizations of more than 100,000 proteins from model organisms with well-characterized proteomes (http://www.cs.ualberta.ca/~bioinfo/PA/GOSUB). Another example is the Protein Research Foundation Database (http://www.prf.or.jp/), which contains information on peptides and unnatural amino acids.

Intellectual property. Patome is a sequence database containing protein sequences disclosed in patents and published applications along with detailed analysis (http://www.patome.org/).

5.2 PROTEIN FAMILIES AND EVOLUTIONARY RELATIONSHIPS

Evolutionary relationships between proteins are based on homology

Proteins with closely related sequences are statistically unlikely to have originated independently and are therefore said to be **homologous**. This means simply that the proteins have arisen during evolution by divergence from a common ancestor and therefore they are said to belong to the same **family**. Two proteins are either homologous or not. Homology is an absolute term and in strict usage should not be quantified. In the same way that it is impossible to be 65% pregnant, it is impossible for two sequences to be 65% homologous. For the quantitation of the degree of relationship between two sequences, the terms **identity** or **similarity** should be used instead (see p.93). Different members of a protein family may show similarities over their

BOX 5.3 CASE STUDY.
Proteogenomics.

The use of proteomic data to improve the annotation of genome sequences is an emerging field known as **proteogenomics**. Although the scope of this approach is wide, the most basic implementation is the use of MS data, which is normally used to search protein databases for matching peptides, to search instead genome databases to find matching gene sequences based on their six-frame translations. The correspondence between a peptide sequence and a genome sequence provides added confidence that the genome sequence represents a gene and can also improve the functional annotation. Discordance between the proteomic and genomic sequences can also identify programmed frameshifts, sites of proteolytic cleavage (for example, signal peptides), intron/exon boundaries (including alternative splice sites), and also sequences that promote post-translational modifications (see Chapter 8).

The analysis of the human genome provides a useful proteogenomics case study. Tanner and colleagues (see Further Reading) created an algorithm that allowed anonymous (that is, non-annotated) genomic sequence data to be searched with MS data. They used 18.5 million publically available MS/MS spectra representing human protein samples to screen the human genome, and in doing so validated the presence of more than 39,000 exons and 11,000 introns. This included evidence for novel or extended exons in 16 genes, and the existence of 224 hypothetical proteins (that is, proteins whose existence was predicted by genome data but which had not yet been confirmed at the protein level). They also validated 40 alternative splicing events, some of which were novel, and 308 single nucleotide polymorphisms.

entire lengths, which suggests the sequences have diverged by the accumulation of point mutations alone (for example, the human α-globin and myoglobin sequences, which are aligned in **Figure 5.1**). In such cases, the degree of relatedness often corresponds to the level of functional conservation. If two protein sequences from two different organisms are highly conserved, it is likely they are functionally equivalent and have accumulated mutations due to **speciation**. Such proteins are known as **orthologs**, for example, the human and mouse β-globin proteins (**Figure 5.2**). Two related

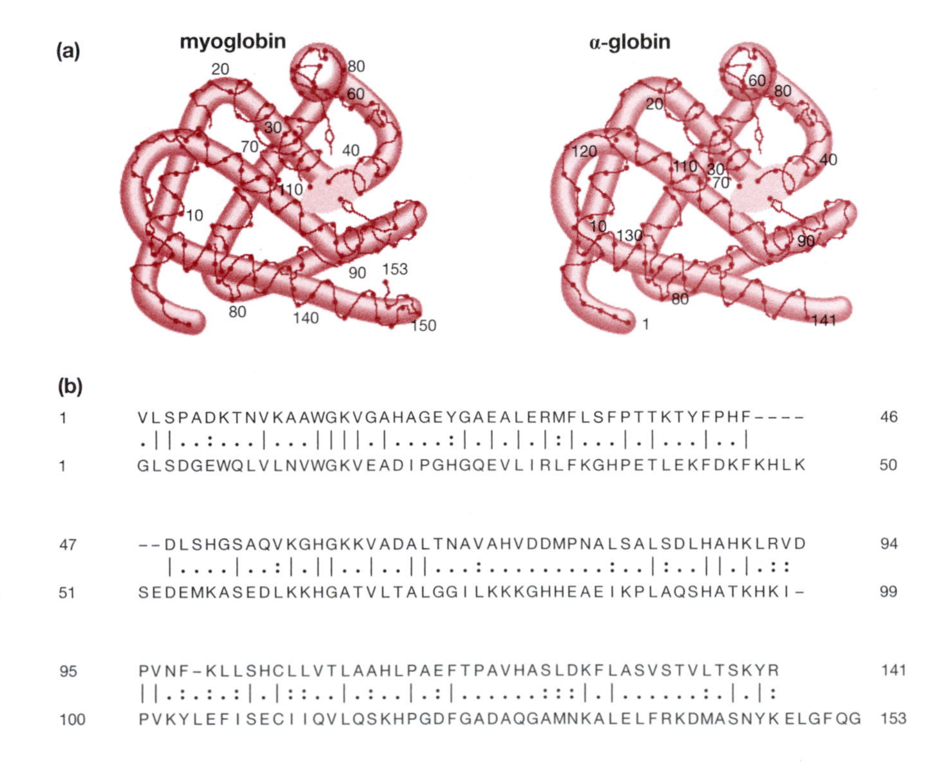

FIGURE 5.1 Comparison of human myoglobin and α-globin. (a) The three-dimensional structures of human myoglobin (left) and α-globin (right) are strongly conserved even though (b) the sequences of α-globin (upper) and myoglobin (lower) have diversified significantly (26% identity, 39% similarity). The sequences were aligned using the EBI EMBOSS-Align program (http://www.ebi.ac.uk/emboss/align/). (From Primrose SB & Twyman RM (2002) Principles of Genome Analysis and Genomics, 3rd ed. With permission from John Wiley & Sons, Ltd.)

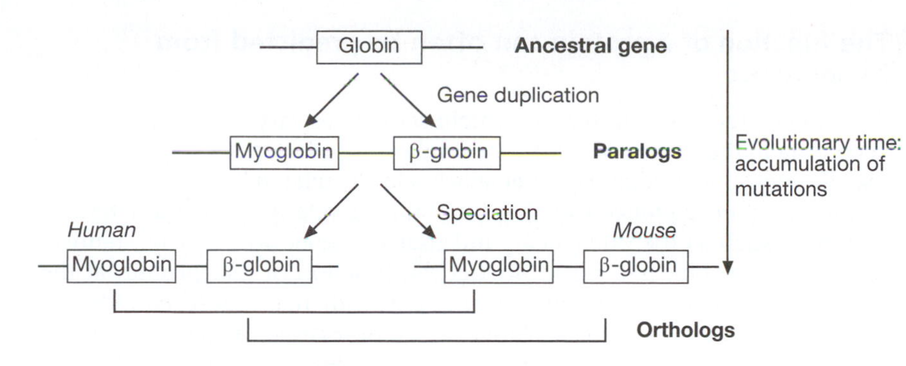

FIGURE 5.2 Evolution by the accumulation of point mutations alone leads to large families of proteins related to each other along their entire primary sequences. Proteins that have arisen by gene duplication within a species are known as paralogs, whereas equivalent proteins in different species are known as orthologs.

proteins in the same organism are known as **paralogs**, and they arise by gene duplication and divergence within a genome. Depending on the age of the duplication event, they may be more or less functionally conserved. Human myoglobin and β-globin are paralogs, as are mouse myoglobin and β-globin (Figure 5.2). However, paralogous relationships are not restricted to within a species, since, for example, the human myoglobin and mouse β-globin proteins are also paralogs.

In other cases, proteins are not related over their entire lengths but show partial alignments corresponding to individual domains. This reflects the modular nature of proteins and the fact that different functions can be carried out by different domains. Such proteins have not diverged simply by the accumulation of point mutations, but also by more complex events such as recombination between genes and gene segments leading to the shuffling and rearrangement of exons. Human proteins involved in the blood-clotting cascade provide a useful example of this process (**Figure 5.3**). Tissue plasminogen activator (TPA) contains four types of domain: a fibronectin type II domain (fnII), an epidermal growth factor domain (EGF), two kringle domains, and a serine protease domain. These domains are shared with a number of other hemostatic proteins, but the organization is different in each case. For example, the fnII domain in TPA is adjacent to the EGF domain, whereas in factor XII, the fnII domain is sandwiched between two EGF domains. In contrast, urokinase lacks a fnII domain and is therefore not activated by fibronectin, but it does contain an EGF domain and a kringle domain.

FIGURE 5.3 Human tissue plasminogen activator is a multidomain protein whose domains are widely shared within the family of hemostatic proteins.

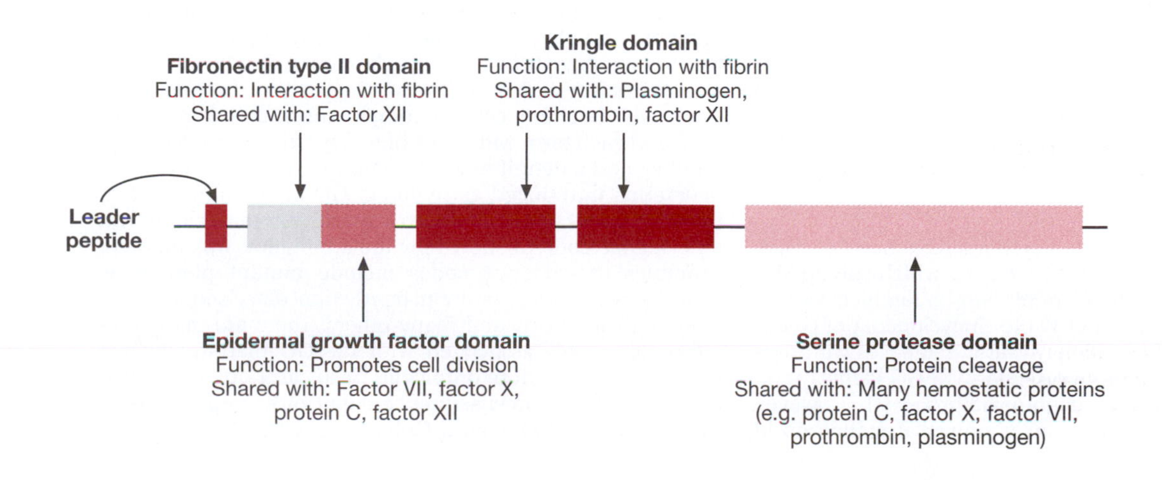

The function of a protein can often be predicted from its sequence

Functional predictions based on protein sequences vary in their usefulness according to the degree of sequence similarity. Orthologous sequences are usually very good predictors, especially in closely related species. For example, if a protein sequence was determined for a relatively uncharacterized mammal, such as the polar bear, and that sequence was nearly identical to human β-globin, one could predict the function of that protein at the biochemical, cellular, and biological levels with reasonable confidence. Paralogous sequences are less reliable indicators of functional conservation, but usually allow at least the biochemical function of a protein to be predicted. For example, a newly identified protein with a globin-like sequence that is not closely related to any known globin is still likely to function at the molecular level as an oxygen-carrier, although a more detailed cellular function would need to be established experimentally. In this context, it is very important to have a standardized nomenclature for the characterization of protein functions, a task which has been developed and refined by the **Gene Ontology Consortium** (**Box 5.4**).

For shuffled proteins, functions can be assigned to individual domains. In the case of TPA, for example, a combined biochemical function can be deduced from the functions of the individual domains: the presence of a fnII domain suggests the protein interacts with fibrin, the serine protease domain indicates that it cleaves other proteins, and the EGF domain indicates that TPA influences cell division. These individual functions are all

BOX 5.4 BACKGROUND ELEMENTS.
The function of a protein.

The high-throughput functional annotation of proteins is dependent on the use of databases to store the sequence, structural, and functional information of homologous proteins in a readily accessible form. The globalization of database resources made possible by the Internet means that a systematic nomenclature for protein function is required so that scientists all over the world can exchange information and understand what that information means. However, standardization is not easy to achieve because scientists work on different systems and organisms, and because the function of a protein can be described at three different levels, namely its molecular/biochemical function, its role within the cell, and its biological role in the whole organism. Molecular and cellular functions are the easiest to standardize because they can be described very precisely (for example, the reaction catalyzed by an enzyme and its role in a given metabolic pathway), but a standard approach for the classification of biological function is more difficult to envisage.

Several functional classification systems have been devised. One of the oldest and most established, but which only applies to enzymes, is the Enzyme Commission hierarchical system for enzyme classification. The Kyoto Encyclopedia of Genes and Genomes (KEGG) contains several databases with useful functional classifications, particularly KEGG Pathway and KEGG Ligand (http://www.genome.ad.jp/kegg/). However, the closest we currently have to a universal standardized platform is Gene Ontology (GO; http://www.geneontology.org/), a flexible system that is not restricted to a hierarchical classification architecture and that allows molecular and biological functions to be assigned independently.

GO was developed by a consortium of researchers studying three different model organisms (the yeast *Saccharomyces cerevisiae*, the fruit fly *Drosophila melanogaster*, and the mouse *Mus musculus*) in order to facilitate the identification of functionally related proteins across species. The platform has expanded to encompass many other organisms, including microbes and plants, and this has not only improved the way researchers share functional data but also helped to refine the classification system itself. The ontology covers three domains (molecular function, cellular component, and biological process) and each term within the ontology has a name, a unique number, and a definition. The definitions can be refined and new terms introduced as required. GO annotation involves the assignment of a GO term to a specific protein, along with an evidence code to show how the annotation was merited. Examples of evidence codes include mutant phenotype, gene overexpression, protein interaction data, sequence or structural similarity, and many others. There are many bioinformatics tools associated with the GO platform, which can be used to explore ontologies or to carry out automated GO annotation of new sequence data (for example, AmiGO BLAST, Blast2GO, GoFigure, Gotcha, and GOPET).

required for its overall biological function in the regulation of blood clotting. As we shall see later, however, similar sequence does not guarantee a similar function, as some very similar proteins can carry out quite distinct roles in the cell or body (p. 113).

5.3 PRINCIPLES OF PROTEIN SEQUENCE COMPARISON

Protein sequences can be compared in terms of identity and similarity

The basis of protein sequence comparison is the ability to align two sequences and determine the number and position of shared residues. The result is an **alignment score** that represents the quality of the alignment and, at the same time, the closeness of the evolutionary relationship. For nucleotide sequences, comparisons are often made on the basis of **sequence identity**, which is the percentage of identical residues in the alignment. For protein sequences, identity can be suitable for the comparison of highly conserved sequences, but a more useful measure is **sequence similarity**, which takes into account **conservative substitutions** between chemically or physically similar amino acids (for example, leucine and isoleucine). When changes occur in protein sequences, they tend to involve substitutions between amino acids with similar properties because such changes are less likely to affect the structure and function of the protein. Therefore, although sequence identity will reveal the total number of differences between two proteins, sequence similarity attaches less significance to those changes that substitute equivalent amino acids and much more significance to changes that are more likely to impact on protein structure and function. **Taylor's Venn diagram** of amino acids, which clusters amino acids on the basis of conserved physical and chemical properties, is shown in **Figure 5.4**. This is one of several approaches to define similarity.

Homologous sequences are found by pairwise similarity searching

The similarity between any two short sequences can be demonstrated by **manual alignment** as shown in **Figure 5.5a**. Most of the amino acids in the top sequence have an equivalent in the bottom sequence, and in evolutionary terms we can presume that any changes between the sequences resulted from the accumulation of point mutations in the corresponding genes.

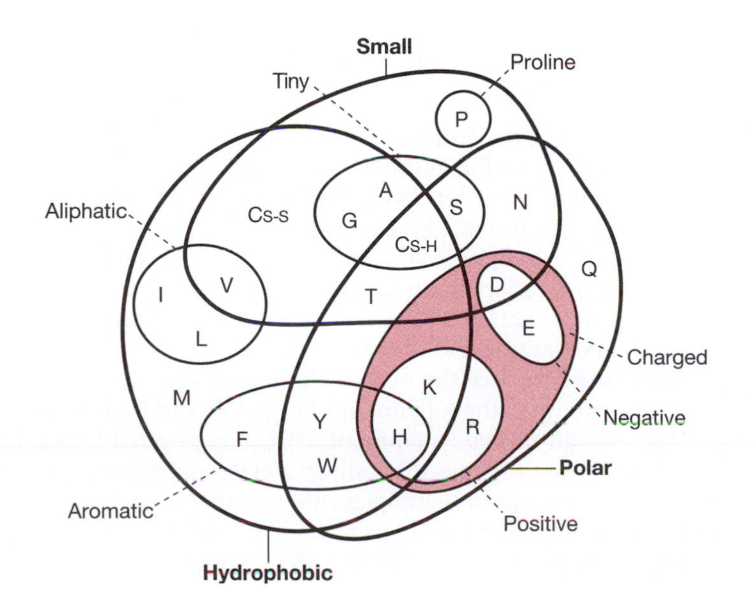

FIGURE 5.4 Taylor's Venn diagram of amino acid properties. $C_{S\text{-}S}$, cysteine in a disulfide bond; $C_{S\text{-}H}$, free cysteine.

(a)

These two peptide sequences appear to have nothing in common when judged on the basis of sequence identity

However, a meaningful alignment can be achieved by introducing a single gap and pairing up amino acids with similar chemical properties

(b)

```
1  ATDPMGVAKLRHHDKYWKKRAIV

2  PVATEEDMPMRGRVIAKDKYIHW
```

These two peptide sequences are unrelated

```
   AT--D-PM-G-V-AKLRHHDKY--WKKRAIV
   ||   | || |  | ||       |||   |
PVATEEDMPMRGRVIAK----DKYIHW
```

But the indiscriminate insertion of gaps can force them to align

FIGURE 5.5 Manual alignment of peptide sequences. (a) Manual alignment of two short peptide sequences to demonstrate the use of amino acid substitution scores where there is little or no sequence identity. A meaningful alignment can still be achieved if conservative substitutions are allowed. (b) Any two sequences can be made to align if enough gaps are introduced, which is why gap penalties are required to generate meaningful alignments.

However, one of the amino acids in the top sequence has no equivalent and a **gap** has been introduced into the bottom sequence to make the alignment more meaningful. We can presume that the gap arose due to a deletion in the bottom sequence or an insertion in the top sequence, although without further information from other protein sequences it is impossible to tell which. For this reason, gaps are sometimes called **indels**.

Real protein sequences are generally much longer than those shown in Figure 5.5 and algorithms are required to find the best alignments. There are two algorithms in common use, known as the **Needleman–Wunsch algorithm** and the **Smith–Waterman algorithm**, and both of them use **dynamic programming** to achieve the best alignment scores. Although the algorithms work on similar principles, the Needleman–Wunsch algorithm looks for **global similarity** between sequences, which works best when sequences are of similar lengths, whereas the Smith–Waterman algorithm focuses on shorter regions of **local similarity** and works best where a short sequence is aligned with a longer one (in such cases, global alignment methods might "stretch" the sequence to introduce gaps to force them into register). The Smith–Waterman algorithm is therefore the most useful for identifying partial sequence alignments such as those found in proteins that share a domain but are dissimilar in other respects. Both algorithms can be used to align sequences over their entire lengths.

In their simplest forms, dynamic programming algorithms find alignments containing the largest possible number of identical and similar amino acids by inserting gaps wherever necessary. The problem with this approach is that the indiscriminate use of gaps can make any two sequences match, no matter how dissimilar (Figure 5.5b). Apart from making alignments meaningless, this does not reflect the true nature of evolution, where insertions and deletions occur much less frequently than substitutions. The problem is addressed by constraining the dynamic programming algorithms with **gap penalties**, which reduce the overall alignment score as more gaps are introduced. For example, the alignment of α-globin and β-globin is shown in **Figure 5.6**. A head-to-head alignment with no gaps provides a relatively low score (Figure 5.6a) whereas the indiscriminate insertion of gaps would produce a higher score but a meaningless alignment. A sensible gap penalty, which reduces the alignment score as more gaps are introduced, produces the optimal alignment shown in Figure 5.6b in which there are three gaps.

(a)
```
VLSPADKTNVKAAWGKVGAHAGEYGAEALERMFLSFPTT
VHLTPEEKSAVTALWGKVNVDEVGGEALGRLLVVYPWTQ

KTYFPHFDLSHGSAQVKGHGKKVADALTNAVAHVDDMPN
RFFESFGDLSTPDAVMGNPKVKAHGKKVLGAFSDGLAHL

ALSALSDLHAHKLRVDPVNFKLLSHCLLVTLAAHLPAEF
DNLKGTFATLSELHCDKLHVDPENDRLLGNVLVCVLAHH

TPAVHASLDKFLASVSTVLTSKYR
FGKEFTPPVQAAYQKVVAGVANALAHKYH
```

(b)
```
-VLSPADKTNVKAAWGKVGAHAGEYGAEALERMFLSFPT
VHLTPEEKSAVTALWGKV--NVDEVGGEALGRLLVVYPW

TKTYFPHF-DLSH-----GSAQVKGHGKKVADALTNAVA
TQRFFESFGDLSTPDAVMGNPKVKAHGKKVLGAFSDGLA

HVDDMPNALSALSDLHAHKLRVDPVNFKLLSHCLLVTLA
HLDNLKGTFATLSELHCDKLHVDPENFRLLGNVLVCVLA

AHLPAEFTPAVHASLDKFLASVSTVLTSKYR
HHFGKEFTPPVQAAYQKVVAGVANALAHKYH
```

FIGURE 5.6 Alignment of two globin sequences. Alignment of α-globin (upper) and β-globin (lower) with only identical residues highlighted. (a) Head-to-head alignment results in a relatively small number of identical residues. (b) Although the indiscriminate insertion of gaps would make the alignment meaningless (not shown), the introduction of a small number of gaps permits a better alignment where a greater number of identical residues are paired.

Although it is possible to apply a constant penalty regardless of gap length, most algorithms employ more complex penalty systems with an initial penalty for opening a gap and then a lower penalty for extending it, described as an **affine gap penalty** (Figure 5.6b).

Dynamic programming algorithms are guaranteed to find the best alignment of two sequences for a given substitution matrix and gap penalty system but they are slow and resource-hungry. Therefore, if they are applied to large sequence databases, the searches can take many hours to perform. Even so, such platforms are available through the SSEARCH (local alignment, based on the Smith–Waterman algorithm) and GGSEARCH (global alignment, based on the Needleman–Wunsch algorithm) facilities provided in the FASTA suite (see below). To allow more rapid searches, alternative heuristic methods have been developed (known as **word methods** or ***k*-tuples methods**) that are not based on dynamic programming, and which are therefore not guaranteed to find an optimal alignment solution but are more efficient for searching large databases. These have been instrumental in the development of Internet-based database search facilities, which otherwise could be rapidly saturated by researchers carrying out similarity searches.

The two principal heuristic algorithms are **BLAST** and **FASTA**, which differ in various aspects, including the fact that BLAST is a **basic local alignment search tool** (that is, it searches for local similarity) whereas FASTA looks for global similarity. There are several variants of each algorithm that are adapted for different types of searches depending on the nature of the query sequence and the database, including variants developed specifically for use with peptide sequences identified by mass spectrometry (**Table 5.1**). Both BLAST and FASTA take into account the fact that high-scoring alignments are likely to contain short stretches of identical or near-identical letters (**words**). In the case of BLAST, the first step is to look for words of a certain fixed word length (W, which is usually equivalent to three amino acids) that score above a given user-determined threshold level, T. In FASTA, this word length is two amino acids and there is no T value because the match must be perfect. Both programs then attempt to extend their matching segments

TABLE 5.1 VARIANTS OF THE BLAST AND FASTA ALGORITHMS FOR SIMILARITY SEARCHING

Program	Query	Database
BLASTN	Nucleotide	Nucleotide
BLASTX	Translated nucleotide	Protein
BLASTP	Protein	Protein
TBLASTN	Protein	Translated nucleotide
TBLASTX	Translated nucleotide	Translated nucleotide
MS-BLAST	Peptide derived from MS data	Protein
MegaBLAST	Concatenated input sequences	User-defined
PSI-BLAST PHI-BLAST CS-BLAST/CSI-BLAST DELTA-BLAST	Protein	Protein (sensitive detection of more distantly related sequences, see p. 100)
FASTA	Nucleotide/protein	Nucleotide/protein
FASTX/FASTY[a]	Translated nucleotide	Protein
TFASTX/TFASTY[a]	Protein	Translated nucleotide
FASTS/FASTF[b]	Peptide derived from MS data (or corresponding nucleotide sequence)	Nucleotide/protein

[a]FASTX and TFASTX allow only frameshifts between codons, whereas FASTY and TFASTY allow substitutions or frameshifts within a codon.

[b]FASTS searches with peptide sequences of unknown order, as obtained from mass spectrometry, whereas FASTF searches with mixed peptide sequences, as generated by the Edman sequencing of unseparated mixtures of peptides.

to produce longer alignments, which in BLAST terminology are called **high-scoring segment pairs**. FASTA is slower but more sensitive than BLAST because the final stage of the alignment process involves alignment of the high-scoring regions using full dynamic programming.

Substitution score matrices rank the importance of different substitutions

The significance of different amino acid substitutions can be incorporated into alignment scores by the use of a **substitution score matrix**. This is simply a table that attaches probabilities (or weights) to all the possible exchanges and applies this weighting when the alignment score is calculated. In the absence of substitution scores, an **identity matrix** is used where identical amino acids in an alignment are given the score 1 and non-identical residues are given the score 0, but this does not reflect what occurs in nature. A genetic code matrix has been proposed based on the number of mutations required to convert different pairs of codons, but the best matrices are based on empirical data. The first empirical substitution matrices were devised by Margaret Dayhoff and colleagues in the 1970s and are sometimes called **Dayhoff matrices** or **mutation data matrices** because they were generated by studying alignments of very similar protein sequences and counting the frequencies with which each type of substitution occurred. Because such matrices are based on the tabulation of actual mutations (mutations that have been "accepted" in an evolutionary sense), they are generally referred to as PAM matrices, with PAM meaning "percentage of accepted point mutations." The result is a set of relative **mutability scores** for each amino acid, based on a defined evolutionary unit of time measured in **PAM units**. One PAM represents the evolutionary time for one residue to change in a sequence of amino acids 100 residues long. The most widely used PAM matrix is the PAM_{250} matrix (**Figure 5.7**), which represents a much longer evolutionary timescale during which a sequence of 100 amino acids is

C	12																			
S	0	2																		
T	−2	1	3																	
P	−1	1	0	6																
A	−2	1	1	1	2															
G	−3	1	0	−1	1	5														
N	−4	1	0	−1	0	0	2													
D	−5	0	0	−1	0	1	2	4												
E	5	0	0	−1	0	0	1	3	4											
Q	−5	−1	−1	0	0	−1	1	2	2	4										
H	−3	−1	−1	0	−1	−2	2	1	4	3	6									
R	−4	0	−1	0	−2	−3	0	−1	−1	1	2	6								
K	−5	0	0	−1	−1	−2	1	0	0	1	0	3	5							
M	−5	−2	−1	−2	−1	−3	−2	−3	−2	−1	−2	0	0	6						
I	−3	−1	0	−2	−1	−3	−2	−2	−2	−2	−2	−2	−2	2	5					
L	−6	−3	−2	−3	−2	−4	−3	−4	−3	−2	−2	−3	−3	4	2	6				
V	−2	−2	0	−1	0	−1	−2	−2	−2	−2	−2	−2	−2	2	4	2	4			
F	−4	−3	−3	−5	−4	−5	−4	−6	−5	−5	−2	−4	−5	0	1	2	−1	9		
Y	0	−3	−3	−5	−3	−5	−2	−4	−4	−4	0	−4	−4	−2	−1	−1	−2	7	10	
W	−8	−2	−5	−6	−6	−7	4	7	7	5	3	2	−3	−4	−5	−2	−6	0	0	17
	C	**S**	**T**	**P**	**A**	**G**	**N**	**D**	**E**	**Q**	**H**	**R**	**K**	**M**	**I**	**L**	**V**	**F**	**Y**	**W**

FIGURE 5.7 The PAM_{250} matrix.

expected to undergo 250 changes (that is, an average of 2.5 mutations per residue). The PAM_{250} matrix gives high scores for very common substitutions (for example, valine/isoleucine substitutions score 4) and low scores for rare ones (for example, proline/tryptophan substitutions score −6). By applying this matrix to the two short sequences in Figure 5.5b, which show almost no identity, a meaningful alignment is achieved in which the basic residues lysine and arginine are aligned, the hydrophobic residues leucine and valine are aligned, and the aspartic and glutamic acid residues are aligned. Note that the scores for alignments of identical amino acids, which are shown along the diagonal of the matrix, are not all the same. This reflects the fact that not all amino acids are equally common in proteins, and that some residues have a more important impact on protein structure than others and are therefore more highly conserved during evolution. For example, two aligned tryptophan residues score 17 because tryptophan is one of the rarest of the amino acids and it also plays a critical structural role in many proteins, while two aligned arginine residues score only 2 because this is quite likely to happen by chance when any two protein sequences are compared.

Several different substitution matrices are in common use, some of which are applied generally whereas others are used for particular types of protein with special features (for example, integral membrane proteins). The PAM series of matrices was derived by comparing the observed changes in closely related protein sequences and extrapolating those results to longer evolutionary distances. **BLOSUM (blocks substitution matrix) matrices**, in contrast, are not based on an explicit evolutionary model, and are thought to outperform PAM matrices in many situations. BLOSUM matrices are based on amino acid substitutions observed in blocks of aligned sequences with a certain level of identity. For example, the $BLOSUM_{62}$ matrix uses aligned blocks showing 62% sequence identity, which means that the higher the number, the closer the evolutionary relationship between sequences (the opposite of PAM). BLOSUM matrices are based on the local alignment of a much more diverse collection of sequences than PAM.

Sequence alignment scores depend on sequence length

Sequence alignment scores are dependent on the length of the aligned sequences. For example, a score of 100% sequence similarity suggests that two proteins are very closely related, but this is meaningless if the alignment is assessed over just three amino acid residues! A 60% similarity over 30 residues is much more worthy of attention, but 60% similarity over 300 residues would be more significant still.

The difference between chance similarity and alignments that have real biological significance is determined by the statistical analysis of alignment scores, particularly the calculation of E values. First, we need the p value of a similarity score S, which is the probability that a score of at least S would have been obtained in a match between any two unrelated protein sequences of similar composition and length. Significant matches are therefore identified by low p values (for example, $p = 0.01$), which indicate that it is very unlikely that the similarity score was obtained by chance, and probably indicates a real evolutionary relationship. The E value is related to p and is the expected frequency of similarity scores of at least S that would occur by chance. E increases in proportion to the size of the database that is searched, so even searches with low p values (for example, $p = 0.0001$) might uncover some spurious matches in a database containing 100,000 sequences ($E = 0.001 \times 100{,}000 = 10$). E can be calculated in different ways according to the search algorithm. In FASTA, $E = Np$, where N is the number of sequences in the database, but other algorithms use different methods.

Multiple alignments provide more information about key sequence elements

Whereas pairwise alignments can be used to search for related proteins, providing identification and perhaps an initial classification of a new protein sequence, the inter-relationships between members of a protein family are better illustrated by **multiple sequence alignments**. This is because the conservation of any two amino acid residues between two protein sequences could occur by chance, but if that same residue is found in five or ten proteins in the family, especially if the proteins are otherwise diverse, the residue is likely to play a key functional role. Multiple alignments are also the basis of phylogenetic trees (see **Box 5.5**).

BOX 5.5 ALTERNATIVE APPLICATIONS.
Phylogenetic analysis of proteins.

Related proteins identified by sequence similarity searches can be used to construct phylogenetic trees to resolve their evolutionary relationships. These are like multiple sequence alignments but incorporate the evolutionary distance between individual sequences based on the number of substitutions. Phylogenetic analysis methods are broadly divided into distance-matrix and tree-searching methods, the first building trees by calculating the percentage distance of all combinations of sequence pairs and the second searching for a tree that best fits the information present in each column of the multiple sequence alignment. Examples of distance-matrix methods include neighborhood joining and examples of tree-searching methods include maximum parsimony and maximum likelihood.

Phylogenetic analysis can be carried out using a range of software packages that embrace different methods (for example, MEGA, PAUP, PHYLIP) or that employ a specific analytical approach (for example, PhyML and RAxML use the maximum likelihood method). Further software is required to visualize the resulting phylogenetic tree (for example, TreeView).

An important component of phylogenetic analysis is bootstrapping, which is used to validate the reliability of a phylogenetic tree. This involves the random sampling of the original dataset followed by the construction of additional trees, which are then scored on the basis of their relation to the full-dataset tree. Jack-knifing is a similar technique but in this case 50% of the original dataset is resampled.

An example of a multiple alignment within the serine protease domains of some of the hemostatic proteins discussed earlier in the chapter is shown in **Figure 5.8**. This alignment shows that some residues are absolutely conserved, some positions are occupied only by similar amino acid residues (those giving the highest substitution scores in the PAM_{250} matrix), and others are more variable. The most strongly conserved residues are those whose physical and chemical properties are absolutely essential to maintain protein function. For example, the histidine residue sixth from the right is part of the catalytic triad of the enzyme, and is essential for the peptidase activity of the protein. As might be expected, it is conserved in all the sequences. The maintenance of tertiary structure is equally important, since this brings all the functionally critical residues into the correct relative spatial positions. In this respect, there are two completely conserved cysteine residues, one adjacent to the aforementioned histidine residue and one just to the right of the sequence gap. These are conserved because they form a disulfide bridge, which is required to hold two parts of the polypeptide backbone in the correct relative positions. There is also a conserved proline residue, proline having an unusual side chain that allows the formation of *cis*-peptide bonds, thereby influencing the way the polypeptide backbone folds. There are also highly conserved residues in the secondary structural elements, and we return to this subject in Chapter 6.

Several software suites have been developed for multiple sequence alignment, the most popular of which employ a method known as **progressive alignment** that basically involves initial pairwise alignments to find the most closely related members of the sequence collection followed by the construction of a **guide tree** based on the order of similarity of the other sequences. This method is implemented with various tweaks and variations in the widely used programs Clustal, ProbCons, and T-Coffee. The advantage is speed, but the main drawback is that information in distant sequence relationships that could improve the overall alignment is lost. In many cases, the multiple alignments have to be adjusted manually, for example, to bring conserved cysteine residues into register when it is known that such residues are involved in disulfide bonds. T-Coffee is slower but generally more accurate than Clustal because it combines the output from Clustal with that from the local alignment program LALIGN (part of the FASTA suite), which finds multiple regions of local alignment between two sequences.

Alternatives to progressive alignment include iterative methods that use progressive alignment but iteratively realign the initial sequences as well as adding new sequences (examples include PRRN/PRRP, CHAOS/DIALIGN, MAFFT, and MUSCLE) and **hidden Markov models** (HMMs), which assign

FIGURE 5.8 Part of a multiple alignment of 15 serine protease sequences. Symbols at the bottom of each column indicate the degree of conservation at that residue position: * = completely conserved (same residue in each sequence); : = highly conserved (conserved residues in each sequence); .= partly conserved (predominantly conservative substitutions). Symbols at the top of each column indicate secondary structure predictions based on structural propensity (see Chapter 6).

```
SecStructure      ......................bBBBBb...----.bBBBBBb.....bBBb.aaa.bba
THRB_HUMAN        LESYIDGRIVEGSDAEIGMSPWQVMLFRKSP----QELLCGASLISDRWVLTAAHCLLYP
THRB_BOVIN        FESYIEGRIVEGQDAEVGLSPWQVMLFRKSP----QELLCGASLISDRWVLTAAHCLLYP
THRB_MOUSE        LDSYIDGRIVEGWDAEKGIAPWQVMLFRKSP----QELLCGASLISDRWVLTAAHCILYP
THRB_RAT          LDSYIDGRIVEGWDAEKGIAPWQVMLFRKSP----QELLCGASLISDRWVLTAAHCILYP
LFC_TACTR         SDSPRSPFIWNGNSTEIGQWPWQAGISRWLADHNMWFLQCGGSLLNEKWIVTAAHCVTYS
FA9_RAT           EPINDFTRVVGGENAKPGQIPWQVILNGEIE------AFCGGAIINEKWIVTAAHCLK--
FA9_RABIT         QSSDDFTRIVGGENAKPGQFPWQVLLNGKVE------AFCGGSIINEKWVVTAAHCIK--
FA9_PIG           QSSDDFIRIVGGENAKPGQFPWQVLLNGKID------AFCGGSIINEKWVVTAAHCIEP-
FA7_BOVIN         NGSKPQGRIVGGHVCPKGECPWQAMLKLNGA------LLCGGTLVGPAWVVSAAHCFER-
FA7_MOUSE         NSSSRQGRIVGGNVCPKGECPWQAVLKINGL------LLCGAVLLDARWIVTAAHCFDN-
FA7_RABIT         GASNPQGRIVGGKVCPKGECPWQAALMNGST------LLCGGSLLDTHWVVSAAHCFDK-
PRTC_HUMAN        QEDQVDPRLIDGKMTRRGDSPWQVVLLDSKK-----KLACGAVLIHPSWVLTAAHCMDE-
PRTC_RAT          EELELGPRIVNGTLTKQGDSPWQAILLDSKK-----KLACGGVLIHTSWVLTAAHCLES-
PRTC_MOUSE        DELEPDPRIVNGTLTKQGDSPWQAILLDSKK-----KLACGGVLIHTSWVLTAAHCVEG-
PSS8_HUMAN        CGVAPQARITGGSSAVAGQWPWQVSITYEGV------HVCGGSLVSEQWVLSAAHCFPS-
                          :  *        ***. :             **. ::   *:::****.
```

likelihoods to all possible combinations of gaps, matches, and mismatches and generate a collection of potential alignments that must be ranked according to their potential biological relevance, often based on structural information, which can help to provide informative alignments (see Chapter 6). Hidden Markov models are regarded as the most efficient and accurate multiple alignment tools, although they are more complex and resource-hungry, especially in large-scale sequence analysis projects. They have been implemented in software suites such as HHsearch, HMMER, and SAM, and are combined with structural information in the MUMMALS program. Structural information can also be layered onto the standard alignment tools as in the cases of PROMALS3D and 3D-Coffee.

5.4 STRATEGIES TO FIND MORE DISTANT RELATIONSHIPS

The sensitivity of the BLAST and FASTA algorithms depends on the length of the query sequence, but there comes a point at which the alignment score between two distantly related proteins is too low to produce a significant score regardless of length. These more distant evolutionary relationships between proteins are important because they help to bridge the gap in the sequence–function relationship by identifying proteins that may have similar overall structures (and therefore functions) even though their sequences have diverged beyond recognition. This problem is approached from the opposite perspective in **structural proteomics** initiatives, as we shall see in Chapter 6. A number of strategies have been developed to improve the sensitivity of sequence matching in an effort to capture more distantly related proteins.

PSI-BLAST uses sequence profiles to carry out iterative searches

PSI-BLAST (**position-specific iterated BLAST**) is an extension of the standard BLAST algorithm that can identify up to three times as many related proteins. The principle of PSI-BLAST is that the first round of hits from a standard BLAST search are collected to form a sequence signature known as a **position-specific score matrix** (**PSSM**) with appropriate weights attached to alternative residues at each position, and this is then used for a second round of searching. The process can be repeated for a defined number of cycles as determined by the user, or it can be repeated indefinitely until no more hits are obtained. The theoretical basis of PSI-BLAST is outlined in **Figure 5.9**. Essentially, a given query sequence A will find any sequences that show a significant degree of similarity (B, C, D) but would be less likely to find sequences that are more distantly related (E, F, G). However, if B, C, and D are used as the search queries, the threshold of detection would be extended to include E and F. Therefore, a PSSM incorporating weighted values from each of the sequences A–D should identify E and F. In the next iteration, the rebuilt profile that includes all the sequences from A to F should identify G.

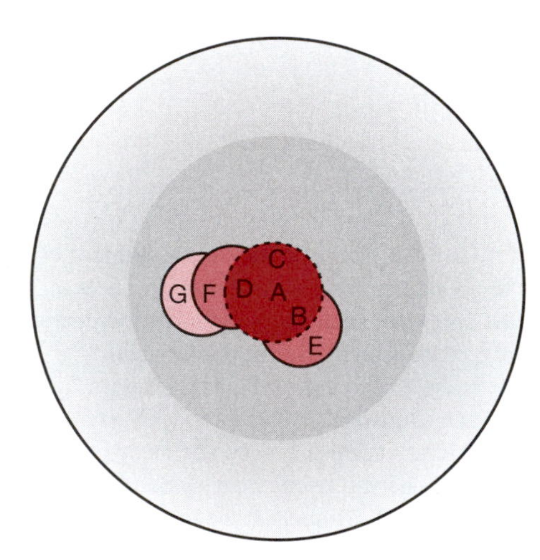

FIGURE 5.9 Theoretical basis of PSI-BLAST. The largest circle represents the whole of sequence space. The smaller *gray* circle represents all sequences that are homologous to the query sequence A, which is placed in the center of the diagram. The *red* inner circle with a broken circumference incorporates all sequences that will be identified when A is used as a query in a standard BLAST search. Sequences E, F, and G, which are more distantly related to A, will not be identified in this initial search. However, sequence F would be identified if D were used as the query and sequence E would be identified if sequence B were used as the query. By combining A to D in a sequence profile, the next search should identify E and F. A further iteration would identify G, because the features of F would be included in the profile.

Although the sensitivity of PSI-BLAST is greater than that of the standard BLAST algorithm, one problem is its tendency to identify spurious matches. This often reflects the incorporation of false-positive matches early in the process, which progressively contaminate further iterations. For example, if the original query sequence identifies a multidomain protein that shares one domain (A) with the query sequence and a second unrelated domain (Z), then the resulting PSSM will incorporate both domains. Further iterations will identify proteins containing the Z domain that are completely unrelated to protein A. Indeed this method, known as the **domain fusion method**, is sometimes used intentionally for the functional annotation of bacterial genomes on the basis that multidomain proteins in one species may be represented by several single-domain proteins in others (p. 136). The problem can be addressed by processing the hits with another BLAST variant known as **PHI-BLAST** (**pattern-hit initiated BLAST**). This algorithm searches for proteins that contain a signature specified by the user and that are similar to the query sequence in the vicinity of the signature, which (using the example above) would allow proteins only containing the Z domain to be excluded from the list of matches.

Pattern recognition methods incorporate conserved sequence signatures

The standard BLAST algorithm considers the probability of matches at the single amino acid level when calculating alignment scores. PSI-BLAST improves the sensitivity of searching by constructing a PSSM, which provides context to the search and therefore achieves a greater probability of finding more distantly related sequences. The principle of **context-based searching** has also been introduced into other variants of BLAST that, rather than building the PSSM from first principles as in the case of PSI-BLAST, instead combine the database search with information from a library of short protein profiles, that is, pre-constructed PSSMs. This is the basis of **context-specific BLAST** (**CS-BLAST**), in which the alignment scores are calculated taking into account the six neighboring amino acids on either side, and **domain enhanced lookup time accelerated BLAST** (**DELTA-BLAST**), which uses a database of conserved protein domains. A hybrid version of CS-BLAST and PSI-BLAST, which uses both a library of protein profiles and *de novo* PSSM generation, has also been developed (**context-specific iterated BLAST**, or CSI-BLAST).

The context-based methods rely on the recognition of sequence signatures to improve their sensitivity, and this is essentially an extension of the multiple alignment strategy for identifying structurally and functionally conserved elements of proteins. The information has been distilled and stored in resources known as **protein signature databases** (**Table 5.2**).There are many different ways of representing the information derived from multiple sequence alignments, and different databases employ different methods:

- **Consensus sequences**. A consensus sequence is a single sequence that represents the most common amino acid residues found at any given position in a multiple alignment. Generally, a lower threshold is set to improve the stringency of the consensus. That is, if at any given position no single amino acid is shared by 60% or more of the sequences, then there is no consensus and the residue is represented by X. The major drawback of this approach is that it does not take into account conservative substitutions (for example, leucine, isoleucine, and valine) that would be informative. In addition, it biases the consensus in favor of any sequence family that predominates in the alignment. A consensus sequence for the last few residues of the protein alignment shown in Figure 5.8, for example, would be W-V-X-T-A-A-H-C. Note that the initial tryptophan and the last four residues are invariant, and the valine and threonine are the consensus residues for the second and fourth positions because the alternative residues (isoleucine and serine, respectively) are in the minority. The big surprise is the X in the third position. This occurs because approximately half the residues are leucine and half are valine, that is, there is no consensus. Due to these disadvantages, consensus sequences are rarely used in protein databases.

- **Sequence patterns**. Sequence patterns are like consensus sequences except that variation is allowed at each position and is shown within brackets. For example, the sequence pattern equivalent to the above consensus sequence would be W-[VI]-[LV]-[ST]-A-A-H-C. Sequence patterns

TABLE 5.2 PROTEIN SIGNATURE DATABASES

Databases	Contents	URL
ASC	Short amino acid sequences with known biological activity	http://bioinformatica.isa.cnr.it/ASC/
CDD	HMMs and multiple sequence alignments derived from other signature databases	http://www.ncbi.nlm.nih.gov/cdd
BLOCKS, PRINTS	Highly conserved regions in multiple alignments of protein families. These are called **blocks** in BLOCKS and **motifs/fingerprints** in PRINTS	http://www.blocks.fhcrc.org http://bioinf.man.ac.uk/dbbrowser/PRINTS
ELMS	Short linear motifs representing regulatory interfaces	http://elm.eu.org
InterPro	Integrated signatures from other databases	www.ebi.ac.uk/interpro/
PROSITE/ProRule, ProTeus, SBASE	**Sequence patterns** associated with protein families and longer **sequence profiles** representing full protein domains. ProTeus contains signatures from protein termini	http://ca.expasy.org/prosite http://www.proteus.cs.huji.ac.il/
Pfam, SMART, ProDom	Collections of protein **domains** as well as HMMs in Pfam and SMART	http://www.sanger.ac.uk/Software/Pfam http://smart.embl-heidelberg.de/ http://prodom.prabi.fr/prodom/current/html/home.php
Superfamily, Gene3D	HMMs based on structural classifications from SCOP (SUPERFAMILY) and CATH (Gene3D)	http://supfam.org/SUPERFAMILY/
SitEx	Projections of protein functional sites on exons	http://www-bionet.sscc.ru/sitex/
PROT-FAM	Protein sequence homology database	http://www.mips.biochem.mpg.de/desc/protfam/
ProClass and iProclass	Protein classifications based on PROSITE patterns and PIR superfamilies	http://pir.georgetown.edu/iproclass/ http://pir.georgetown.edu/gfserver/proclass.html
ProtoMap	Automatic hierarchical classification of all SWISS-PROT and TrEMBL sequences	http://protomap.cornell.edu/
SYSTERS	Protein families database	http://systers.molgen.mpg.de/

Protein signature databases are resources containing sequences relating to conserved protein sequences. The table lists general databases only and excludes those representing specific protein families or functions (for example, catalytic sites, metal-binding sites, and modification sites).

are found in the database PROSITE, although the actual pattern for the above protein family would be shown as W-[LIVM]-[ST]-A-[STAG]-H-C, representing further variations found in other sequences that are not listed in Figure 5.8. Although variation is allowed, probabilities are not shown. Therefore, the fact that valine and isoleucine are equally represented at the second position but methionine is comparatively rare is not evident. PROSITE sequence patterns are generally shorter than consensus sequences and can therefore be useful in assigning distant homologs to protein families when only the most conserved regions remain. However, their very shortness can lead to false assignments, even when common patterns such as those involved in post-translational modification are taken into account.

- **Blocks**. These are not individual sequences but multiply aligned ungapped segments derived from the most highly conserved regions in protein families. They are found in two databases: PRINTS (where they are called **motifs**) and **BLOCKS** (where they are called, eponymously, **blocks**). The use of "motifs" to describe conserved regions of protein families is now unusual, and is generally restricted to the terminology used with the PRINTS database. Motifs are usually defined as functionally relevant short conserved sequences that have been defined experimentally or are assumed to be SLiMs/MiniMotifs (see below). Motifs can also be defined in a three-dimensional structural context, either generally (for example, helix-turn-helix) or specifically (for example, a specific catalytic triad in an enzyme). This is discussed in more detail in Chapter 6. In PRINTS, individual motifs from a single protein family are grouped together as **fingerprints** (**Figure 5.10**). Because the fingerprints are larger than PROSITE patterns and the search process uses an amino acid substitution matrix, it is possible to identify more distant relationship in PRINTS than in PROSITE. The PSSMs generated by improved BLAST variants are essentially representations of motifs/blocks that incorporate weighting information and can be grouped into **sequence profiles** (**gapped weight matrices**) that describe larger conserved sequence fragments, which are also stored in the PROSITE database. The PSSM models themselves are stored in the Conserved Domains Database (CCD).

- **Short Linear Motifs** (**SLiMs/MiniMotifs**). These are regulatory protein modules characterized by compact interaction interfaces (3–11 contiguous amino acids in length) that are enriched in natively unstructured regions of proteins. They bind with relatively low affinity to their targets and because few mutations are necessary to generate a novel motif, they often arise by convergent evolution to introduce novel interaction interfaces to proteins, particularly those involving signal transduction, protein trafficking, and post-translational modification. These are stored in ELM, the database of eukaryotic linear motifs.
- **Domains**. A **protein domain** is an independent unit of structure or function, which can often be found in the context of otherwise unrelated sequences. A number of databases have been established to catalog protein domains, including ProDom, which lists the sequences of known protein domains created automatically by searching protein primary sequence databases. Other databases also contain elements of protein domains (such as the sequence profiles stored in PROSITE). Pfam and SMART contain multiple domain alignments and **hidden Markov models**, which are among the statistically most sophisticated tools for representing protein domains.

Each of the above protein signature databases has its strengths and weaknesses, which can make the comparative interpretation of results from different databases challenging (**Figure 5.11**). To resolve this problem, an integrated cross-referencing tool called **InterPro** was established in 1999 so that query sequences could be screened against different protein signature databases and presented in a clear format. Currently, InterPro integrates the data from PROSITE, Pfam, PRINTS, ProDom, SMART, TIGRFAMs, PIR SuperFamily, and SUPERFAMILY, with the integration of CATH and PANTHER HMMs underway. InterPro covers nearly 80% of UniProt and also matches its entries against GO terms and structural data from MSD, CATH, and SCOP, which are important in the context of archiving and classifying protein structures. We will consider their importance in Chapter 6.

FIGURE 5.10 Example sequences of the four conserved motifs that define the SH3 domain, as shown in the PRINTS database. These represent the most conserved regions from the multiple alignment of many SH3 domains. Only five examples are shown, and many further examples can be found in the database itself.

Motif 1	Motif 2	Motif 3	Motif 4
GYVSALYDYDA	DELSFDKDDIISVLGR	EYDWWEARSL	KDGFIPKNYIEMK
YTAVALYDYQA	GDLSFHAGDRIEVVSR	EGDWWLANSL	YKGLFPENFTRHL
RWARALYDFEA	EEISFRKGDTIAVLKL	DGDWWYARSL	YKGLFPENFTRRL
PSAKALYDFDA	DELSFDPDDVITDIEM	EGYWWLAHSL	YKGLFPENFTRRL
EKVVAIYDYTK	DELGFRSGEVVEVLDS	EGNWWLAHSV	VTGYFPSMYLQKS

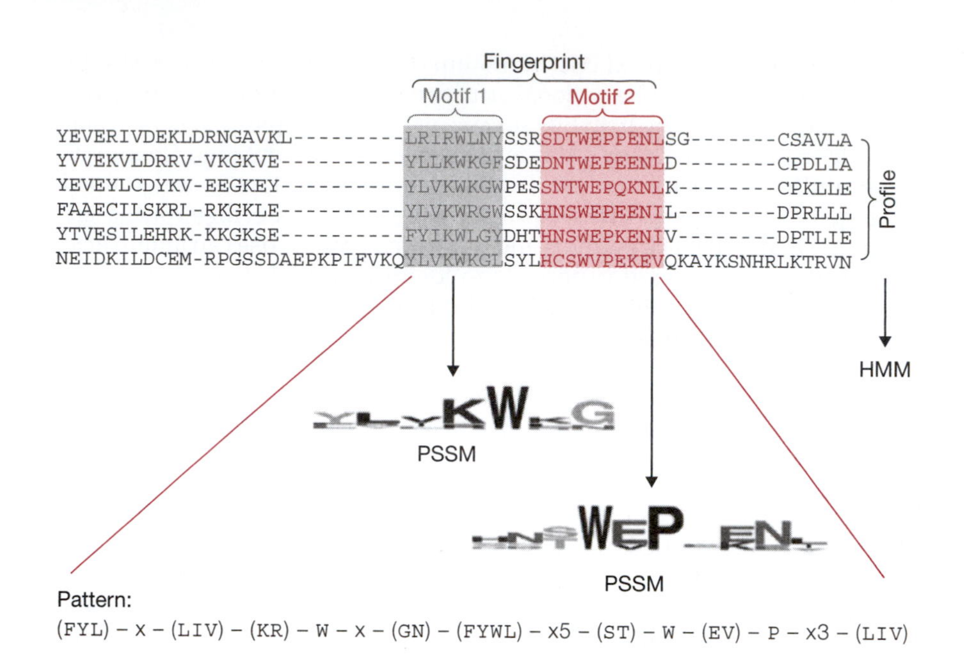

FIGURE 5.11 Comparison of different types of protein signatures. A motif is a single conserved region. A group of motifs forms a fingerprint. PSSMs are generated by weighting motifs based on the amino acid residue frequencies at each position (frequency being denoted by height in this example). Patterns represent the core functional and structural features of the sequence indicating alternative amino acids at each position. The protein depicted is a member of the chromodomain family and the invariant tyrosine (W) is a key catalytic residue explaining the absence of alternative residues at this position. Profiles represent the complete conserved regions, including gaps that are used to generate hidden Markov models for comparative profiling. (From Pavlopoulou A & Michalopoulos I (2011) *Int. J. Mol. Med.* 28, 295. With permission from Spandidos Publications.)

5.5 THE RISK OF FALSE-POSITIVE ANNOTATIONS

Standard similarity searches, recursive methods and pattern or profile searching can all identify sequences that are more or less related to a particular query. However, these methods are not foolproof and all have the potential to come up with spurious matches or annotations. One of the dangers is **database pollution**. Databases contain errors, so annotating a new sequence on the basis of database information alone can sometimes serve only to reinforce and propagate misinformation. Some databases contain better-quality data than others, primarily reflecting the degree of manual curation (for example, the distinction between UniProtKB/Swiss-Prot and UniProtKB/TrEMBL, see Box 5.1) and the quality of the evidence required for functional assignment (see Box 5.4). The progressive integration of databases and the standardization of nomenclature should reduce the amount of pollution in the future.

Errors can also be introduced by the user if search algorithms are not understood properly, for example, if a BLAST search is carried out with an *E* value cutoff that is not stringent enough. This is particularly relevant in iterative searches because low-quality alignments are incorporated into the PSSM used for the next round of searching. The bad seed can grow, resulting in a large collection of false positives that bear no relationship to the original query sequence.

As stated earlier, another drawback to similarity searching is that sequence conservation does not always predict functional conservation. Although many bioinformatics tools work on the basis that sequence/structure/function are closely related, sequences also diverge and become functionally distinct. There are many examples of proteins that show strong sequence conservation but perform quite different functions, for example, the enzyme lysozyme and the regulatory protein α-lactalbumin. There are also proteins that have entirely different sequences but perform essentially the same function, such as the diverse collection of metabolic enzymes that have been recruited as crystallins in the lens of the vertebrate eye. It is also necessary to consider the impact of **low-complexity sequences**, that is, sequences such as transmembrane domains that are present in many proteins with extremely diverse functions. These are often masked out prior to similarity searches to prevent false positives. Additional evidence for functional annotation can be gathered by investigating protein functions and interactions, as discussed in Chapters 6 and 7.

FURTHER READING

Agrawal A, Brendel VP & Huang X (2008) Pairwise statistical significance and empirical determination of effective gap opening penalties for protein local sequence alignment. *Int. J. Comput. Biol. Drug Des.* 1, 347–367.

Altschul SF, Wootton JC, Gertz EM et al. (2005) Protein database searches using compositionally adjusted substitution matrices. *FEBS J.* 272, 5101–5109.

Ansong C, Purvine SO, Adkins JN et al. (2008) Proteogenomics: needs and roles to be filled by proteomics in genome annotation. *Brief. Funct. Genomic. Proteomic.* 7, 50–62.

Kumar S & Filipski A (2007) Multiple sequence alignment: in pursuit of homologous DNA positions. *Genome Res.* 17, 127–135.

Lan N, Montellione GT & Gerstein M (2003) Ontologies for proteomics: towards a systematic definition of structure and function that scales to the genome level. *Curr. Opin. Chem. Biol.* 7, 44–54.

Lesk A (1998) Introduction to Bioinformatics. Oxford University Press.

Löytynoja A (2012) Alignment methods: strategies, challenges, benchmarking, and comparative overview. *Methods Mol. Biol.* 855, 203–235.

Mulder NJ, Kersey P, Pruess M & Apweiler R (2008) *In silico* characterization of proteins: UniProt, InterPro and Integr8. *Mol. Biotechnol.* 38, 165–177.

Nash PD (2012) Why modules matter. *FEBS Lett.* 586, 2572–2574.

Orengo CA & Thornton JM (2005) Protein families and their evolution - a structural perspective. *Annu. Rev. Biochem.* 74, 867–900.

Pavlopoulou A & Michalopoulos I (2011) State-of-the-art bioinformatics protein structure prediction tools. *Int. J. Mol. Med.* 28, 295–310.

Stastna M & Van Eyk JE (2012) Analysis of protein isoforms: can we do it better? *Proteomics* 12, 2937–2948.

Stein L (2001) Genome annotation: from sequence to biology. *Nat. Rev. Genet.* 2, 493–503.

Tanner S, Shen Z, Ng J et al. (2007) Improving gene annotation using peptide mass spectrometry. *Genome Res.* 17, 231–239.

The UniProt Consortium (2013) Update on activities at the Universal Protein Resource (UniProt) in 2013. *Nucleic Acids Res.* 41(D1), D43–D47.

Valdar WSJ & Jones DT (2003) Amino acid residue conservation. In Advanced Text—Bioinformatics (CA Orengo, DT Jones, JM Thornton eds), pp 49–64. BIOS Scientific Publishers.

Weatheritt RJ & Gibson TJ (2012) Linear motifs: lost in (pre) translation. *Trends Biochem. Sci.* 37, 333–341.

CHAPTER 6

The analysis of protein structures

6

6.1 INTRODUCTION

We discussed in Chapter 5 the intimate relationship between the sequence, structure, and function of proteins. This relationship exists because the primary amino acid sequence of a protein determines how the polypeptide chain folds up in space, and the structure of the folded protein determines how it interacts with other molecules in the environment. These interactions then constitute the basis of protein function. For example, a folded protein often contains clefts and cavities that complement particular ligands, substrates, and indeed other proteins (allowing the formation of complexes, see Chapter 7); and the manner in which a protein folds up determines the distribution of charges over its surface and the positioning of key amino acid residues, hence defining its physicochemical properties, its potential interaction partners, and (in the case of enzymes) its ability to catalyze particular reactions. Proteins do not usually adopt their final structure spontaneously and instantaneously but become structured at different levels in the form of local secondary structures and quasi-independent domains. A brief overview of the principles of protein structure is provided in **Box 6.1**.

Structures, like sequences, have a predictive value in the assignment of protein functions because two proteins with similar structures are likely to interact with common substrates or ligands and therefore may carry out similar activities in the cell. However, structures can be even more useful than sequences in a comparative sense because structures tend to be more strongly conserved over evolutionary timescales. Two proteins with sequences that can no longer be recognized as homologous based on the comparison methods discussed in Chapter 5 may nevertheless have similar structures, indicating they are homologous but more distantly related than proteins with matching sequences. Therefore, solving the structure of a protein may provide some information about its function even if no related sequences can be found in any of the databases.

In the past, protein structural analysis was undertaken only when the function of a protein was already well understood, but the predictive value of the structure/function relationship has highlighted the merits of bringing structural analysis to the beginning of the investigative process as a way to bridge the current sequence-to-function knowledge gap. The main difficulty with this approach is that the methods used to solve protein structures have traditionally required laborious, expensive, and time-consuming work that is unsuitable for automation. One of the greatest breakthroughs in proteomics over the last decade has been the development of high-throughput automated

BOX 6.1 BACKGROUND ELEMENTS.
An overview of protein structure.

Proteins are macromolecules comprising one or more polypeptides, each of which is a linear chain of amino acids. There are 20 standard amino acids specified by the genetic code plus at least two modified derivatives discovered thus far, selenocysteine and pyrrolysine, which are inserted in a context-dependent manner (**Figure 1**). Amino acids have a standard structure (**Figure 2**) but possess chemically diverse residual groups or side chains, allowing the synthesis of proteins with a wide range of physicochemical properties. Further diversity is generated by over 400 different types of post-translational modification (Chapter 8).

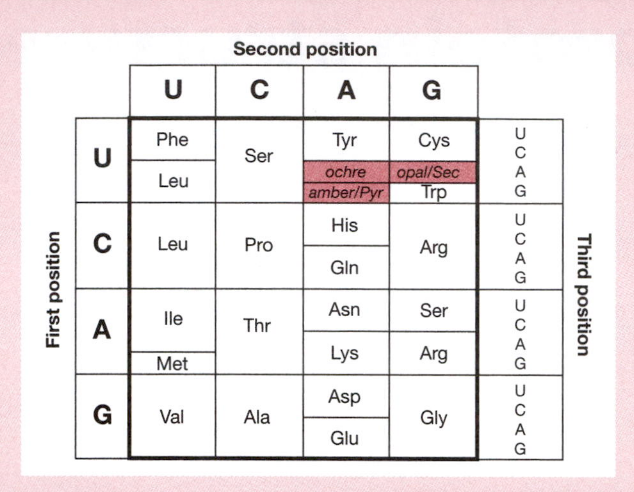
Second position

First position	U	C	A	G	Third position
U	Phe	Ser	Tyr	Cys	U
	Phe	Ser	Tyr	Cys	C
	Leu	Ser	ochre	opal/Sec	A
	Leu	Ser	amber/Pyr	Trp	G
C	Leu	Pro	His	Arg	U
	Leu	Pro	His	Arg	C
	Leu	Pro	Gln	Arg	A
	Leu	Pro	Gln	Arg	G
A	Ile	Thr	Asn	Ser	U
	Ile	Thr	Asn	Ser	C
	Ile	Thr	Lys	Arg	A
	Met	Thr	Lys	Arg	G
G	Val	Ala	Asp	Gly	U
	Val	Ala	Asp	Gly	C
	Val	Ala	Glu	Gly	A
	Val	Ala	Glu	Gly	G

FIGURE 1 The universal genetic code. Amino acids are identified by their three-letter designation. Nonsense codons are identified by name. Note the dual functions of codons UAG and UGA. Sec, selenocysteine; Pyr, pyrrolysine.

Primary structure

The sequence of amino acids in a polypeptide is known as the **primary structure**. The amino acids are joined together by peptide bonds, which usually adopt the *trans* configuration such that the carbonyl oxygen and amide hydrogen of adjacent amino acids point away from each other (**Figure 3**). The peptide bond itself is rigid, but the other bonds are quite flexible and allow the polypeptide backbone to fold in space. Exceptionally, proline residues have limited conformational freedom because the residual group is bonded to the main polypeptide backbone (indeed, strictly speaking, proline is an **imino acid** rather than an amino acid). Such residues are therefore also able to form *cis* peptide bonds so the carbonyl oxygen and amide hydrogen of adjacent residues project in the same direction, although the *trans* configuration is still preferred. This has a major influence on the folding of the peptide backbone and the substitution of proline residues with other amino acids inevitably has a significant effect on the overall structure. For this reason proline residues are often highly conserved in protein folds. Similarly, glycine residues are important because their small residual group (a single hydrogen) allows a much greater degree of flexibility than other residues. Cysteine residues are also highly conserved because they have the ability to form **disulfide bridges**, which help to stabilize the three-dimensional structure of individual polypeptide chains as well as joining discrete polypeptides together.

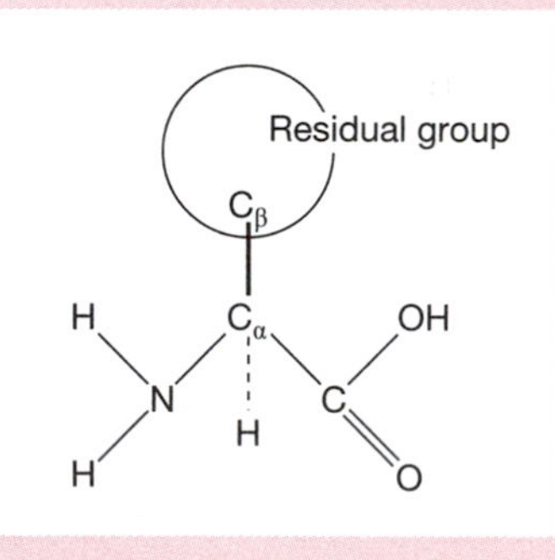

FIGURE 2 General chemical structure of an amino acid.

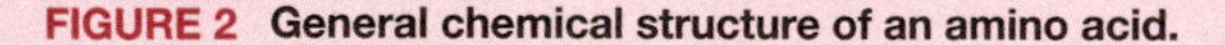

Secondary structure

Secondary structures in proteins are regular and repeating local configurations generated by intramolecular hydrogen bonds. These sometimes involve polar side chains (such as those of serine and threonine residues), but the polypeptide backbone itself is polar because the NH group can act as a hydrogen donor whereas the C=O group can act as a hydrogen acceptor. The regular spacing of peptide bonds throughout the polypeptide chain allows regular, ordered structures to form. The two most common structures are the α-helix and the β-sheet, which are often depicted as the only secondary structures in proteins, interspersed with nonstructured regions known as coils (**Figure 4**). Typically, α-helices are right-handed and range in size from 4–40 residues, corresponding to 1–12 turns of helix. They occur when hydrogen bonds form between peptide units four residues apart, aligning them and giving the entire structure a significant dipole moment although the bond angles are acute. Other helical secondary structures (the 3_{10}-helix and the π-helix) are rarely seen except at the end of more typical α-helices because of

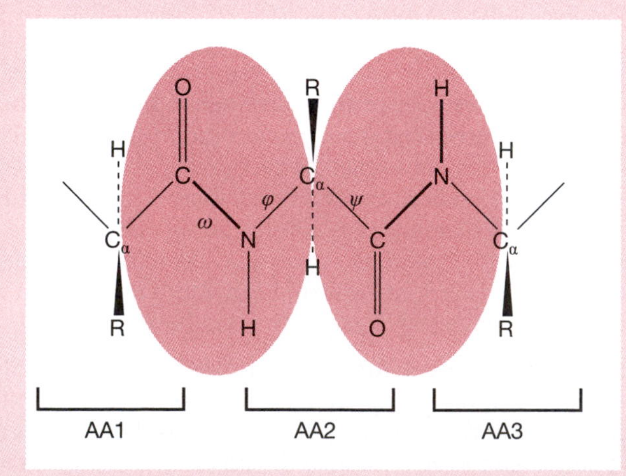

FIGURE 3 Structure of the *trans* peptide bond. AA, amino acid.

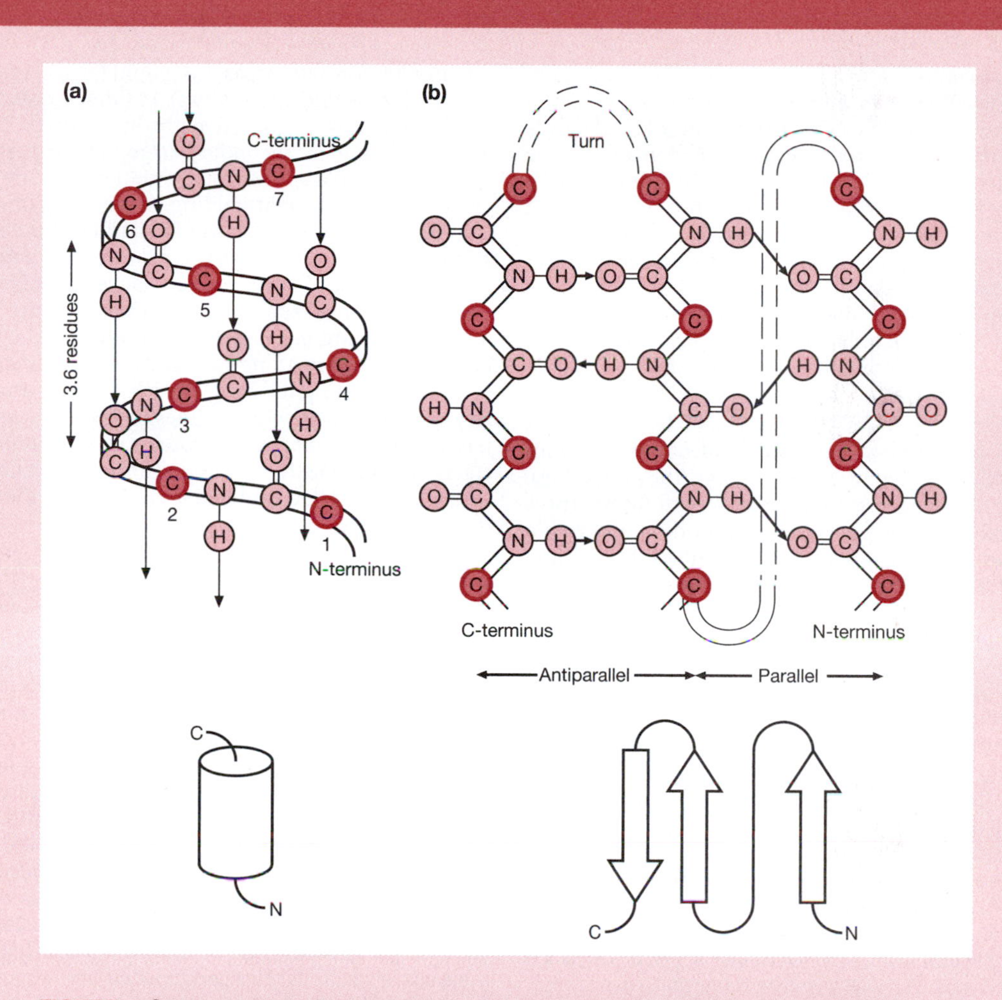

FIGURE 4 Common secondary structures in proteins. (a) The α-helix. (b) The β-sheet. Detailed molecular structures are shown above; arrows show direction of hydrogen bonds. Topological structures are shown below.

unfavorable packing constraints. Unlike helices, β-sheets form from regions of the polypeptide chain where the bond angles are fully extended (these are known as β-strands). Several β-strands can align in parallel, antiparallel, or mixed arrays, and hydrogen bonds form between peptide units in adjacent strands. Both α-helices and β-sheets may be joined together by linker regions that adopt their own secondary structures, which can be defined as bends or **turns**. For example, a β-turn is formed when a hydrogen bond forms between peptide units three residues apart. Where no hydrogen bonds are present, linker regions are known as **loops**. The core of a protein is often rich in secondary structures, because this allows energy-efficient packing, whereas loops are generally found on the surface where interactions can occur with the solvent. Loops are generally much more mutable than the core regions because they do not interfere with the way the protein is packaged and act more like "decorations" on the protein surface to control its interactions.

Motifs (supersecondary structure)

Proteins are usually classified as being predominantly α-helical, predominantly β-sheet, or mixed, and this often depends on the type of motifs that are present. A motif, at the sequence level, is generally defined as a functionally relevant sequence module that has been experimentally determined (although see Chapter 5 for more discussion). In structural biology, a motif is a group of secondary structures that are found connected together and perform a common function. Simple examples include the helix–turn–helix, which is often found in DNA-binding proteins, the helix–loop–helix, which acts as a dimerization interface, and the coiled coil, which is often found in fibrous proteins such as keratin. More complex examples, which contain more secondary structures, include the globin fold (eight α-helices), the Greek key (a four-strand antiparallel β-sheet), and the αβ-barrel (in which several β-α-β motifs roll up into a cylinder).

Continued on next page

BOX 6.1 BACKGROUND ELEMENTS (*Continued*). An overview of protein structure.

Tertiary structure (fold)

The **tertiary structure** or **fold** of a polypeptide is its overall shape, reflecting the way the secondary structures and motifs pack together to form compact domains. A domain can be regarded as a part of a polypeptide chain that can fold independently into a stable tertiary structure, but domains can also be defined as units of protein function. A protein may contain a single domain or multiple domains, and in the latter case the different domains can carry out individual functions in the context of the overall biological function of the protein. As stated above, disulfide linkages between cysteine residues are often required to maintain tertiary structures.

Quaternary structure

Many proteins are single polypeptides but others are composed of multiple polypeptide subunits. The way these subunits assemble determines the quaternary structure. There is no functional difference between a multidomain protein and a protein with several different polypeptide subunits, and many proteins can exist in both forms. For example, most transcription factors are single polypeptides with DNA-binding and transcriptional activation domains, but others assemble from independent subunits. Indeed, the assembly of a transcription factor from interacting subunits is the basis of the two-hybrid system for detecting binary protein interactions (see p. 145). Protein subunits may interact noncovalently, or may be joined together by inter-polypeptide disulfide bridges.

Protein folding

Like all physical and chemical reactions, protein folding is driven by the need to attain a state of minimum thermodynamic free energy with respect to the surrounding solvent molecules. This ideal state is known as the **native conformation** of the protein and is generally the state in which the protein is functional. For every native state, there are an infinite number of **denatured conformations**. Protein folding therefore cannot involve a random search through all these possible conformations, as this would take an infinite amount of time (the **Levinthal paradox**). In other words, protein folding must follow a defined pathway, perhaps by forming a framework of local secondary structures or perhaps by condensing around a specific nucleation point. One of the major determinants of protein folding, at least in globular proteins, is **hydrophobic collapse**: the formation of a central core of hydrophobic residues excluded from contact with the solvent. Experiments with some proteins have identified intermediate folding states, such as the **molten globule**, which lacks tertiary structure but is rich in secondary structures, providing support for the framework and collapse models. However, several small proteins have been shown to undergo single-step global folding, which agrees with the nucleation model. It is also notable that many proteins cannot attain their native states spontaneously, and require the assistance of specialized enzymes called **molecular chaperones** that catalyze the folding process.

Intrinsically unstructured proteins

Intrinsically unstructured proteins (also known as **naturally unfolded proteins** or **intrinsically disordered proteins**) lack any stable tertiary structure under physiological conditions, thus challenging the paradigm that proteins must be well ordered to function correctly. Many such proteins only adopt stable structures when they form complexes with other proteins or alternative ligands (**coupled folding and binding**). Other proteins have intrinsically disordered domains alongside compact structured domains. Intrinsically unstructured proteins often lack the bulky hydrophobic amino acids that, in globular proteins, collapse to form the hydrophobic core, or feature low-complexity sequences such as repeats of the same few amino acid residues. They tend to be less dense than globular proteins and can be detected using density-sensitive methods such as NMR spectroscopy and small-angle X-ray scattering. Their lack of secondary structure can also be measured by far-UV circular dichroism spectroscopy.

methods for structural analysis, mirroring the earlier revolution in genomics brought about by high-throughput automated DNA sequencing. These techniques form the basis of what is now known as structural genomics.

Although **structural genomics** and **structural proteomics** were initially regarded as synonymous, the former is now considered to reflect the mission to determine the structure of a representative member of every protein fold family (this Chapter), whereas the latter also includes the structure of protein complexes and macromolecular assemblies (see Chapter 7).

6.2 STRUCTURAL GENOMICS AND STRUCTURE SPACE

Coverage of structure space is currently uneven

To understand the value of structural genomics, it is necessary to understand the concepts of **sequence space** and **structure space**. These can be envisaged on a theoretical level as the sum of all possible protein sequences that could potentially exist, and the sum of all the structures that those sequences could possibly generate while obeying the laws of physics and geometry. In reality, not all possible sequences and structures exist in nature, because existing sequences have been selected over millions of years of evolution. Therefore, sequence and structure space can be defined on a practical level as the sum of all sequences and structures in existence.

The important difference between sequence space and structure space is that structure space is much smaller, that is, there are fewer structures than there are sequences. This is because many sequences can give rise to the same structure. Amino acids with similar chemical properties are often interchangeable, but even when substitutions occur between dissimilar amino acids the overall effect on the structure of a large protein can be marginal. Many proteins contain functionally critical residues (such as those in the active site in an enzyme) and structurally critical residues (such as cysteine residues required for the formation of disulfide bonds), but the importance of other residues is often additive, so cumulative changes may be required to make radical differences to the overall structure of the protein.

The practical consequence of the above, as already stated, is that proteins with dissimilar sequences can have the same overall structure (or **fold**) and that solving the structure of a protein can provide an indication of its function and its evolutionary relationship to other proteins even if there are no matching sequences. For example, structural analysis of the uncharacterized secreted protein AdipoQ revealed a structural relationship to the tumor necrosis factor (TNF) family of chemokines even though the level of sequence identity between AdipoQ and TNFα is just 9%, a relationship too distant to be picked up by a BLAST search. With the evidence of a structural relationship in hand, it became possible to align multiple sequences and identify conserved residues and secondary structures in these proteins (**Figure 6.1** and color plates).

(a)

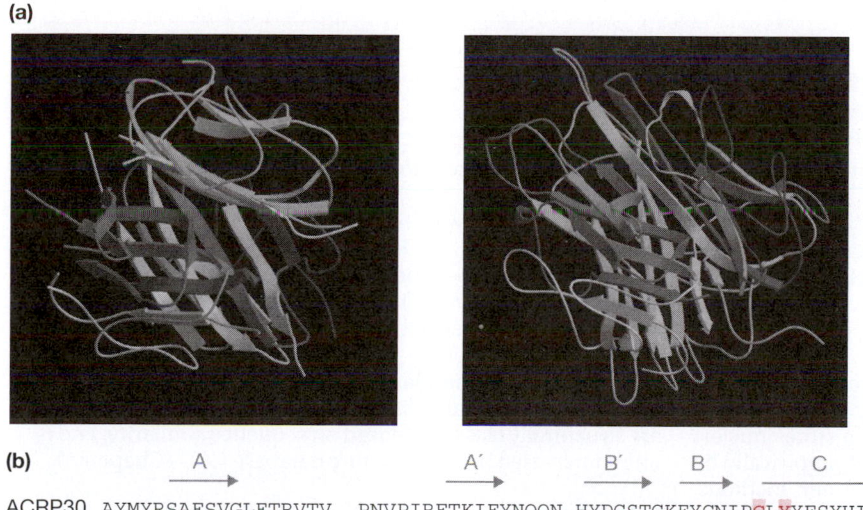

(b)

```
              A                    A′             B′       B        C                        D
ACRP30  AYMYRSAFSVGLETRVTV   PNVPIRFTKIFYNQON HYDGSTGKFYCNIPGLYYFSYHITV                YMKDVKVSLFKK
C1qA    GATQKVAFSALRTINSPLR PNQVIRFEKVITNANE NYEPRNGKFTCKVPGLYYFTYHASS                 RGNLCVNLVRK
TNFα RTPSDKP   VAHVVANPQAEGQ   LQWLNRRANALLANGV ELRD NQLVVPSEGLYLIYSQVLFKGQGCP         STHVLLTHTISRIAV
TNFβ    TLKP   AAHLIGDPSKQNS   LLWRANTDRAFLQDGF SLSN NSLLVPTSGIYFYYSQVVFSGKAYSPKATSSPLYLAHEVQLFSS
CD40L GDQNPQIAAHVISEASSKTTSVLQW AEKGYYTMSNNLVTLENGKQLTVKROGLYYIYAQVTF    CSNREASSQAPFIASLCLKSPG
              A                      A′      B′     B           C                          D

           E                         F            G                           H
ACRP30   DKAVLFTYDQYQ              EKNVDQASGSVLLHLEVGDQVWLQVYGDGDHNGLYADNVNDSTFTGFLLYHDTN
C1qA     GRDSMQKVVTFC              DYAQNTFQVTTGGVVLKLEQEEVVHLQATDKNSLLGIEGANSIFTGFLLFPDMDA
TNFα SYQTKVNLLSAIKSPCQRETPEGAEAKPWYEPIYLGGVFQLEKGDRLSAEINRPDYLDFAESGQV   YFGIIAL
TNFβ QYPFHVPLLSSQKMVY         PGLQEPWLHSMYHGAAFQLTQGDQLSTHTDGIPHLVLSPSTV   FFGAFAL
CD40L RFERILLRAANTHS           SAKPCGQQSIHLGGVFELQPGASVFVNVTDPSQVSHGTGFT   SFGLLKL
             E                        F             G                         H
```

FIGURE 6.1 Identification of related proteins by structural comparison. (a) A ribbon diagram comparison of AdipoQ (left) and TNFα (right). The structural similarity is equivalent to that within the TNF family. (b) Structure-based sequence alignment between several members of the TNF family (CD40L, TNFα, and TNFβ) and two members of the C1q family (C1qA and AdipoQ, the latter labeled ACRP30). Highly conserved residues (present in at least four of the proteins) are shaded, and arrows indicate β-strand regions in the proteins. There is little sequence conservation between AdipoQ and the TNF proteins (for example, 9% identity between AdipoQ and TNFα), so BLAST searches would not identify a relationship. See also color plates. (Adapted from Shapiro L & Harris T (2000) *Curr. Opin. Biotechnol.* 11, 31. With permission from Elsevier. Images courtesy of Protein Data Bank.)

If we consider the above example and extrapolate it to the entirety of structure space, it becomes clear that solving a certain number of structures will allow the functional annotation of all sequences, because the level of degeneracy in the sequence-to-structure relationship means that each structure can represent a large number of sequences. Estimates as to the total number of protein folds in existence vary from 4000 to 10,000, but it is clear that we have discovered only a fraction of them so far. The major repository for protein structures is the **Protein Data Bank** (PDB), which is discussed in more detail in **Box 6.2**. At the time of writing, the PDB contains more than 80,000 structures, but many of them are redundant, that is, they represent closely related proteins and variants of the same protein created by point mutations. There are only approximately 1000 unique folds, representing the proteins that have been easiest to prepare under the conditions suitable for structural analysis. Our current coverage of protein structure space is therefore highly uneven and biased. The goal of structural genomics (**Figure 6.2**) is to find representative members of every protein fold family and solve their structures to provide coarse (that is, punctuated) but even coverage of structure space. These representative proteins can then be used as templates for the structural annotation of related proteins using the bioinformatics-based approaches discussed later in the chapter, providing fine coverage of the gaps between these initial coarse targets. This approach should go a long way to establishing the functions and evolutionary relationships among all orphan genes.

Although structural comparisons and the identification of homologous relationships are perhaps the most useful applications of protein structures in proteomics, there are many other benefits to the direct analysis of structures.

BOX 6.2 BACKGROUND ELEMENTS.
Important protein structure databases and resources.

Protein Data Bank. The most important protein structure database is the Protein Data Bank (PDB) which is today maintained by an international consortium comprising the Research Collaboratory for Structural Bioinformatics Protein Database (RCSB PDB), the Protein Data Bank in Europe (PDBe), the Protein Data Bank Japan (PDBj, who founded the collaboration in 2003), and the Biological Magnetic Resonance Data Bank (BMRB, which joined in 2006). Together, these organizations are known as the **Worldwide Protein Data Bank (wwPDB)**. The PDB is the universal repository for all protein three-dimensional structural data (it also stores nucleic acid structures), including structures derived empirically by X-ray diffraction, NMR spectroscopy, and other methods, and those predicted by modeling. PDB data comprise three-dimensional coordinates of the protein structures and notes regarding the methods used for structural determination, the primary amino acid sequences, related literature, and the chemical structures of cofactors and prosthetic groups.

Databases for protein structure classification. As discussed in the main text, structural comparison requires a rigorous definition of the different structures to allow the correct description of novel proteins. There are three major databases of protein structures that focus on such definitions: **SCOP (structural classification of proteins)** and **CATH (class, architecture, topology, homologous superfamily)**, both of which use hierarchical classification schemes that are integrated with the protein signature databases discussed in Chapter 5 to describe overall fold structures; and the Dictionary of Secondary Structures in Proteins, which provides a definitive guide the classification of secondary structures. There is also the **FSSP** database (families of structurally similar proteins), which uses the DALI algorithm to align proteins with similar structures and define fold families on this basis. ProtClustDB is a NCBI database where proteins are structurally classified based on sequence similarity, and is one source used by the signature database CCD (Chapter 5).

Other protein structure databases. Additional protein structure databases have been developed to present more detailed information about modeled protein structures (for example, ModBase) or to investigate ways to visualize and disseminate protein structural data more effectively, including several (PDPWiki and Proteopedia) using the collaborative wiki model. Further sites extend the information available in the PDB by showing how proteins interact with cellular components such as the lipid bilayer (for example, the Orientations of Proteins in Membranes database) or by focusing on particular protein groups (for example, PDBTM, the Protein Data Bank of Transmembrane Proteins).

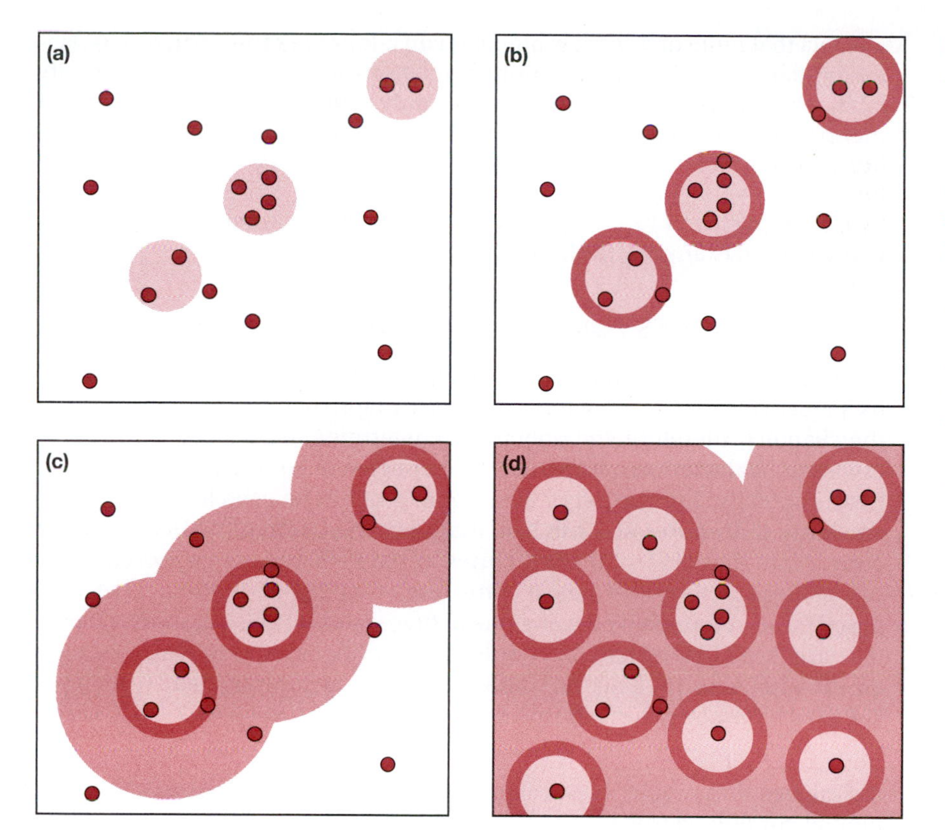

FIGURE 6.2 The goal of structural genomics. Dots represent individual proteins. (a) In current sequence space, many proteins are orphans because their relationship to other proteins cannot be determined at the sequence level. Circles show proteins linked by sequence relationships. (b) Pattern and profile matching algorithms can extend the range of sequence analysis and discover new homologous relationships. (c) Known protein structures can extend these relationships even further, because structures are much more highly conserved than sequences. (d) Structural genomics aims to solve enough structures so that all proteins can be related to other proteins.

First, structural analysis may reveal features of proteins that have obvious functional significance but that cannot be identified by studying the underlying sequence. Examples include the overall shape (which may reveal clefts and crevices that could function as ligand-binding pockets or active sites), the juxtaposition of particular amino acid side chains (which could reveal potential catalytic sites), the electrostatic composition of the surface (which may suggest possible interactions with other molecules), and the crystal packing of the protein (which may reveal possible interacting surfaces and biologically relevant multimeric assemblies). Additionally, the unexpected presence of a ligand, co-factor, or substrate in an experimentally determined protein structure can provide the basis for a functional hypothesis. Some real examples from current structural genomics initiatives are discussed later in the chapter. Finally, a direct application of structural genomics is in the rational design of drugs, a topic to which we return in Chapter 10.

Structure and function are not always related

Although protein structure is helpful in predicting function, it should be emphasized that there is no simple, one-to-one relationship between structure and function. There are many examples of proteins with similar structures that have evolved to perform a myriad of different functions, such as the α/β hydrolase fold that is associated with at least six different enzymatic activities as well as appearing in a cell adhesion molecule (**Table 6.1**). It is generally accepted that 30% sequence identity is required to confidently predict that two proteins will share the same fold (although structures can be conserved with much less sequence identity than this, as shown by the AdipoQ/TNF example above) but that 60% sequence identity is required to confidently predict that two proteins have the same function.

TABLE 6.1 KNOWN FUNCTIONS ASSOCIATED WITH THE α/β HYDROLASE FOLD

Cholesterol esterase
Dienelactone hydrolase
Haloalkane dehalogenase
Neurotactin (cell adhesion molecule)
Non-heme chloroperoxidase
Serine carboxypeptidase
Triacylglycerol lipase

TABLE 6.2 KNOWN STRUCTURES ASSOCIATED WITH GLYCOSYL HYDROLASE ACTIVITY

α/α-toroid
Cellulase-like β/α-barrel
Concanavalin A-like two-layer β-sandwich
Double-psi β-barrel
Orthogonal β-bundle
Six-bladed β-propeller
TIM barrel

TIM, triose phosphate isomerase

The fact that quite divergent sequences can adopt the same structure is useful for identifying distant evolutionary relationships, but it can also identify false relationships where functionally equivalent structures have evolved independently. Such structures are described as **analogous** rather than homologous, since they are not related by descent. Proteins can also be functionally analogous without any obvious homology. The enzyme glycosyl hydrolase, which is represented by at least seven distinct structures, provides a useful example (**Table 6.2**).

6.3 TECHNIQUES FOR SOLVING PROTEIN STRUCTURES

It is not yet possible to predict the tertiary structure of a protein *de novo* from its sequence without some form of preexisting structural data (see later in the chapter). Therefore, the only way to determine the structure of an otherwise uncharacterized protein with any degree of confidence is to solve it experimentally. The two major techniques that can be used for this purpose are **X-ray diffraction** and **nuclear magnetic resonance (NMR) spectroscopy**. More than 98% of the structures in the PDB have been solved using one of these methods, and most of the remaining 2% are theoretical models based on structures solved using one of these methods. Fewer than 100 protein structures in total have been solved using other methods, which include electron microscopy, electron diffraction, electron paramagnetic resonance (EPR) spectroscopy, and neutron scattering.

Both X-ray diffraction and NMR spectroscopy are notoriously demanding techniques that require painstaking work to determine the precise experimental conditions ideal for the structural analysis of a given protein. The preparation of protein crystals for diffraction is regarded almost as an art form, and many attempts to determine protein structures fail because suitable crystals are unavailable. In the case of NMR spectroscopy, which requires ultrapure protein solutions, the proteins must be stable and soluble at high concentrations and must not aggregate or denature under these conditions. The protein sample usually requires single (^{15}N or ^{13}C), double (^{15}N plus ^{13}C), or even triple (^{15}N, ^{13}C, and ^{2}H) isotopic labeling to increase spectral resolution and reduce spectral overlaps. Each technique involves the collection and processing of large amounts of data and the assembly of a model or models of atomic coordinates that agree with the empirical results. Neither method, at a first glance, appears suitable for the type of high-throughput investigations undertaken in proteomics.

Like other analytical techniques, however, both have benefited from advances in technology and the development of highly parallel assay formats that allow many different conditions to be tested simultaneously. Advances in bioinformatics, which allow structural data to be processed and modeled more quickly than ever before, have also made an invaluable contribution to the structural genomics field. We discuss the principles of X-ray diffraction and NMR spectroscopy below and summarize the recent advances that have brought structural biology into the proteomics era. We then briefly consider some additional methods that are used to investigate protein structure.

X-ray diffraction requires well-ordered protein crystals

X-ray diffraction exploits the fact that X-rays are scattered in a predictable manner when they pass through a protein crystal. X-rays are diffracted when they encounter electrons, so the nature of the scattering depends on the number of electrons present in each atom and the organization of the atoms in space. Like other waves, diffracted X-rays can positively or negatively interfere with each other. Therefore, when protein molecules are regularly arranged in a crystal, the interaction between X-rays scattered in the same direction by equivalent atoms in different molecules generates a pattern of spots known as **reflections** on a detector (**Figure 6.3**). These **diffraction patterns** can be used to build a three-dimensional image of the electron clouds of the molecule, which is known as an **electron density map**. The structural model of the protein is built within this map.

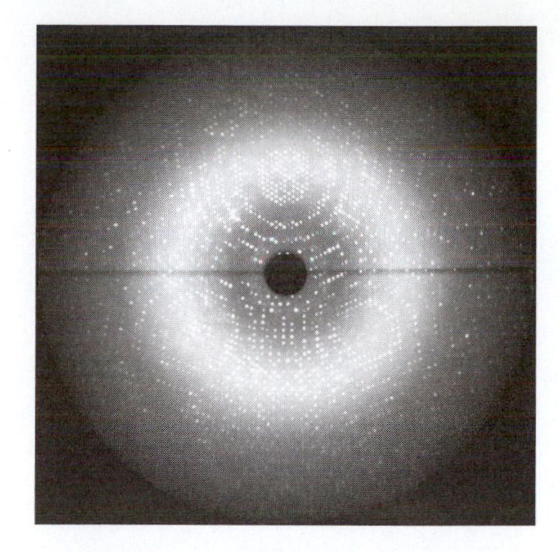

FIGURE 6.3 Pattern of reflections generated by a protein phosphatase crystal. (Courtesy of Daniela Stock, MRC Laboratory of Molecular Biology, Cambridge.)

Accurate structural determination requires a well-ordered crystal that diffracts X-rays strongly. This has been a significant bottleneck because, as discussed above, protein crystals can be notoriously difficult to grow. Hydrophobic proteins or proteins with hydrophobic domains are the most difficult to crystallize, and for this reason the PDB contains relatively few structures of complete membrane proteins. Some proteins also have unstructured regions that tend to reduce the density of electron maps. One major development that has increased the throughput of X-ray diffraction is the use of **automated crystallization workstations**, which allow thousands of different parameters such as different protein concentrations, salt concentrations, solvents, temperatures, and pH to be tested in parallel in order to identify the best crystallization conditions. Smaller sample volumes can also be used, allowing crystallization studies with non-abundant proteins, as long as the solution has a protein concentration in the range 5–25 mg/ml and is more than 95% pure. The small volumes also tend to reduce the time required for equilibration and increase the success rate.

The success of crystallization has also been increased by improving the methods for target selection, that is, identifying and excluding proteins that are likely to generate poor crystals. There are several bioinformatics tools that can identify poor candidates based on their sequence, and also practical solutions such as testing multiple orthologs of the same protein and truncated variants that lack unstructured regions. High-resolution methods based on **hydrogen–deuterium exchange mass spectrometry** (**DXMS**) can be used to identify such regions, allowing them to be removed from expression constructs prior to the preparation of recombinant protein crystals, or even directly by ***in situ* proteolysis** during crystallization. Other **salvage pathways** that can increase the likelihood of achieving useful crystals representing recalcitrant proteins include **surface entropy reduction**, which involves the replacement of high-conformational-entropy solvent-exposed residues with smaller residues that support crystal packing, and the reduction of surface lysine residues by methylation, which promotes the crystallization of recalcitrant proteins and may also increase the resolution of diffraction data from proteins that do form good crystals.

The next problem encountered in X-ray diffraction is the derivation of an electron density map from the diffraction patterns. This process requires three pieces of information: the wavelength of the incident X-rays (which is already known), the amplitude of the scattered X-rays (which can be determined by the intensity of the reflections), and the **phase** of diffraction. The phase cannot be determined from the pattern of reflections, and this has come to be known as the phase problem. Sometimes, it is possible to use phases from related solved structures already in the PDB, an approach known as **molecular replacement**. In standard (low-throughput) structural determinations, further experiments are carried out to determine the diffraction phases by producing heavy-atom-containing **isomorphous crystals**, that is, crystals of the same overall structure incorporating heavier atoms that produce alternative diffraction patterns. For example, the crystals can be immersed in a heavy metal salt solution so that heavy metal atoms diffuse into the spaces originally occupied by the solvent and bind to defined sites in the protein. Metal atoms diffract X-rays more strongly than the atoms normally found in proteins because they contain more electrons. By comparing the reflections generated by several different isomorphous crystals (a process termed **multiple isomorphous replacement**, **MIR**) the positions of the heavy atoms can be worked out and this allows the phase of diffraction in the unsubstituted crystal to be deduced. A complete description of each reflection (wavelength, amplitude, and phase) is known as a **structure factor**.

To increase the throughput of X-ray diffraction, this rather laborious process for determining structure factors has been superseded by methods that rely on the phenomenon of **anomalous scattering**. This occurs when heavy metal atoms in a protein crystal are struck by X-rays of a wavelength close to their natural absorption edge, causing them to re-emit some of the energy as further X-rays. The magnitude of anomalous scattering varies with the wavelength of the incident X-rays, so one type of metal-containing crystal can be bombarded at several different wavelengths and different diffraction patterns obtained from which the phase of scattering can be calculated. This is the basis of techniques such as **SIRAS** (**single isomorphous replacement with anomalous scattering**), **SAD** (**single-wavelength anomalous dispersion**), and **MAD** (**multiple-wavelength anomalous dispersion**). Anomalous scattering requires the use of **synchrotron radiation sources** (which produce high-intensity X-ray beams that can be tuned precisely) and sensitive detectors, such as kappa-geometry goniometers and area detectors. The rapid progress in structural genomics over the last 10 years owes much to the increasing availability of third-generation synchrotron facilities, more than 120 of which are now on-line, each capable of solving 100–300 structures per year. A streamlined approach is to express each protein in bacteria or yeast and incorporate a metal-substituted amino acid derivative, such as **selenomethionine**. This is a routine method in structural proteomics because it is highly compatible with ultra-fast MAD/SAD data collection and processing, and avoids the need to soak crystals in heavy metal solutions.

Finally, a structural model is built into the electron density map. This requires one more crucial piece of information, namely, the amino acid sequence, because carbon, oxygen, and nitrogen atoms cannot be distinguished with certainty by X-ray diffraction so amino acid side chains are difficult to identify. The resulting model is a set of *xyz* atomic coordinates assigned to all atoms except hydrogen (conventional X-ray diffraction cannot resolve the positions of hydrogen atoms and where these are present in crystal structures they have been added by modeling after the structure is determined). The more data used to create the electron density map, the greater the degree of certainty about the atomic positions and the higher the resolution of the model. Even so, there may be areas of the protein for which atomic positions cannot be determined precisely. Each atom is assigned a so-called **temperature factor**, which is a measure of certainty. The higher the temperature factor, the lower the certainty. High temperature factors indicate either a degree of disorder (that is, the particular atom was in different relative positions in different protein molecules within the crystal) or dynamism (that is, the particular atom had the tendency to vibrate around its rest position).

A recent development that may yet revolutionize the application of X-ray diffraction in structural genomics is the use of **X-ray mini-beams**, as narrow as 5 μm in diameter, which produce useful data from protein crystals that are too small for conventional synchrotrons. Furthermore, mini-beams can pinpoint the best-ordered areas of larger crystals to reduce scatter from inhomogeneous regions, increasing the signal-to-noise ratio from protein crystals without causing so much radiation damage.

NMR spectroscopy exploits the magnetic properties of certain atomic nuclei

Nuclear magnetic resonance (**NMR**) is a phenomenon that occurs because some atomic nuclei have magnetic properties. In NMR spectroscopy, these properties are used to obtain data about the relative position of atoms in a molecule, which can be refined into a set of structural models. Subatomic particles can be thought of as spinning on their axes, and in many atoms these spins balance each other out so that the nucleus itself has no overall spin. In hydrogen (^{1}H) and some naturally occurring isotopes of carbon and nitrogen (^{13}C, ^{15}N), the spins do not balance out and the nucleus possesses what is termed a **magnetic moment**. Quantum mechanics tells us that such nuclei can have one of two possible orientations because they are spin-½ particles, and in the absence of a magnetic field the spin polarization is random. However, in an applied magnetic field, the energy levels split because in one orientation the magnetic moment of the nucleus is aligned with the magnetic field and in the other it is not (**Figure 6.4**). Where such energy separations exist, nuclei can be induced to jump from the lower-energy magnetic spin state to the less favorable higher-energy state when exposed to radio waves of a certain frequency. This absorption is called resonance because the frequency of the radio waves coincides with the frequency at which the nucleus spins. When the nuclei flip back to their original orientations, they emit radio waves that can be measured. Protons (^{1}H) give the strongest signals, and this is the basis of protein structural analysis by **NMR spectroscopy**.

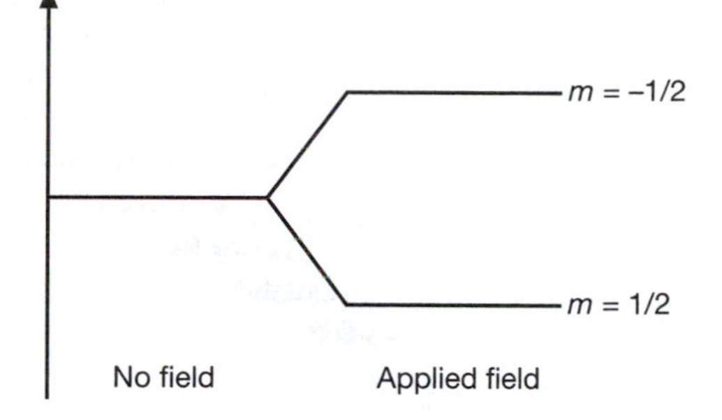

FIGURE 6.4 Energy levels in a nucleus with noninteger spin. In the absence of a magnetic field, the nucleus can exist in one of two orientations, each of which has the same energy. In an applied magnetic field, the energy levels split because in one orientation the magnetic moment of the nucleus is aligned with the field and in the other it is not. *m* is the magnetic quantum number.

Structures can be determined by NMR spectroscopy because the magnetic resonance frequency of each nucleus is influenced by nearby electrons in bonded atoms, which generate their own magnetic fields. The external magnetic field strength must be increased to overcome this opposition (**shielding**), and the degree of perturbation (**chemical shift**) depends on the chemical environment of each nucleus. In this way, it is possible to discriminate between hydrogen atoms in, for example, methyl and aromatic groups. The same principle applies to hydrogen atoms in proteins, but one-dimensional NMR experiments are insufficient for the analysis of such complex molecules because the resulting spectra contain large numbers of overlapping peaks.

One way to address this problem is to use a sequence of radio pulses separated by different time intervals to generate higher-dimensional NMR spectra with additional peaks indicating pairs of interacting nuclei. Several types of interactions can be measured by using different pulse sequences:

- **COSY** (**correlation spectroscopy**) detects, through two or three chemical bonds, the interaction between two protons that are bonded to carbon and/or nitrogen atoms. This allows the experimenter to trace a network of protons linked to bonded atoms.

- **TOCSY** (**total correlation spectroscopy**) detects groups of protons interacting through a coupled network, not just those joined to adjacent bonded pairs of carbon or nitrogen atoms. TOCSY can often identify all the protons associated with a particular amino acid, but cannot spread to adjacent residues because there are no protons in the carbonyl portion of the peptide bond.

- **NOESY** (**NOE spectroscopy**) takes advantage of the **nuclear Overhauser effect** (**NOE**), that is, signals produced by magnetic interactions between nuclei that are close together in space but not associated by bonds. This is most useful for determining protein structures because interactions can be identified between protons in separate amino acid residues.

The combination of COSY/TOCSY and NOESY is useful because COSY/TOCSY assigns protons to different spin systems representing individual amino acid residues and their side chains, whereas NOESY can help to determine the spatial relationships between the spin systems, thus determining the order of residues and the folding of the polypeptide backbone. The above methods are described as **homonuclear** techniques because only the ^{1}H signal is detected. For larger proteins, even the peaks on multidimensional spectra can become crowded, so homonuclear methods are best suited to small proteins and peptides. A higher resolution can be achieved using **heteronuclear** methods, which require the protein to be labeled with ^{13}C and ^{15}N. The extra information derived from heteronuclear NMR is important because magnetization can be transferred through the peptide bonds allowing the different spin systems (residues) to be connected. Even so, NMR spectroscopy is generally only suitable for the analysis of proteins in the 30–40 kDa range (although see below for exceptions).

When all the effects of chemical shifts and coupling are taken into account, the result of NMR analysis is a set of **distance restraints**, which are estimated distances between particular pairs of atoms (either bonded or nonbonded). If enough distance restraints are calculated, the number of protein structures that fit the data becomes finite. Therefore, NMR spectroscopy produces not a precise structure but an ensemble of models called a family that fit the data. The quality of NMR structural models increases with the number of restraints, and typically it is possible to achieve up to 15 restraints per residue with homonuclear NMR and up to 25 restraints per residue with heteronuclear NMR. This will produce models that are equivalent in resolution to a 0.2 nm model produced by X-ray diffraction. The resolution of NMR models can also be increased by incorporating **angle restraints** (restraints on the torsion angles of the chemical bonds), which can be derived from coupling constants, and also **orientation restraints** resulting from residual dipolar coupling (see below).

Like X-ray diffraction, there have been significant advances in NMR-based techniques for protein analysis that have accelerated the rate at which structures can be solved and have improved the accuracy of the resulting models. Currently, approximately 15% of the structures in the PDB are based on NMR data. Two major limitations of NMR spectroscopy in proteomics were the poor performance with large proteins and the restriction to proteins that can be easily solubilized. The problem with large proteins is twofold: first the tendency to produce overlapping signals (addressed in part using heteronuclear and multidimensional techniques) but also the faster relaxation of magnetization, which broadens and weakens the resulting signals. This has been addressed by the development of a novel procedure that reduces the rate of relaxation (**transverse relaxation optimized spectroscopy, TROSY**) combined with the deuteration of proteins, allowing the analysis of proteins of up to 900 kDa. The need for protein solutions has advantages as well as drawbacks, because solution NMR allows the investigation of protein dynamic behavior and interaction with ligands. Indeed, the noninvasive nature of NMR spectroscopy means that it has recently become possible to analyze protein structures in living cells (**in-cell NMR**). However, to analyze the structures of proteins that cannot be dissolved and are not suitable for crystallization (for example, fibrous proteins, membrane proteins), it is now possible to carry out **solid-state NMR**, which relies on the collection of distance restraints that are enhanced by anisotropic interactions in the solid state. In many cases, these broaden NMR spectra and reduce the resolution, but some enhance the structural information that can be obtained, especially internuclear dipolar coupling. Solid-state NMR has already been used to solve the structures of many crystalline proteins, fibrillar proteins, and transmembrane peptides.

Additional methods for structural analysis mainly provide supporting data

Although X-ray diffraction and NMR spectroscopy are regarded as the gold standards for structural determination, various other methods have been used either to provide additional information about solved structures or to look at the structures of recalcitrant proteins.

Far-UV circular dichroism spectrophotometry (CDS) is used to determine protein secondary structures. **Circular dichroism (CD)** is an optical phenomenon that occurs when molecules in solution are exposed to circularly polarized light. Asymmetric molecules such as proteins show different absorption spectra in left and right circularly polarized light, and this allows their secondary structures to be characterized. CDS using light between 160 and 240 nm generates distinct and characteristic spectra for proteins rich in α-helices and β-sheets, respectively (**Figure 6.5**). Although CD spectrophotometry cannot determine protein tertiary structures, the technique is a useful complement to XRC and NMR spectroscopy in structural biology. **Synchrotron radiation CD (SRCD)** allows the rapid structural classification of large numbers of proteins.

Alternative methods for the analysis of protein structure include neutron diffraction, electron diffraction, and electron microscopy. **Neutron diffraction** is used much less frequently than X-ray diffraction because neutron sources are less widely available and the flux of neutron beams is about ten orders of magnitude lower than that of X-ray beams. Neutrons are scattered by the nuclei of atoms in a protein crystal, rather than the electrons. The advantage of neutron scattering is that neutrons are scattered by hydrogen atoms, which cannot be "seen" by X-rays. Neutron diffraction is therefore used to determine the positions of important hydrogen atoms, such as those in critical hydrogen bonds or catalytic sites. **Electron diffraction** is used to study proteins that crystallize or naturally assemble into two-dimensional arrays but do not form orderly three-dimensional crystals. An example is tubulin, whose structure was solved by electron diffraction in 1998 because this protein forms large flat sheets in the presence of zinc ions. Electron microscopy is advantageous because single molecules can be analyzed in the same way as crystalline arrays, allowing the structures of large protein complexes to be determined without crystallization. Although three-dimensional information is lost in an electron microscope image, this can be reconstructed by repeating the analysis at many tilt angles. An interesting recent development is **electron tomography**, which can be used to study the structures of proteins and protein complexes inside the cell.

6.4 PROTEIN STRUCTURE PREDICTION

Structural predictions can bridge the gap between sequence and structure

Despite the technological advances in structural genomics that have increased the rate at which solved structures are deposited in the PDB, this is still a relatively slow and expensive process compared with the accelerating rate of sequence discovery and the falling cost of high-throughput sequencing.

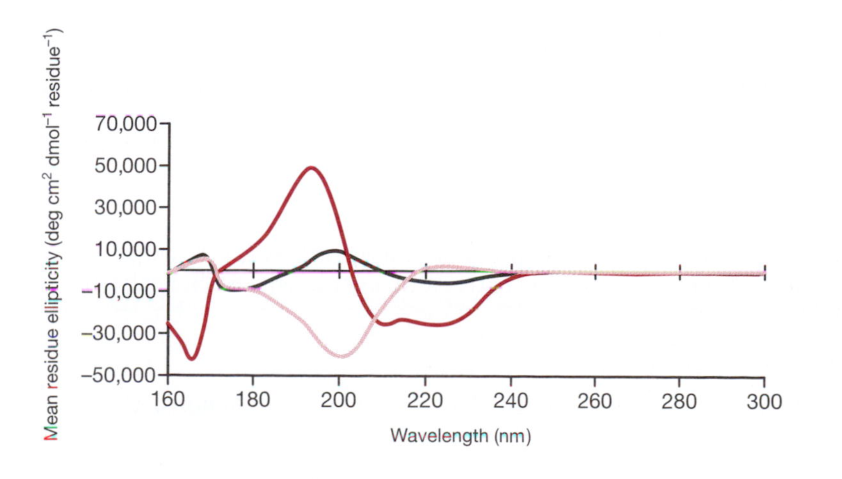

FIGURE 6.5 Circular dichroism spectra of diverse proteins. Spectra representing primarily α-helical (myoglobin, *red*), β-sheet (concanavalin A, *black*), and polyproline II helical (type VI collagen, *pink*) secondary structures, showing that substantial differences are present in the low-wavelength region. (From Wallace BA & Janes RW (2001) *Curr. Opin. Chem. Biol.* 5, 567. With permission from Elsevier.)

Therefore, the targets for structural genomics are chosen carefully to ensure they produce the best coverage of structure space in the smallest number of structures. Once a structural model of a protein is available, there are many bioinformatics-based methods that can be used to predict the structures of proteins with related sequences. There are also certain aspects of protein structure that can be predicted from sequences *de novo* without preexisting structural data. These two approaches, working from opposite ends of the challenge, will hopefully meet up eventually to complete the coverage of structure space. Although not as precise as empirical methods, structural prediction can be carried out in a largely automated manner.

Protein secondary structures can be predicted from sequence data

Secondary structure predictions represent a useful first step toward the structure of an entire fold. They are often known as **three-state predictions** because each residue in a protein sequence can generally be assigned as part of an α-helix (H), an extended β-strand (E), or an unstructured coil (C), with the other five recognized secondary structures playing comparatively minor roles (see Box 6.1 for additional information).

The **Chou–Fasman method** was one of the first empirical techniques for the prediction of secondary structures, and was based on the statistical likelihood of individual amino acid residues appearing in a given type of structure based on the analysis of known protein structures already solved by X-ray diffraction. Some amino acids, such as glutamate, have a **helical propensity** (that is, they are more likely to occur in α-helices than elsewhere in the protein) whereas others, such as valine, have a **strand propensity** (that is, they are more abundant in β-strands and β-sheets). Some amino acids, such as leucine, are equally likely to appear in helices and strands. Glycine and proline residues, due to their unusual residual groups and the effects these have on the flexibility of the polypeptide backbone, are rarely found in secondary structures at all. Indeed they are often found at the ends of helices and strands and thus act as secondary structure breakers (**Table 6.3**). The **GOR** (**Garnier–Osguthorpe–Robson**) **method** was developed at the same time and was based on similar principles (comparison with solved proteins) but also incorporated Bayesian statistics and pairwise interactions as conditional probabilities based on neighboring residues to improve the predictions. Even so, predictions using these methods are only accurate approximately 50% of the time.

Secondary structure predictions based on single proteins are unreliable because there are individual examples of all amino acids appearing in all types of secondary structure. Multiple alignments can remove much of this uncertainty by identifying conserved blocks of residues that favor the formation of helices or strands, and this is the basis of tools such as PSI-PRED, PORTER, and PHD, all of which use neural nets trained with sequence profiles to increase the accuracy of their predictions to over 75%. Other machine learning techniques such as **support vector machines** can also be included (for example, SSpro) and other programs incorporate the concept of relative solvent accessibility, that is, the degree to which different amino acid residues are accessible to the surrounding solvent (for example, SABLE, Jpred).

Accuracy in secondary structure predictions is generally expressed as the Q_3 score, the arithmetic mean of the correlation coefficients for helical, strand, and coil predictions:

$$Q_3 = (C_h + C_s + C_c)/3$$

where C_h is the correlation coefficient for helical predictions, C_s is the

TABLE 6.3 HELICAL AND STRAND PROPENSITIES OF THE AMINO ACIDS

Amino acid	Helical (α) propensity	Strand (β) propensity
Glu	1.59	0.52
Ala	1.41	0.72
Leu	1.34	1.22
Met	1.30	1.14
Gln	1.27	0.98
Lys	1.23	0.69
Arg	1.21	0.84
His	1.05	0.80
Val	0.90	1.87
Ile	1.09	1.67
Tyr	0.74	1.45
Cys	0.66	1.40
Trp	1.02	1.35
Phe	1.16	1.33
Thr	0.76	1.17
Gly	0.43	0.58
Asn	0.76	0.48
Pro	0.34	0.31
Ser	0.57	0.96
Asp	0.99	0.39

A value of 1.0 signifies that the propensity of an amino acid for the particular secondary structure is equal to that of the average amino acid, values greater than one indicate a higher propensity than the average, and values less than one indicate a lower propensity than the average. The values are calculated by dividing the frequency with which the particular residue is observed in the relevant secondary structure by the frequency for all residues in that secondary structure.

correlation coefficient for strand predictions, and C_c is the correlation coefficient for coil predictions.

The helical correlation coefficient C_h is calculated as follows, and the same principles are used to calculate C_s and C_c:

$$C_h = \frac{ab - cd}{\sqrt{(a+c)(a+d)(b+c)(b+d)}}$$

where a is the number of residues assigned correctly as helix, b is the number of residues assigned correctly as non-helix, c is the number of residues assigned incorrectly as helix, and d is the number of residues assigned incorrectly as non-helix.

Another relatively simple way to predict the occurrence of α-helices in proteins is to construct a **helical wheel**, a diagram in which the positions of amino acids are plotted on a circle corresponding to the pitch of an ideal α-helix (**Figure 6.6**). In globular proteins, α-helices tend to exhibit the clustering of hydrophobic residues on one face of the helix and the clustering of polar residues on the other. However, the transmembrane domains of membrane-spanning proteins often contain α-helices composed predominantly of hydrophobic residues. **Transmembrane helices** can therefore be identified by scanning the protein sequence with a moving window of about

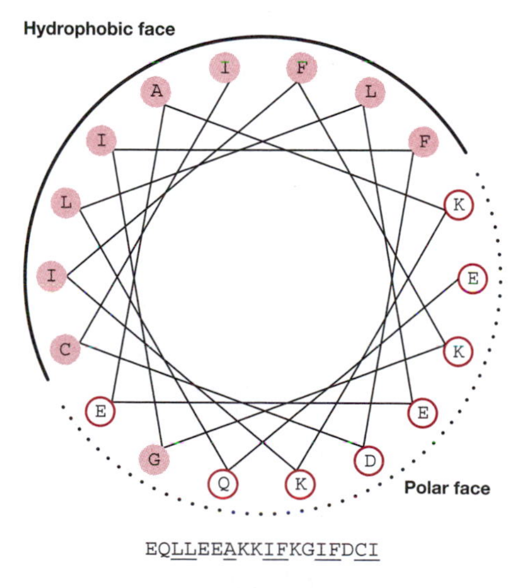

FIGURE 6.6 A short amino acid sequence plotted in the form of a helical wheel. The hydrophobic amino acids are underlined in the sequence and shown shaded on the helical wheel. Glycine can be thought of as either hydrophobic or polar as it has no side chain.

20 residues to identify highly hydrophobic segments, using tools such as SOUSI or PRED-TMR. Other methods for predicting transmembrane helices include hidden Markov models (for example, TMMOD) and the dense alignment surface method (DAS-TMfilter). Not all transmembrane structures are helices, and other programs have been developed to identify **transmembrane barrels** using a combination of hidden Markov models (for example, PRED-TMMB), neural nets (for example, TMBETA-NET), and clusters of membrane proteins (for example, TSEG).

Tertiary structures can be predicted by comparative modeling if a template structure is available

The tertiary structure of a protein can be predicted from its sequence with reasonable accuracy if the structure of a closely related protein is available and can be used as a template. This approach is known as **comparative modeling** or **homology modeling** and generally works well if the two sequences show >30% identity over 80 or more residues.

The first step in comparative modeling is to find suitable templates, which is achieved by searching for homologous protein sequences and identifying those with solved structures. This is not always possible, and the main limitation to comparative modeling as a structural prediction method is the lack of template structures, hence the overarching purpose of structural genomics to cover structure space by punctuating it with representative structures. When a suitable template can been found, the second step of the process is to align the query protein sequence on the structural template. Accurate alignment is critical for the quality of the final structural model, which is built in the third step of the process. Better models are obtained with more closely related proteins. If the template and query proteins are >70% identical, automatic alignment methods are suitable, but for less conserved proteins more human intervention becomes necessary. As discussed earlier, any residue known to be required for protein function (such as key residues in an active site) or to maintain structure (especially cysteine, glycine, and proline residues, see Box 6.1) are likely to have conserved positions in the protein structure and should be aligned in the sequence. If multiple template structures are available, it is often appropriate to superimpose the structures and use the average atomic positions in the modeling template.

Various software suites are available for homology modeling and they focus on different aspects of the process. CPHmodels and Domain Fishing focus on template selection, the former by searching sequence databases to build a PSSM before using this to select templates, and the latter by splitting the query protein into shorter domains that can each be used to select optimal templates. ESyPred3D and Geno3D focus on sequence alignment using a range of different multiple sequence alignment programs, helping to identify residues in the template that are part of the protein structural core and those forming surface loops. Generally, the positions of residues in the structural core are highly conserved because the core is rich in secondary structure. The loops are more variable and these often correspond to gaps in the sequence alignment, that is, places where the template and query proteins have different numbers of residues. Structural prediction in the loop regions is therefore more difficult. A simple method is to use a so-called **spare parts algorithm**, which searches through databases of loop structures from other proteins. The query protein may not necessarily be homologous to any of the proteins in this database, but particular loops of up to four residues in length may be analogous in sequence allowing their structures to be predicted. These structures are fitted into the model. The use of a spare parts algorithm may be combined with other methods that attempt to calculate the folding behavior of particular loops from first principles.

Once the path of the polypeptide backbone has been mapped, further algorithms are employed to predict the positions of the amino acid side chains. Where particular residues are conserved in the template and query sequence, the side-chain positions can be based on those in the template. For non-conserved residues, algorithms are employed that attempt to fill space in the protein interior in the most energetically favorable manner. The initial model may be refined through the use of energy-minimization software that makes minor adjustments to atomic positions to reduce the overall potential energy of the structure. The integrated software suite **SWISS-MODEL** provides resources to build and modify models and validate the different steps.

Ab initio prediction methods attempt to construct structures from first principles

The disadvantage of comparative modeling methods is that only structures with suitable templates can be modeled. In contrast, *ab initio* methods aim to predict protein tertiary structures from first principles, that is, in the absence of any structural information. A typical procedure would be to define a mathematical representation of a polypeptide chain and the surrounding solvent, define an energy function that accurately represents the physicochemical properties of proteins, and use an algorithm to search for the chain conformation that possesses the minimum free energy.

The problem with *ab initio* methods is that even short polypeptide chains can fold into a potentially infinite number of different structures. If enough solvent molecules are incorporated into the model to make it realistic, the system becomes too complex to study without applying some knowledge of the behavior of known proteins. For this reason, *ab initio* methods are impractical as approaches to structural prediction for polypeptides greater than about 200 residues. In the case of shorter polypeptides, recent results have been encouraging. For about a third of all polypeptides less than 150 residues in length that have been analyzed by such methods, one of the resulting models was close enough to the true structure to identify it in the PDB. However, the resolution of each model was poor, and the practical applications of *ab initio* prediction remain limited.

The most popular algorithm for *ab initio* prediction is Rosetta, which models structurally variable regions based on known structures, approximated non-local interactions between them and, the Monte Carlo method to minimize free energy. The huge computer resources required for this process have been mitigated by developing a **distributed computing project rosetta@home**, which works in a similar mode to the much more widely recognized project **folding@home**, namely, by recruiting the idle computer processing resources of thousands of volunteers and sending and receiving packages of data for analysis over the Internet.

Fold recognition (threading) is based on similarities between nonhomologous folds

Although in theory a given polypeptide chain could adopt an almost infinite number of different conformations, logic dictates that most of these would be energetically unfavorable and would never exist in nature. As discussed in Box 6.1, the way a protein chain folds—either by condensation around a nucleation site or through intermediate stages rich in secondary structure—also limits the total number of conformations that are possible. Finally, we know that the total number of protein folds in structure space is limited to a few thousand by the total number of sequences.

These observations and deductions suggest that searching the whole of conformational space looking for energy-efficient ways to fold a polypeptide chain is probably wasteful when only a few thousand energetically stable folds actually exist. Many hypothetical proteins are likely to have homologous structures in the PDB, but without sequence homology or empirical structural data such relationships cannot be recognized. **Fold recognition** (or **threading**) methods address this problem by detecting folds that can be used for structural modeling without homology at the sequence level.

The principle of fold recognition is the identification of folds that are compatible with a given query sequence, which can be achieved by using multiple sequence alignments to construct profiles and/or searching through a database of known protein structures, known as a **fold library**, scoring the folds and identifying candidates that fit the sequence, and aligning the query and best-scoring proteins. Once such a template has been identified, the remainder of the process is the same as comparative modeling. Fold recognition methods are generally based on both sequence similarity searches and structural information. For example, the 3D-PSSM method (three- dimensional position-specific scoring matrix) employs the PSI-BLAST algorithm to find sequences that are distantly related to the query protein and supplements this with secondary structure predictions and information concerning the tendency of hydrophobic amino acids to reside in the protein's structural core. This is the basis of tools such as pGenTHREADER and PHYRE, whereas M-TASSER uses a structure-based method for fold recognition to build a template followed by model assembly and refinement. As with homology modeling, threading methods are generally able to detect distantly related sequences but the accuracy of structural prediction is limited by errors in sequence alignment.

6.5 COMPARISON OF PROTEIN STRUCTURES

Once the tertiary structure of a protein has been determined by X-ray diffraction or NMR spectroscopy, or modeled by any of the techniques discussed above, it is deposited in the PDB and can be accessed by other researchers. As discussed at the beginning of the chapter, the key benefit of structural data in proteomics is the ability to compare protein structures and predict functions on the basis of conserved structural features. There are two requirements to fulfill this aim: an objective method for comparing protein structures and a system of structural classification that can be applied to all proteins, so that protein scientists in different parts of the world use the same descriptive language.

Several programs are available, many free over the Internet, which convert PDB files into three-dimensional models (for example, Rasmol, MolScript, Chime). Furthermore, a large number of algorithms have been written to allow protein structures to be compared. Generally, these work on one of two principles, although some of the more recent programs employ elements of both. The first method is **intermolecular comparison**, where the structures of two proteins are superimposed and the algorithm attempts to minimize the distance between superimposed atoms (**Figure 6.7a**). The function used to measure the similarity between structures is generally the **root mean square deviation** (**RMSD**), which is the square root of the average squared distance between equivalent atoms. The RMSD is smaller for structures that are more similar, and is zero if two identical structures are superimposed. Examples of such algorithms include Comp-3D and ProSup. The second method is **intramolecular comparison**, where the structures of two proteins are compared side by side, and the algorithm measures the internal distances between equivalent atoms within each structure and identifies alignments in which these internal distances are most closely matched (Figure 6.7b). An example of such an algorithm is DALI. Algorithms that employ both methods include COMPARER and VAST.

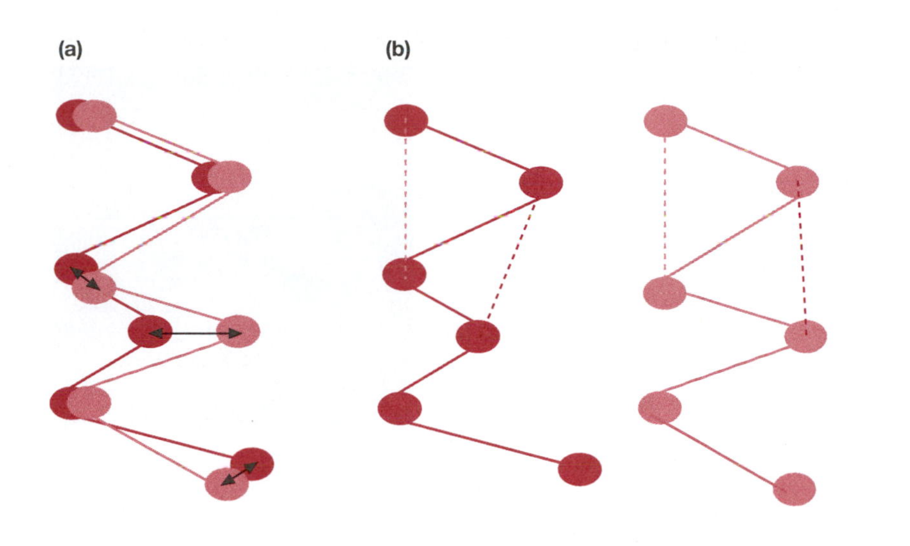

FIGURE 6.7 Comparison of protein structures. Circles represent Cα atoms of each amino acid residue and lines represent the path of the polypeptide backbone in space. (a) Intermolecular comparison involves the superposition of protein structures and the calculation of distances between equivalent atoms in the superimposed structures (shown as bidirectional arrows). These distances are used to calculate the root mean square deviation (RMSD), with the following formula

$$\text{RMSD} = \sqrt{\frac{1}{N}\sum_i d_i^2}$$

where d_i is the distance between the ith pair of superimposed Cα atoms and N is the total number of atoms aligned. A small RMSD value computed over many residues is evidence of significantly conserved tertiary structure. (b) Intramolecular comparison involves side-by-side analysis based on comparative distances between equivalent atoms within each structure (shown as color-coded dotted lines).

Similar methods are used to gauge the accuracy of structural models when the actual structures become available. When alignments are good, as is generally the case with comparative modeling, then very accurate models are possible. RMSDs of less than 0.1 nm represent very good predictions, because this is similar to the degrees of difference between two separate experimental determinations of the same protein structure. When the percentage sequence identity between template structures and target sequence exceeds 70% it is reasonable to expect that the model should be accurate to an RMSD of less than 0.2–0.3 nm even using completely automated methods. When the percentage identity drops below 40% then getting a good alignment, often with manual intervention, becomes more critical.

The **global distance test** (**GDT**) is a more accurate way to measure the similarity between protein structures, and is used when comparing models that have been solved independently by both X-ray diffraction and NMR spectroscopy. It is less sensitive to the impact of outlier regions created by the poor modeling of surface loops when the structure of the protein core is accurate, and is therefore the preferred metric for the annual benchmarking projects CASP (Critical Assessment of Techniques for Protein Structure Prediction) and CAFASP (Critical Assessment of Fully Automated Structure Prediction), which focus on expert prediction systems, as well as LiveBench and EVA, which run continually and focus on publically accessible prediction software.

6.6 STRUCTURAL CLASSIFICATION OF PROTEINS

Functional annotation on the basis of protein structure requires a rigorous and standardized system for the classification of different structures. Several different hierarchical classification schemes have been established, which divide proteins first into general classes based on the proportion of various secondary structures they contain (predominantly α-helix, predominantly β-strand, and mixed), then into successively more specialized groups based on how those structures are arranged. These schemes are implemented in databases such as SCOP, CATH, and FSSP (Box 6.2).

These databases differ in the way classifications are achieved. For example, the FSSP system is implemented through fully automated structural comparisons using the DALI program. CATH is semi-automatic, with comparisons carried out using the program SSAP, but the results of comparisons are manually curated. SCOP is a manual classification scheme and is based

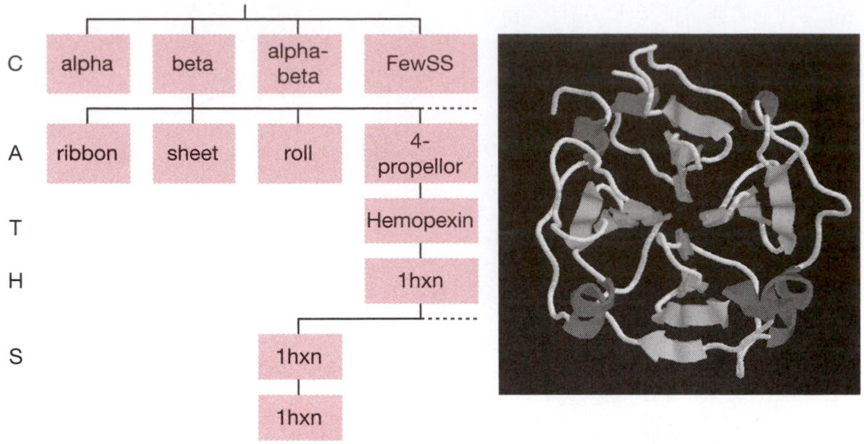

FIGURE 6.8 Structural classification of proteins using the CATH database. The protein shown is hemopexin, a protein rich in β-sheets with few α-helices. See also color plates. (Courtesy of Christine Orengo.)

on evolutionary relationships as well as geometric criteria. Not surprisingly, the same protein may be classified differently when the alternative schemes are used. There is broad general agreement in the upper levels of the hierarchy, but problems are encountered when more detailed classifications are sought because these depend on the thresholds used to recognize fold groups in the different classification schemes. An example CATH classification, showing the structural classification hierarchy, is shown in **Figure 6.8** (see also color plates).

Additional problems that lead to confusion in the structural classification of proteins include the existence of so-called **superfolds**, such as the TIM (triose phosphate isomerase) barrel, which are found in many proteins with diverse tertiary structures and functions. It is necessary to distinguish between homologous structures (which are derived from a common evolutionary ancestor) and analogous structures (which evolved separately but have converged). Similarly, variations in the fold structure between diverse members of the same protein family can result in a failure to recognize homologous relationships. In its most extreme form, this can be seen as the **Russian doll effect**, which describes the continuous variation of structures between fold groups (**Figure 6.9** and color plates).

6.7 GLOBAL STRUCTURAL GENOMICS INITIATIVES

Structural genomics initiatives have been set up all over the world, some comprising dispersed laboratories working toward a common goal and some focused at particular centralized sites. In America, the National Institute of General Medical Sciences (NIGMS) funded nine structural genomics pilot centers, and several additional academic and industrial consortia have been established in America, Europe, and Japan (see Further Reading). Although the overall goal of structural genomics is to provide representative structures for all protein families, various different approaches have been used to select an initial set of target proteins. Research has focused on microbes, which have smaller genomes (and thus smaller proteomes) than higher eukaryotes, but a fundamentally similar basic set of protein structures. Several groups chose thermophilic bacteria such as *Methanocaldococcus jannaschii* for their pilot studies, on the basis that proteins from these organisms should be easy to express in *Escherichia coli* in a form suitable for crystallography and/or NMR spectroscopy. A favorable strategy in model eukaryotes is to focus on proteins that are implicated in human diseases, for example, the Tuberculosis Structural Genomics Consortium focuses on *Mycobacterium tuberculosis*, and is examined as a case study in **Box 6.3**. Overall, the idea

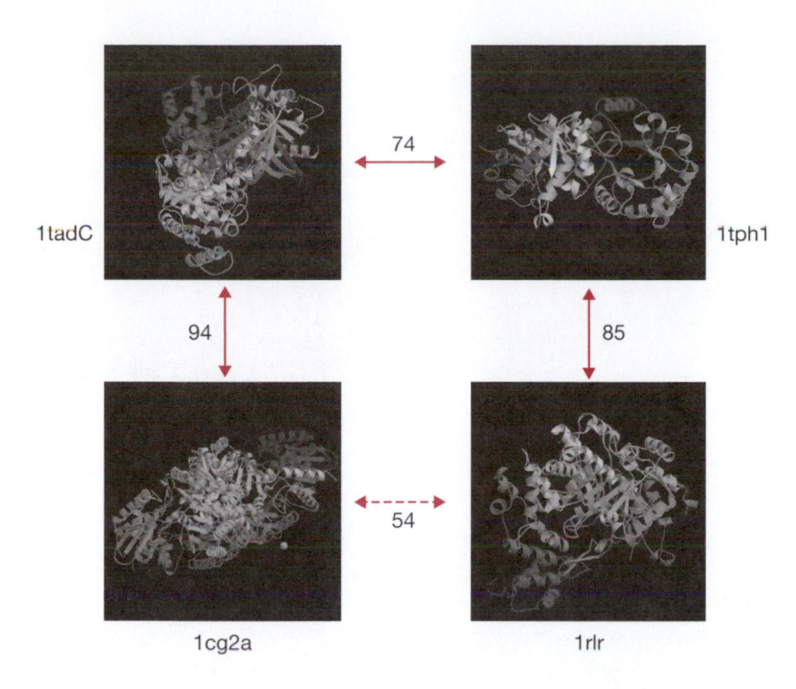

FIGURE 6.9 The Russian doll effect. Four proteins are illustrated that show continuous structural variation over fold space. Each of the proteins shares at least 74 structurally equivalent residues with its nearest neighbor, but the two extreme proteins show only 54 structurally equivalent residues when compared directly. Key: 1cg2a, carboxypeptidase G2; 1tadC, transducin-K; 1tph1, triose phosphate isomerase; 1rlr, ribonucleotide reductase protein R1. See also color plates. (From Domingues FS, Koppensteiner WA & Sippl MJ (2000) *FEBS Lett.* 476, 98. With permission from Elsevier. Images courtesy of Protein Data Bank.)

has been to choose structures that maximize the amount of information returned from each structural genomics program (**Figure 6.10**).

Much of the progress described above in terms of technological advances in structural determination and modeling, the use of novel cloning systems that facilitate high-throughput analysis (for example, ligation-independent cloning and PIPE—polymerase incomplete primer extension), small-angle X-ray scattering, improved access to synchrotron facilities, improved salvage strategies for recalcitrant proteins, solid-state NMR, and better software for

BOX 6.3 CASE STUDY.
The TB Structural Genomics Initiative.

The TB Structural Genomics Consortium was launched in 2000 and currently includes nearly 150 principal investigators representing 94 research organizations in 15 countries. The overall aim of the consortium is to solve the structures of *Mycobacterium tuberculosis* proteins, focusing on those that will improve the diagnosis and treatment of tuberculosis.

The core of the consortium is a pipeline comprising several facilities that respectively carry out cloning, protein expression, protein purification, and X-ray data collection. The cloning facility at Texas A&M University has prepared a proteomewide library of 3600 vectors, and several tiers of genes have been submitted for protein expression based on their perceived therapeutic value. The initial set (Top 100) were selected as the most promising drug targets whereas the second set (Target 600) were selected based on two criteria: their necessity for bacterial survival and their coverage of structure space. This selection strategy therefore demonstrates a mixture of medically relevant structures and those required to meet the general aims of structural genomics initiatives.

To increase the success rate and reduce candidate attrition at each stage of the pipeline, the consortium has developed a range of expression vectors that accommodate different tags, and has embraced a number of salvaging pathways to reduce the loss of candidates at the crystallization stage (for example, surface entropy reduction, p. 115). The data collection facility at the Lawrence Berkley National Laboratory offers a third-generation synchrotron source with small-angle X-ray scattering (SAXS) to identify protein aggregates in solution, which can be used as a quality indicator for protein crystallization.

By 2012, the consortium had solved the structures of approximately 250 proteins (accounting for more than one-third of the *M. tuberculosis* proteins in the PDB). Many were solved as complexes with their natural substrates and cofactors, providing additional data relevant for the rational design of drugs. In addition to the protein structures, the consortium has provided extensive protein expression and interaction data. The structural data in the PDB are complemented by a gene expression correlation database hosted by the consortium (gecGrid), algorithms to predict functional interactions (Prolinks), and the ProKnow database, which links the protein structures, functions, and interactions with Gene Ontology terms.

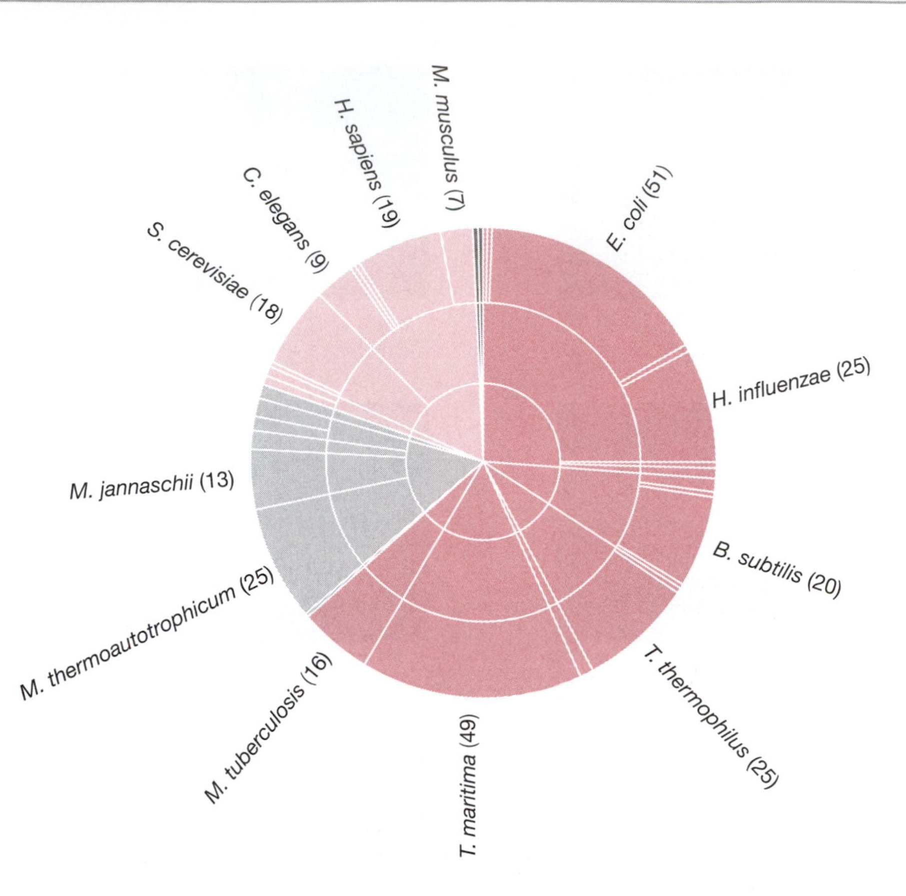

FIGURE 6.10 Species distribution of 316 structures solved by major structural genomics initiatives. The colored sectors represent superkingdoms (dark pink = bacteria, light gray = archaea, light pink = eukaryotes, dark gray = viruses) and the concentric circles represent descending levels of the NCBI Taxonomy Database. Labels on the outer ring identify species with the largest number of solved structures. (From Todd AE, Marsden RL, Thornton JM & Orengo CA (2005) *J. Mol. Biol.* 384, 1235. With permission from Elsevier.)

model building and comparison, has been driven by these programs. The benefits of these large-scale collaborations are becoming clear. An analysis of the international structural genomics consortia in 2005 found that they contributed more than half of all novel structurally categorized protein families and more than five times the number of unique novel folds as the rest of the community, despite accounting for only 20% of the structures annually deposited in the PDB (**Figure 6.11**).

In principle, the value of the structural genomics approach has been validated by the functional annotation of many of the initial hypothetical proteins chosen for structural analysis. For example, of the first 10 proteins analyzed in the *Methanobacterium thermoautotrophicum* project, 7 could be assigned a function due to structural similarity with known protein folds

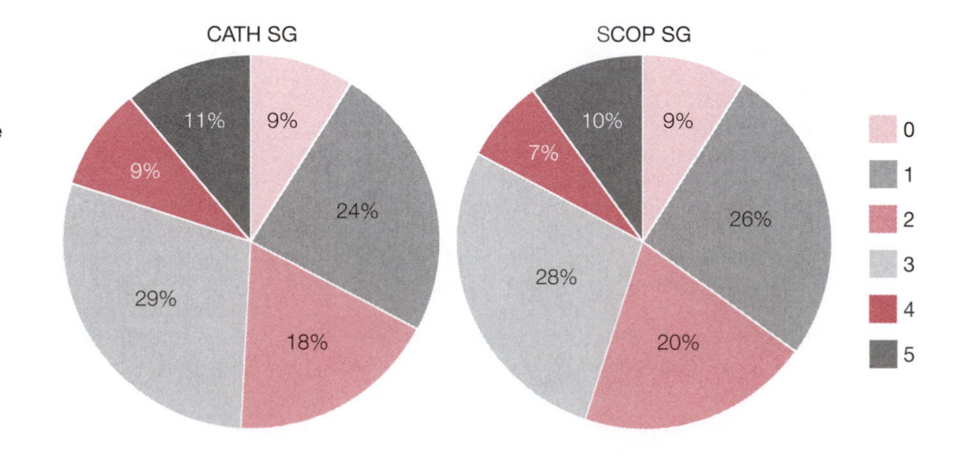

FIGURE 6.11 Novelty of folds discovered by structural genomics projects. Ratings are based on CATH and SCOP classifications and use the following criteria: 0 = identical or near-identical to PDB protein; 1 = more than 30% identity to PDP protein; 2 = distant homolog detectable by HMM methods; 3 = distant homolog detected by structure comparison only; 4 = new superfamily from known fold; 5 = new fold. (From Todd AE, Marsden RL, Thornton JM & Orengo CA (2005) *J. Mol. Biol.* 384, 1235. With permission from Elsevier.)

or other structural criteria, including the presence of bound ligands in the crystal (**Table 6.4**). The presence of a ligand or cofactor can often be helpful, and was instrumental in the functional annotation of *M. jannaschii* hypothetical protein MJ0577, the first structure to be generated in a structural genomics initiative. In this example, the crystal contained ATP, suggesting a role in ATP hydrolysis that was later confirmed by biochemical experiments. Therefore, even when the structure of a protein does not match any other in the database, structural analysis may still provide functional information that can be followed up with other experiments. Another interesting example is hypothetical protein TM0423, from *Thermotoga maritime*, which co-purified and co-crystallized with a molecule of Tris buffer. In this case, the position of the buffer suggested that the protein would be able to bind to glycerol, and identified it as a glycerol hydrogenase.

The main goal of structural genomics is to assign functions to hypothetical proteins on the basis of their relationship to known folds, but another is to discover new folds and assemble a comprehensive directory of protein space. It appears that about 35% of the structures emerging from current structural genomics initiatives contain novel folds, which confirms the hypothesis that protein space is finite and probably comprises at most a few thousand distinct structures. Every time a new fold is discovered, a little bit more of that protein space is filled. Furthermore, many of the new folds can

TABLE 6.4 SOME OF THE EARLY STRUCTURES DETERMINED BY STRUCTURAL PROTEOMICS INITIATIVES

Target ID	Organisms	Technique	Fold family	Function
HI1434	*H. influenzae*	X-ray	Novel	Unknown
Maf	*B. subtilis*	X-ray	MJ0226-like	Nucleoside triphosphate binding
MJ0226	*M. jannaschii*	X-ray	Novel	Nucleoside triphosphate hydrolysis
MJ0541	*M. jannaschii*	X-ray	Unknown	Nicotinamide mononucleotide adenylyltransferase
MJ0577	*M. jannaschii*	X-ray	Adenine nucleotide-binding domain-like	ATP hydrolysis
MJ0882	*M. jannaschii*	X-ray	Rossmann fold	Unknown
MTH1048	*M. thermoautotrophicum*	NMR	Novel	Subunit in RNA polymerase II
MTH1175	*M. thermoautotrophicum*	NMR	Ribonuclease H-like	Unknown
MTH1184	*M. thermoautotrophicum*	NMR	Novel	Unknown
MTH0129	*M. thermoautotrophicum*	X-ray	TIM-barrel	Orotidine 5′-monophosphate decarboxylase
MTH0150	*M. thermoautotrophicum*	X-ray	Nucleotide-binding	NAD^+ binding
MTH0152	*M. thermoautotrophicum*	X-ray	Novel	Ni^{2+} and FMN binding
MTH1615	*M. thermoautotrophicum*	NMR	Armadillo repeat	DNA binding, transcription factor
MTH1699	*M. thermoautotrophicum*	NMR	Ferredoxin-like	Transcription elongation factor
MTH0040	*M. thermoautotrophicum*	NMR	Three-helix bundle	Zn^{2+} binding, scaffold in RNA polymerase II
MTH0538	*M. thermoautotrophicum*	NMR	Rossmann fold	Mg^{2+} binding, putative ATPase
Ybl036C	*S. cerevisiae*	X-ray	TIM barrel	PLP binding
Ycih	*E. coli*	NMR	eIF1-like	Translation initiation factor
YjgF	*E. coli*	X-ray	*Bacillus* chorismate mutase-like	Unknown
Yrdc	*E. coli*	X-ray	Novel	RNA binding

TIM, triose phosphate isomerase; eIF, eukaryotic initiation factor; FMN, flavin mononucleotide; NAD^+, nicotinamide adenine dinucleotide; PLP, pyridoxal phosphate.

be assigned functions because they bind particular ligands or have other properties, and this reveals new structure–function relationships that can be applied more widely. Sequence analysis and structural comparisons with these novel folds can identify previously unanticipated evolutionary relationships. At some point in the future, we may reach the stage where there is no such thing as an orphan gene or a hypothetical protein.

FURTHER READING

Aloy P, Oliva B, Querol E et al. (2002) Structural similarity to link sequence space: new potential superfamilies and implications for structural genomics. *Protein Sci.* 11, 1101–1116.

Banci L, Bertini I, Luchinat C & Mori M (2010) NMR in structural proteomics and beyond. *Prog. Nucl. Magn. Reson. Spectrosc.* 56, 247–266.

Bork P & Koonin EV (1998) Predicting functions from protein sequences—where are the bottlenecks? *Nat. Genet.* 18, 313–318.

Brenner SE (2001) A tour of structural genomics. *Nat. Rev. Genet.* 2, 801–809.

Burley SK, Almo SC, Bonanno JB et al. (1999) Structural genomics: beyond the Human Genome Project. *Nat. Genet.* 23, 151–157.

Campbell ID (2002) Timeline: the march of structural biology. *Nat. Rev. Mol. Cell Biol.* 3, 377–381.

Chim N, Habel JE, Johnston JM et al. (2011) The TB Structural Genomics Consortium: a decade of progress. *Tuberculosis* 91, 155–172.

Cim N, McMath LM, Beeby M & Goulding CW (2009) Advances in *Mycobacterium tuberculosis* structural genomics: investigating potential chinks in the armor of a deadly pathogen. *Infect. Disord. Drug Targets* 9, 475–492.

Heinemann U (2000) Structural genomics in Europe: slow start, strong finish? *Nat. Struct. Biol.* 7, 940–942.

Joachimiak A (2009) High-throughput crystallography for structural genomics. *Curr. Opin. Struct. Biol.* 19, 573–584.

Kim Y, Babnigg G, Jedrzejczak R et al. (2011) High-throughput protein purification and quality assessment for crystallization. *Methods* 55, 12–28.

Lan N, Montelione GT & Gerstein M (2003) Ontologies for proteomics: towards a systematic definition of structure and function that scales to the genomic level. *Curr. Opin. Chem. Biol.* 7, 44–54.

Lesley SA, Kuhn P, Godzick A et al. (2002) Structural genomics of the *Thermotoga maritima* proteome implemented in a high-throughput structure determination pipeline. *Proc. Natl Acad. Sci. USA* 99, 11664–11669.

Levitt M (2009) Nature of the protein universe. *Proc. Natl Acad. Sci. USA* 106, 11079–11084.

Mao B, Guan R & Montelione GT (2011) Improved technologies now routinely provide protein NMR structures useful for molecular replacement. *Structure* 19, 757–766.

Marsden ML, Lewis TA & Orengo CA (2007) Towards a comprehensive structural coverage of completed genomes: a structural genomics viewpoint. *BMC Bioinformatics* 8, 86.

Mittl PRE & Grutter MG (2001) Structural genomics: opportunities and challenges. *Curr. Opin. Chem. Biol.* 5, 402–408.

Montelione GT, Arrowsmith C, Girvin ME et al. (2009) Unique opportunities for NMR methods in structural genomics. *J. Struct. Funct. Genomics* 10, 101–106.

Murillo AC, Li HY, Alber T et al. (2007) High throughput crystallography of TB drug targets. *Infect. Disord. Drug Targets* 7, 127–139.

Murzin AG (1998) How divergent evolution goes into proteins. *Curr. Opin. Struct. Biol.* 8, 380–387.

Orengo CA & Thornton JM (2005) Protein families and their evolution—a structural perspective. *Annu. Rev. Biochem.* 74, 867–900.

Pavlopoulou A & Michalopoulos I (2011) State-of-the-art bioinformatics protein structure prediction tools. *Int. J. Mol. Med.* 28, 295–310.

Sali A, Glaeser R, Earnes, T & Baumeister W (2003) From words to literature in structural proteomics. *Nature* 422, 216–225.

Slabinski L, Jaroszewski L, Rodrigues AP et al. (2007) The challenge of protein structure determination—lessons from structural genomics. *Protein Sci.* 16, 2472–2482.

Stevens RC, Yokohoma S & Wilson IA (2001) Global efforts in structural genomics. *Science* 294, 89–92.

Structural Genomics Consortium, China Structural Genomics Consortium, Northeast Structural Genomics Consortium et al. (2008) Protein production and purification. *Nat. Methods* 5, 135–146.

Terwilliger TC (2000) Structural genomics in North America. *Nat. Struct. Biol.* 7, 935–939.

Terwilliger TC, Park MS, Waldo GS et al. (2003) The TB structural genomics consortium: a resource for *Mycobacterium tuberculosis* biology. *Tuberculosis* 83, 223–249.

Yokoyama S, Hirota H, Kigawa T et al. (2000) Structural genomics projects in Japan. *Prog. Biophys. Mol. Biol.* 73, 363–376.

Zhang C & Kim SH (2003) Overview of structural genomics: from structure to function. *Curr. Opin. Chem. Biol.* 7, 28–32.

CHAPTER 7

Interaction proteomics

7

7.1 INTRODUCTION

In Chapters 5 and 6, we explored the link between protein structure and function, and showed that the tertiary structure of a protein determines the overall shape of the molecule, the distribution of surface charges and the juxtaposition of critical functional residues. Such residues might constitute the active site of an enzyme, the ligand-binding site of a receptor, or the antigen-recognition domain of an antibody. The structure of a protein therefore influences its function by determining the other molecules with which it can interact and the consequences of those interactions. Proteins interact with small molecules, nucleic acids, and/or other proteins, and such interactions lie at the heart of every biological process. Almost all proteins are gregarious, functioning as parts of larger complexes rather than working in isolation. Within such complexes, the interactions between proteins may be static or transient, the latter occurring, for example, in signaling pathways, often as a consequence of reversible post-translational modifications (Chapter 8). From the above, it is clear that protein interactions and functions are intimately related, and it follows that the investigation of protein interactions can help in the functional annotation of uncharacterized proteins and their grouping into functional networks.

Interaction proteomics or **interactomics** is the investigation of protein interactions at multiple levels (**Figure 7.1**). The highest-resolution methods are those we considered in Chapter 6, namely, X-ray diffraction and nuclear magnetic resonance spectroscopy, which can help to characterize interactions on the atomic scale, producing detailed data that show the precise structural relationships between interacting chemical groups. These methods reveal the precise configuration of the **interfaces** between interacting proteins and are also useful for revealing interactions between proteins and small-molecule ligands, substrates, and co-factors. As discussed in Chapter 6, the presence of a bound ligand or co-factor may help to reveal the function of a protein whose structure has been resolved but whose activity is unknown (p. 129).

The key interaction proteomics platforms allow the high-throughput analysis of protein interactions on the molecular scale. These are typically used to study **protein–protein interactions**, although variant methods have been developed to probe interactions between proteins and nucleic acids or small molecules. Some methods detect **binary interactions** (that is, interactions between pairs of proteins) and others detect **complex interactions** (that is, interactions between multiple proteins to form complexes). These methods do not reveal the precise chemical nature of

Method	Atomic scale	Molecular scale: Binary interactions	Molecular scale: Complex interactions	Cellular scale
X-ray and NMR	■	■		
Competition binding		■		
Gel-retardation assays		■		
ELISA		■	■	
Two-hybrid test		■	■	
Affinity purification		■	■	
BIAcore sensor chip/plasmon resonance		■	■	
Electron paramagnetic resonance	■	■	■	
Gel-filtration chromatography		■	■	
Mass-spectrometric screening		■	■	
Cross-linking		■	■	
Co-immunoprecipitation		■	■	
Co-sedimentation		■	■	
Sizing column		■	■	
Sucrose-gradient sedimentation			■	
Co-purification			■	
Electron microscopy			■	
Native gel			■	
Immunoprecipitation				■
Immunofluorescence				■
Immunolocalization				■
Immunostaining	■	■	■	■
FRET analysis			■	■
Monoclonal antibody blockage				■
Interaction adhesion assay				■
Knock-out				■
Antisense				■
Transient co-expression				■

FIGURE 7.1 Methods to detect protein interactions have different resolutions from atomic, through molecular to cellular. The direct analysis of interactions on a molecular scale can focus on binary interactions (only two proteins) or complex interactions (more than two proteins). (From Xenarios I & Eisenberg D (2001) *Curr. Opin. Biotechnol.* 12, 334. With permission from Elsevier.)

the interactions but simply report that such interactions take place. Many different techniques can be used to detect molecular interactions on a small scale, but the major high-throughput technologies are **two-hybrid systems** and their derivatives for binary interactions and **systematic affinity purification–mass spectrometry** techniques for the characterization of protein complexes. Protein microarrays are also emerging as useful tools for the characterization of protein interactions, but we defer the discussion of these miniature devices until Chapter 9.

In reductionist terms, molecular interaction analysis is useful for the functional characterization of proteins because proteins that interact are likely to be engaged in the same activity (**guilt by association**). The principle is that if an uncharacterized protein X interacts with proteins Y and Z, both of which are already known to be required for mRNA splicing, then it is likely that protein X is also a splicing component of some description. Interactions in the context of an assay do not necessarily represent a definitive functional association *in vivo*. Indeed, interaction assays are renowned for their tendency to generate false positives, so further experiments are necessary to validate the results. Furthermore, some proteins interact promiscuously with many different partners so even genuine interactions detected in an assay may not be functionally relevant. However, molecular interaction analysis is useful in global terms because it can be applied on a massive scale, and can therefore be used to construct **interaction maps** of the entire proteome, that is, diagrams that show proteins or protein complexes as nodes and interactions between them as links (see Section 7.11). As we shall see later, such diagrams not only provide a holistic view of the functioning cell but also show which proteins or complexes are vital hubs in the system and which are redundant. This can help in the selection of compounds that bind to critical interaction interfaces (**interaction hotspots**), which are often the most effective drug targets (Chapter 10).

Finally, interaction studies can be carried out at the cellular level by determining where proteins are localized. This important but often-overlooked component of interaction data can support molecular interaction studies by placing two proteins in the same place at the same time, and can provide evidence against spurious functional interactions suggested by molecular assays (if two proteins that apparently interact *in vitro* never actually encounter each other *in vivo*). It may also be possible to predict the function of a protein directly from its localization, for example, a nuclear pore protein, a component of a flagellum, or part of the actin cytoskeleton (**Box 7.1**).

BOX 7.1 ALTERNATIVE APPLICATIONS.
Protein localization and organelle proteomics.

Protein localization can provide important evidence either to support or challenge the data from interaction screens. For example, the co-localization of proteins in the same compartment supports the potential for interactions, whereas the demonstration that two proteins never exist in the same compartment refutes it, even if the proteins do happen to interact when present in the same solution. Some of the techniques described in this chapter are suitable for the analysis of protein interactions *in vivo* and can reveal their compartmentalization by the localization of fluorescence, but it is important to exclude artifacts caused, for example, by protein overexpression, which can result in proteins escaping from their normal compartments and contaminating others.

As well as helping to confirm or refute putative interactions, protein localization data can be useful in their own right to provide functional annotations (for example, proteins located in the thylakoid membrane of a chloroplast are probably involved in photosynthesis) and to define targets (for example, secreted proteins and cell surface proteins are often useful drug targets, and in the case of pathogens may also be useful for vaccine development). For this reason, many investigators have carried out studies of subcellular compartments or organelles (**organelle proteomics**) by isolating the appropriate fraction before analysis. Unsurprisingly, this can increase the resolution of proteomic data. For example, when identical amounts of protein from a total macrophage lysate or purified phagosomes were analyzed by 2DGE-MS, several hundred spots were identified in each gel even though the phagosome proteome should be a subset of the total cell lysate. Instead, only about 20 of the phagosome proteins were detected in the total cell lysate, which was dominated by actin. More than 90% of the organelle proteins were undetected in the total lysate analysis.

Several large-scale screens have been carried out to determine the localization of proteins, one of which involved the systematic replacement of yeast proteins with tagged versions that could be detected using antibodies. This revealed that about half of the yeast proteome is cytosolic, about 25% is nuclear, 10–15% is mitochondrial, and 10–15% is found in the secretory pathway. About 20% of proteins overall were localized in membranes of various organelles or the plasma membrane. Similar studies have been conducted with fluorescent protein tags allowing real-time analysis and the assignment of proteins into more than 20 compartment-specific categories in yeast, mammalian, and plant cells. Protein localization can also be predicted directly from sequence data using a number of algorithms that have been refined for particular species, some of which are shown in **Table 1**.

TABLE 1

Predictor name	Website URL	Organism	Number of subcellular locations covered
Hum-mPLoc	http://chou.med.harvard.edu/bioinf/hum-multi	Human	14 (including those for proteins with multiple sites)
Plant-PLoc	http://chou.med.harvard.edu/bioinf/plant	Plant	11
Euk-mPLoc	http://chou.med.harvard.edu/bioinf/euk-multi	Eukaryotic	22 (including those for proteins with multiple sites)
Gneg-PLoc	http://chou.med.harvard.edu/bioinf/Gneg	Gram negative	8
Gpos-PLoc	http://chou.med.harvard.edu/bioinf/Gpos	Gram positive	5
Virus-PLoc	http://chou.med.harvard.edu/bioinf/virus	Virus	7

From Chou KC & Shen HB (2008) *Nat. Protoc.* 3, 153. With permission from Macmillan Publishers Ltd.

This chapter begins by introducing the principles of interaction analysis and showing how these principles have been developed into high-throughput technologies. The accomplishments of proteome-scale interaction analysis are discussed and we conclude by considering the bioinformatics strategies for dealing with interaction data.

7.2 METHODS TO STUDY PROTEIN–PROTEIN INTERACTIONS

Protein–protein interactions are central to virtually every biological process and many different analytical methods have been developed to study them. Most of these methods are suitable for studying interactions within a small group of proteins and cannot be employed on a proteomic scale, but the principles are similar to the higher-throughput technologies discussed later on. Small-scale methods are often used to corroborate the data produced in large-scale studies and can thus help to eliminate false positives.

Genetic methods suggest interactions from the combined effects of two mutations in the same cell or organism

Classical genetics can be used to investigate protein interactions by combining different mutations in the same cell or organism and observing the resulting phenotype. This approach has been widely used in genetically amenable microbial species and model metazoans species such as the fruit fly *Drosophila melanogaster*, the nematode *Caenorhabditis elegans*, the mouse *Mus musculis*, and the small flowering plant *Arabidopsis thaliana*.

A straightforward example is a screen for **suppressor mutants**, that is, secondary mutations that correct the phenotype of a primary mutation. As shown in **Figure 7.2**, the principle is that a mutation in the gene for protein X that prevents its interaction with protein Y will result in a loss of function that generates a mutant phenotype. However, a second mutation in the gene for protein Y could introduce a compensatory change that restores the interaction. Suppressor mutants identified in the screen are then mapped and the corresponding genes and proteins identified.

The advantage of genetic screens is that they help to identify functionally significant interactions, sifting through the proteome for those interactions that have a recognizable effect on the overall phenotype. However, it is important to remember that genetic screens only provide indirect evidence for interactions and further direct molecular evidence is necessary. One potential problem is that the suppressor mutation may map to the same gene as the primary mutation, because a second mutation in the same gene can suppress the primary mutant phenotype by introducing a compensatory conformational change within the same protein. Even if the suppressor maps to a different gene, the two gene products might not actually interact. For example, a mutation that abolishes the activity of an enzyme required for amino acid biosynthesis could be suppressed by a gain-of-function mutation in a transport protein that increases the uptake of that amino acid from the environment.

Another genetic approach is a screen for **enhancer mutations**, that is, those that worsen the phenotype generated by a primary mutation. One example

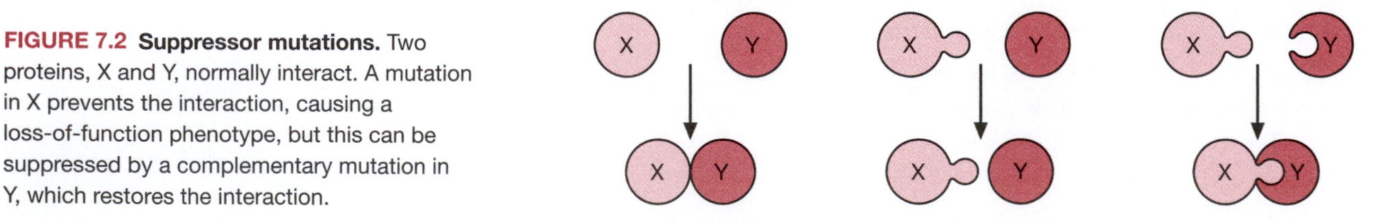

FIGURE 7.2 Suppressor mutations. Two proteins, X and Y, normally interact. A mutation in X prevents the interaction, causing a loss-of-function phenotype, but this can be suppressed by a complementary mutation in Y, which restores the interaction.

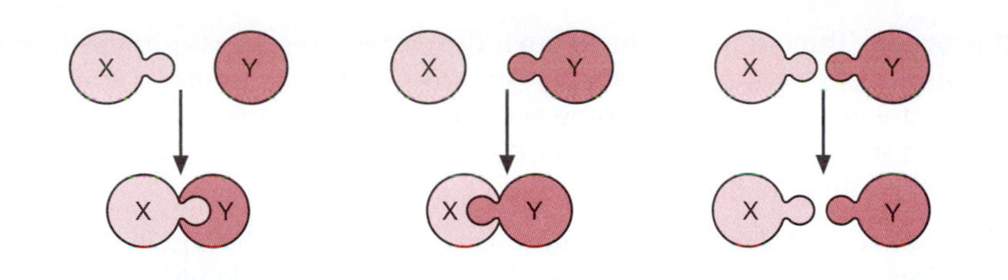

FIGURE 7.3 Synthetic lethal effect. The same two proteins can still interact if there is a mutation in either X or Y that does not drastically affect the interaction between them. However, if the mutations are combined, protein interaction is abolished and a lethal loss-of-function phenotype is generated.

of this strategy is the **synthetic lethal screen**, where individual mutations in the genes for proteins X and Y do not prevent interaction and are therefore viable, but simultaneous mutations in both genes prevent the interaction and result in a lethal phenotype (**Figure 7.3**). The *Saccharomyces cerevisiae* **synthetic genetic array** (**SGA**) is a high-throughput example of this principle in which yeast strains with mutations in a single gene are crossed against an array of 5000 viable deletion mutants representing much of the genome, allowing synthetic lethal interactions to be mapped systematically.

Mutations in different genes that generate similar phenotypes often indicate that the corresponding proteins are part of the same complex or the same biochemical or signaling pathway. For pathways, the order of protein function can often be established by experiments to determine **epistasis**, a genetic phenomenon in which a mutation in one gene masks the phenotype of a mutation at another locus. In this type of experiment, loss-of-function and gain-of-function mutations (with opposite phenotypes) are combined in the same cell or organism. If a loss-of-function mutation in gene X overrides (is epistatic to) the phenotype of a gain-of-function mutation in gene Y, this suggests that protein X acts downstream of protein Y in the pathway and the two corresponding proteins may interact (**Figure 7.4**). Analogous to the SGA described above, **epistatic miniarray profiles** (**E-MAPs**) can be generated to systematically test panels of proteins for epistatic interactions.

Like suppressor mutants, synthetic lethal/enhancer mutants and epistatic interactions only suggest that two gene products interact. There are many other plausible explanations for such genetic effects, and candidate protein interactions must be confirmed at the biochemical level.

Protein interactions can be suggested by comparative genomics and homology transfer

The availability of complete genome sequences for many different organisms allows **comparative genomics** to be used for the functional annotation

FIGURE 7.4 Establishing gene order in a pathway by epistasis. A loss-of-function mutation in either gene X or gene Y causes a plant to become stunted. A gain-of-function mutation in gene Y causes a growth burst. If the phenotype of the double mutant is stunted, then X acts downstream of Y, but if its phenotype is tall, then the converse is true.

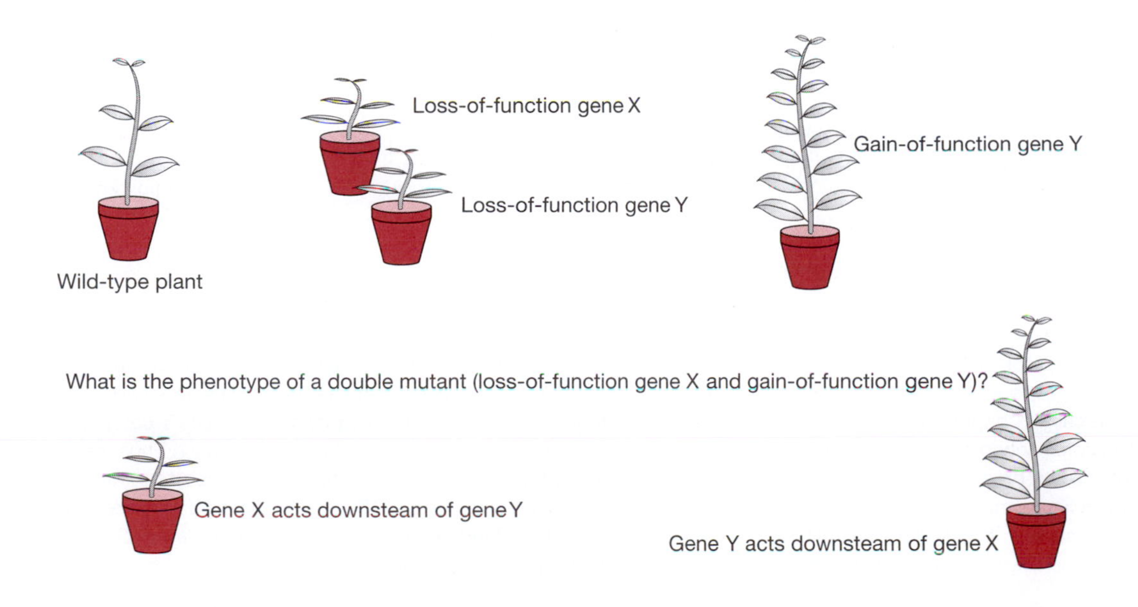

of proteins. Three methods have been developed to suggest protein interactions directly from genomic data. These work best in bacteria and microbial eukaryotes because more genome sequences are available for comparison and both the genes and genomes tend to be simpler in structure and organization than those of higher eukaryotes.

The first is called the domain fusion method (**Figure 7.5**) and is based on the principle that protein domains are structurally and functionally independent units that can therefore operate either as discrete but interacting subunits or as part of the same polypeptide chain (see Chapter 6). Therefore, multidomain proteins in one species may be represented by two or more interacting subunits in another, reflecting the impact of deletions, inversions, or translocations at the DNA level that separate or bring together functionally independent protein segments. A good example is the enzyme topoisomerase II, which is a single polypeptide (with two domains) in the yeast *S. cerevisiae* but two separate polypeptides in the bacterium *Escherichia coli*. The principle of the domain fusion method is that the sequence of protein X, a single-domain protein from one species, is used as a search query against other genomes. This identifies any single-domain proteins related to protein X and also any multidomain proteins such as protein X-Y in other species. The sequence of protein X-Y can in turn be used to search for the individual gene for protein Y in the original genome. If this single protein Y gene exists and is uncharacterized, then the domain fusion method infers that protein X interacts with protein Y and provides a tentative biological function. Both these inferences can then be tested in further experiments. The protein X-Y sequence may also match further domain fusions, such as protein Y-Z, linking all three proteins (X, Y, and Z) into a potential interacting complex. This principle can be extended to reconstruct entire complexes, pathways, and networks in the cell.

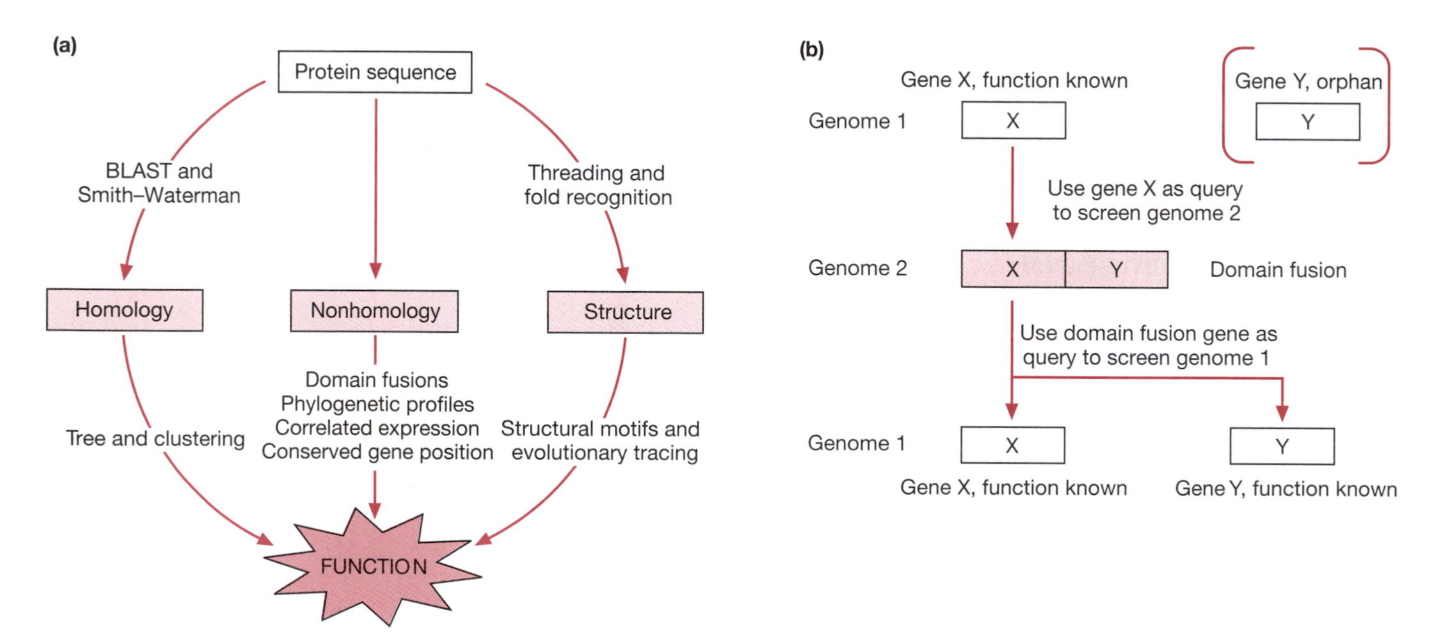

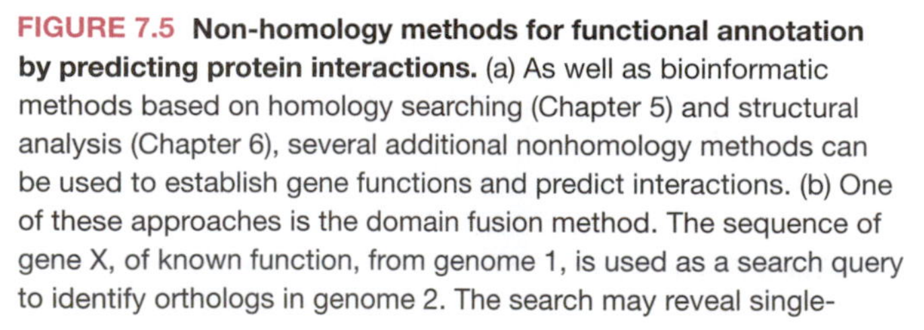

FIGURE 7.5 **Non-homology methods for functional annotation by predicting protein interactions.** (a) As well as bioinformatic methods based on homology searching (Chapter 5) and structural analysis (Chapter 6), several additional nonhomology methods can be used to establish gene functions and predict interactions. (b) One of these approaches is the domain fusion method. The sequence of gene X, of known function, from genome 1, is used as a search query to identify orthologs in genome 2. The search may reveal single-domain orthologs of gene X, but may also reveal domain fusion genes such as X-Y. As part of the same protein, domains X and Y are likely to be functionally related. The sequence of domain Y can then be used to identify single-domain orthologs in genome 1. Thus, gene Y, formerly an orphan with no known function, becomes annotated due to its association with gene X. The two proteins are also likely to interact. (Part (a) from Marcotte EM (2000) *Curr. Opin. Struct. Biol.* 10, 359. With permission from Elsevier.)

The second comparative genomics method is based on the knowledge that bacterial genes are often organized into operons and that such genes are often functionally related even if their sequences are diverse. Therefore, if two genes are neighbors in a series of bacterial genomes, it suggests they are functionally related and that their products may interact. Caution is required in expanding this principle of conservation of gene position to all bacterial genomes because functionally unrelated genes may also be organized into operons. The converse applies in eukaryotes, where functional clusters of structurally unrelated genes are rare, although there are examples in fungi and plants.

The final method is based on **phylogenetic profiling** and exploits the evolutionary conservation of genes involved in the same function. For example, the conservation of three or four uncharacterized genes in 20 aerobic bacteria and their absence in 20 anaerobes might indicate that the products are required for aerobic metabolism. Because proteins usually function as complexes, the loss of one component would render the entire complex nonfunctional, and would tend to lead to the loss of the other components over evolutionary time because mutations in the corresponding genes would have no further detrimental effect. The use of phylogenetic profiling to assign a function to the yeast hypothetical protein YPL207W is shown as an example in **Figure 7.6**.

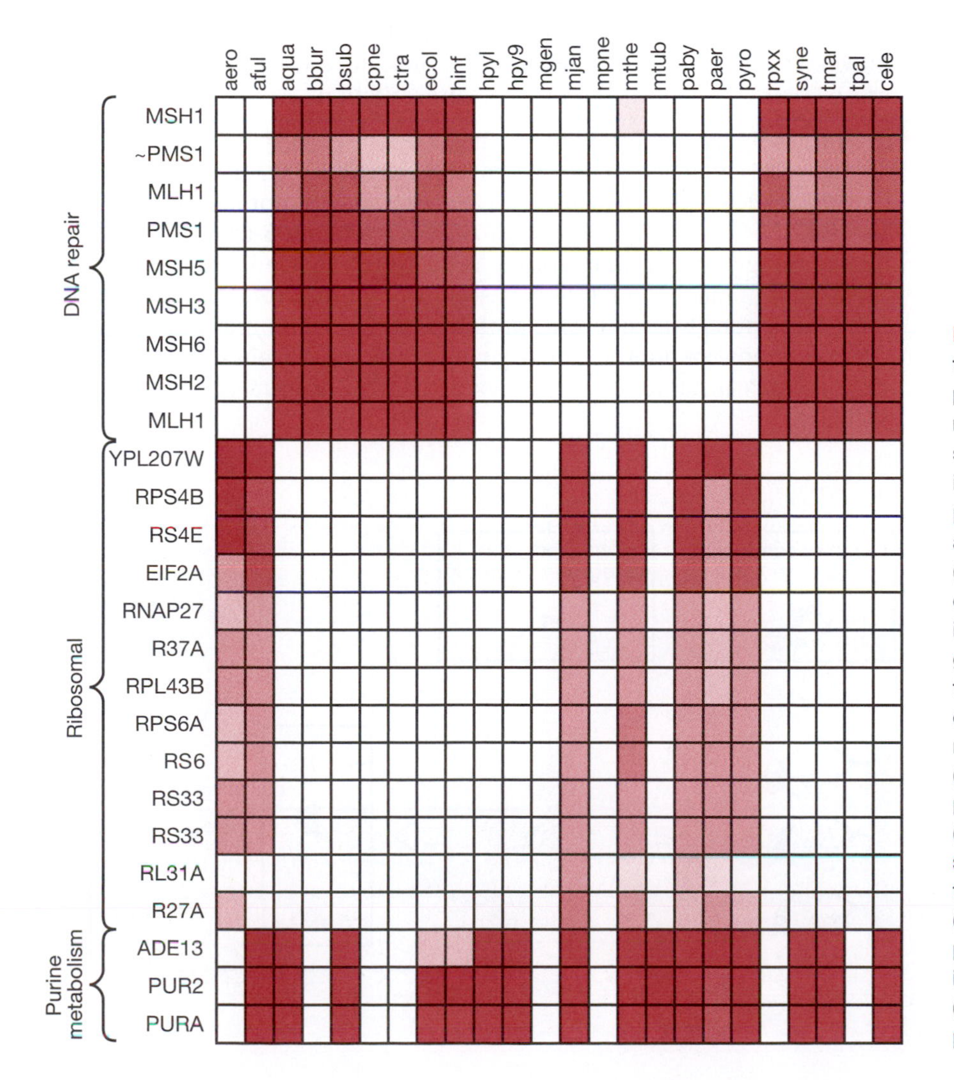

FIGURE 7.6 Phylogenetic profiles for three groups of yeast proteins (ribosomal proteins and proteins involved in DNA repair and purine metabolism) sharing similar co-inheritance patterns. Each row is a graphical representation of a protein phylogenetic profile, with elements colored according to whether a homolog is absent (*white box*) or present (*colored box*) in each of 24 genomes (*columns*). When homology is present, the elements are shaded on a gradient from *light red* (low level of identity) to *dark red* (high level of identity). In this case, homologs are considered absent when no BLAST hits are found with expectation (*E*) values $< 1 \times 10^{-5}$. When homologs are present, the profile receives a score ($-1/\log E$) that describes the degree of sequence similarity with the best match in that genome. The uncharacterized protein (YPL207W) clusters with the ribosomal proteins and can be assigned a function in protein synthesis. (From Marcotte EM (2000) *Curr. Opin. Struct. Biol.* 10, 359. With permission from Elsevier.)

In a more general sense, protein interactions can be inferred by **homology transfer**. This means that interactions confirmed in one organism can be used to predict interactions among homologous proteins in another organism (see Section 7.11).

Affinity-based biochemical methods provide direct evidence that proteins can interact

Genetic and genomic methods suggest protein interactions from genetic evidence but other methods are needed to confirm that the interactions take place. In contrast, biochemical methods demonstrate conclusively that interactions can take place but other methods are required to confirm that they do take place *in vivo*. This is because biochemical methods take the interacting components out of their natural context and may bring into contact proteins that are normally separated by membranes inside cells.

Biochemical methods are based on the principle of affinity, that is, that two proteins interacting with each other *in vivo* will also interact *ex vivo* in a comparable environment because they have an intrinsic affinity for each other and therefore bind to each other selectively. This can be exploited in two major approaches, known as **affinity capture** and **affinity pull-down**. Both approaches use one of the interacting components as a **bait** to purify interacting partners, which are known as **prey**. In affinity capture methods, the bait protein is immobilized on a solid support, such as the resin in a chromatography column (affinity chromatography). In affinity pull-down methods, the bait is added to a complex solution in the liquid phase and is then recovered once it has captured the prey, for example, by using magnetic beads coupled to ligands that bind the bait or through the use of cross-linking reagents that encourage precipitation.

A typical affinity capture platform is shown in Figure 7.7. Here, the bait (protein X) is attached to the chromatography resin via an antibody that is immobilized on the resin material. An alternative approach is to express protein X as a fusion with the protein glutathione-*S*-transferase, which binds strongly to chromatography resin functionalized with glutathione. After equilibration, cell lysate is passed through the column and any proteins that interact directly with protein X will be captured. Importantly, if the conditions are appropriate to maintain native protein complexes, then this method can capture the complexes intact. After washing with a solution

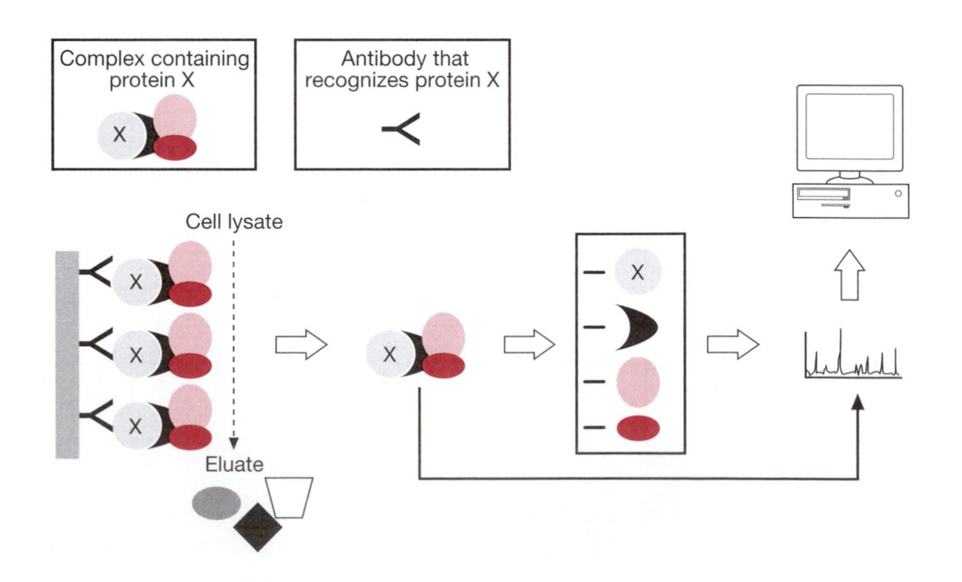

FIGURE 7.7 Affinity chromatography can be used to trap interacting proteins. If Protein X is immobilized on Sepharose beads (for example, using specific antibodies), then proteins (and other molecules) interacting with protein X can be captured from a cell lysate passed through the column. After washing away unbound proteins, the bound proteins can be eluted, separated by SDS-PAGE (optional) and analyzed by mass spectrometry.

that preserves affinity contacts but washes out nonspecific binders, the interacting proteins (prey) can be eluted selectively and in a pure form by increasing the salt concentration or by adding SDS to the buffer, both of which disrupt the affinity contacts. Proteins in the eluate can then be separated by electrophoresis, digested into peptides individually, and analyzed by mass spectrometry as shown in Figure 7.7, or the entire complex can be digested into peptides, separated by reversed-phase HPLC, and analyzed by mass spectroscopy with the aim of identifying the components (Chapter 3). Stepwise increments in the salt or SDS concentration can be used to discriminate between proteins that bind with high or low affinity to protein X, and controls are required to eliminate proteins that bind to irrelevant components of the experiment, such as the antibody or the glutathione-*S*-transferase. This fundamental "capture and analyze" strategy is widely exploited in the high-throughput interaction technologies discussed later in the chapter, and is one of the underpinning principles of large-scale interaction analysis.

Two affinity pull-down approaches are shown in **Figure 7.8**. The first such method to be developed was **co-immunoprecipitation** (Figure 7.8a), which is based on the principle that the addition of antibodies specific for protein X to a cell lysate will result in the precipitation of the antibody–antigen complex when additional reagents are used to cross-link the antibodies, allowing them to be recovered by centrifugation. Once purified, the recovered complexes can be fractionated and analyzed by mass spectrometry as above, with steps taken to remove the signals produced by the antibody. More recently, pull-down techniques have been developed that involve the use of microscopic beads coated with capture agents specific for a particular tag, which has the advantage that standard protocols can be used and spurious interactions can be avoided, resulting in its development for large-scale applications

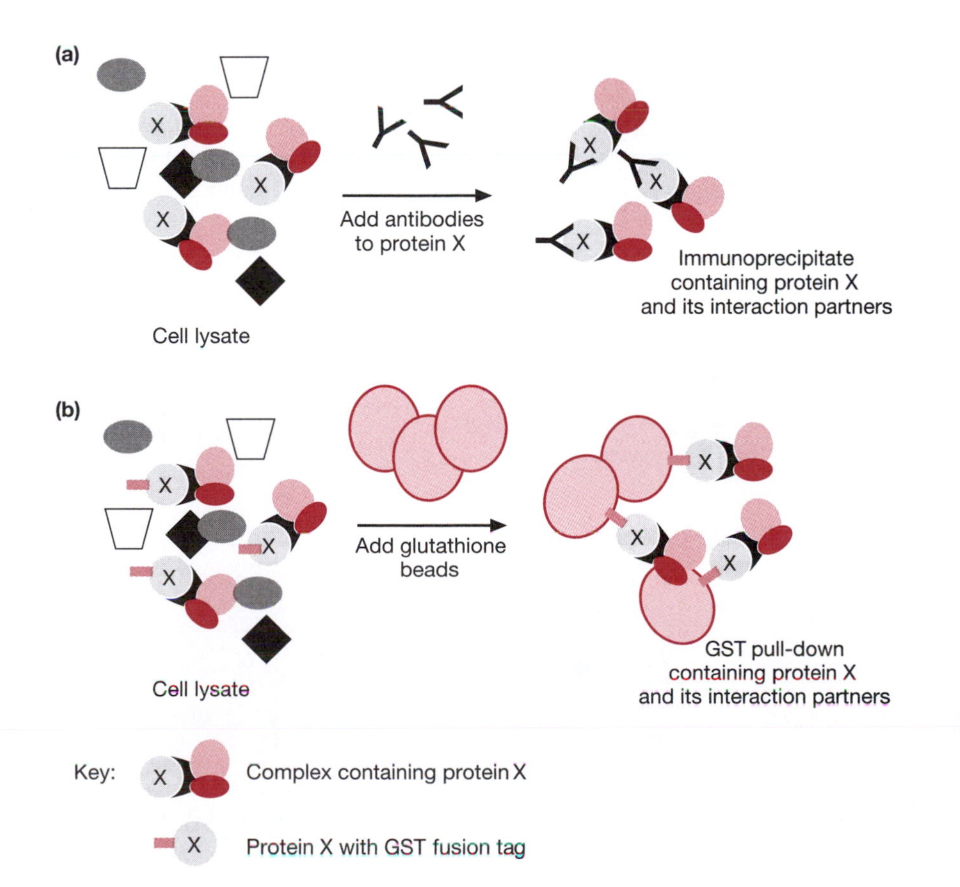

FIGURE 7.8 Two affinity pull-down strategies. (a) Immunoprecipitation can be used to isolate proteins interacting with protein X if a specific antibody is available. (b) Alternatively, protein X can be expressed as a glutathione-*S*-transferase (GST) fusion protein and captured using glutathione beads.

(Section 7.9). **GST pull-down** is shown as an example in Figure 7.8b. The bait (protein X) is expressed as a fusion to GST and is attached to beads functionalized with glutathione. These beads are mixed with the cell lysate, directly capturing the proteins (or protein complexes) that interact with protein X. The beads can be recovered by centrifugation, although there are also variants of the approach where the beads have paramagnetic properties to facilitate their separation from the cell lysate. Analogous techniques can be used to capture proteins containing oligohistidine tags (immobilized metal ion affinity chromatography, IMAC) or antibodies specific for epitope tags such as FLAG and c-Myc. The antibody can be used as a direct way to capture complexes, but alternatively the antibody can be pre-loaded with the bait and the bait used to capture interacting proteins, including complexes. One key difference between the direct and indirect use of antibodies is that direct capture is sympathetic to the physiological conditions in the cell, specifically the abundance of the bait. Pre-loading the antibody with bait usually increases the amount of bait beyond its normal physiological concentrations and may result in the capture of complexes that would not usually form under normal conditions.

If an entire protein complex can be captured, direct interactions among its components can be characterized by cross-linking, where interacting proteins are covalently joined together. A useful strategy is shown in **Figure 7.9**. This involves the use of a photoactivated cross-linking reagent, which contains a radioactive label. If this reagent is covalently joined to purified protein X *in vitro*, then the conjugate can be added to a cell lysate and cross-linking can be induced by exposure to light. However, because the label is on the photoactivated moiety of the cross-linking reagent, it is transferred to the interacting partner, allowing it to be purified and characterized. Cross-linking followed by two-dimensional gel electrophoresis has been widely used to study the architecture of protein complexes (Figure 7.9b). Note that cross-linking is also a useful method to study protein interactions with nucleic acids (**Box 7.2**). Other strategies include cross-linking with formaldehyde and the direct analysis of digested complexes by mass spectrometry in solution or after an in-gel digest.

A final note about affinity trapping methods is that they are not limited to single bait proteins. By switching from homogeneous (in solution) assay formats to solid-phase assays (where the bait is immobilized on a solid surface

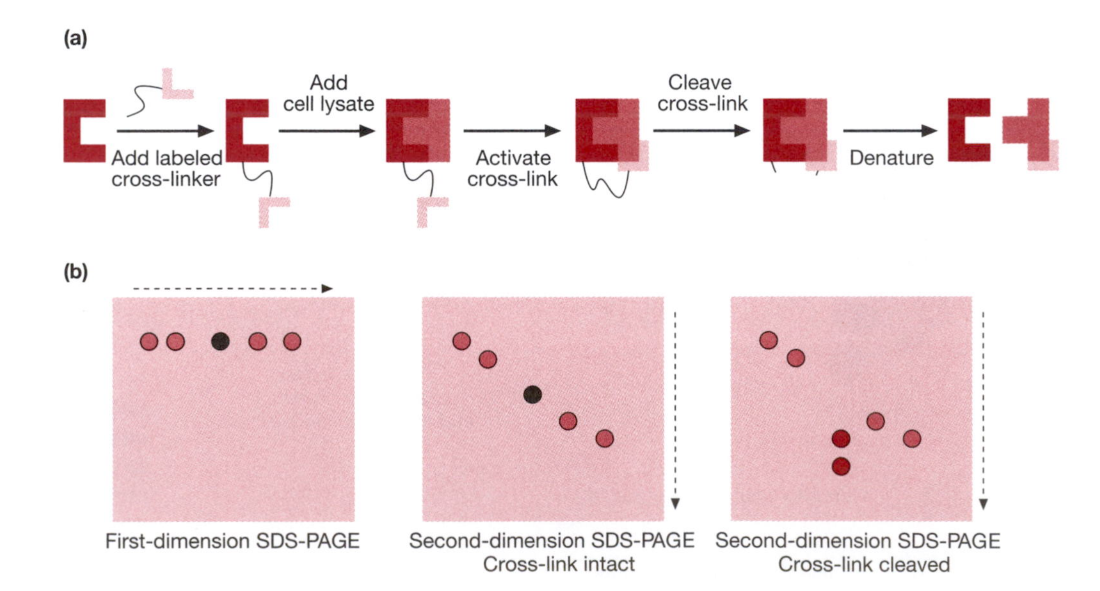

FIGURE 7.9 Interacting proteins can be identified by cross-linking. (a) A labeled cross-linker is added to protein X *in vitro* and the cell lysate is added so that interactions can occur. If the cross-link is activated at this stage, interacting proteins become covalently attached to the bait. After purification, the cross-link can be cleaved and the interacting proteins separated. The label remains on the interaction partner. (b) Mapping complex architecture by 2D-PAGE. Interacting proteins are cross-linked and separated by SDS-PAGE in two dimensions. If the cross-link remains intact, proteins will form a diagonal pattern because the smallest proteins move farthest in both dimensions. However, if the cross-link is cleaved between the two gel runs, formerly cross-linked proteins can be identified because they move off the diagonal.

BOX 7.2 RELATED TECHNOLOGIES.
Detection of protein–nucleic acid interactions.

Protein–nucleic acid interactions underlie some of the most fundamental biochemical processes, including DNA replication, DNA repair, recombination, transcription, mRNA processing, and translation. They are also important for packaging nucleic acids (for example, histones in chromatin) and transporting them around the cell (for example, chromosome segregation, RNA export from the nucleus, and RNA localization). All nucleic acids associate with proteins at some stage and often exist as permanent or semi-permanent nucleoprotein complexes. Some proteins interact nonspecifically with DNA and/or RNA whereas others only bind to particular sequences. The latter are the most interesting because they often have a regulatory function (see Lee et al., Further Reading). The functions of proteins can also be determined by promoter analysis (see Werner, Further Reading).

Biochemical techniques for the investigation of protein-nucleic acid interactions can be divided into two major categories:

Affinity-dependent purification and screening methods. Nucleic acid-binding proteins can be purified by exploiting their affinity for DNA or RNA. A successful early method for the isolation of RNA-binding proteins was simply to filter cell lysates through nitrocellulose. The RNA, and any associated proteins, would bind to the nitrocellulose while other proteins would be washed through. Slightly more sophisticated methods are required to isolate sequence-specific binding proteins. First, the cell lysate must be mixed with an excess of total genomic DNA or tRNA (as appropriate) in order to block the nonspecific binding proteins. Sequence-specific binding proteins can then be isolated by affinity chromatography in which a particular oligonucleotide is used as the affinity matrix. The affinity of proteins for nucleic acids can also be exploited to identify DNA- and RNA-binding proteins on membranes or in expression libraries. After blocking with nonspecific DNA, a labeled DNA or RNA probe is applied to the membrane and will only bind to those proteins with affinity for that specific sequence. One disadvantage of this method, known as southwestern screening for DNA-binding proteins and northwestern screening for RNA-binding proteins, is that nucleic acid-binding proteins made up of several different subunits will not be detected, because the components will be present as separate clones.

A variant of the yeast two-hybrid system, known as the yeast one-hybrid system, is useful for the identification of transcription factors. Essentially, this involves the transformation of yeast with a construct comprising a minimal promoter and reporter gene, with several tandem copies of a candidate transcription factor-binding motif placed upstream. A cDNA expression library is then prepared in which all proteins are expressed as transactivation domain hybrids. These will activate the target gene only if they contain a DNA-binding domain that interacts with the chosen promoter sequence. This system can only identify proteins that bind to DNA autonomously.

The one-and-a-half hybrid system is similar, but can detect proteins that bind DNA as heterodimers with a second, accessory protein.

The one-two hybrid system can search for both autonomous binders and proteins that bind only as heterodimers.

Another variant of the two-hybrid system, known as bait-and-hook or three-hybrid, is useful for the identification of RNA-binding proteins or protein interactions with small ligands. In this system, one of the components (the hook) comprises the DNA-binding domain of a transcription factor and a sequence-specific RNA-binding protein that attaches to one end of a synthetic RNA molecule. The other end of the RNA molecule contains the sequence for which candidate interactors are sought. A prey library is constructed as normal, with each protein expressed as a fusion to a transactivation domain. Only in cells where the prey interacts with the RNA sequence attached to the hook will the transcription factor be assembled and the reporter gene activated.

Methods for the precise characterization of protein–nucleic acid interactions. These methods are diverse and are usually designed to identify the sequence to which a particular protein binds. The gel retardation assay is used to demonstrate protein–DNA interactions but it can also identify the approximate location of protein-binding sites in DNA or RNA when DNA fragments with a putative binding site are used. It is based on the fact nucleic acid/protein complexes move through electrophoretic gels more slowly than naked DNA or RNA. DNase footprinting can identify the exact nucleotides covered by a protein, because these will be protected from nuclease digestion. Methylation interference and methylation protection are techniques that identify the specific bases that make contact with a binding protein, either because these are protected from methylation when the protein is bound or because they interfere with the normal interaction if they are already modified. RNA–protein interactions in complexes are often studied by chemical cross-linking or treatment with nucleases. A very useful technique is the hybridization of short DNA oligonucleotides to RNA molecules in an RNA–protein complex followed by digestion with RNaseH, which is specific for DNA/RNA hybrids. This allows the systematic functional testing of parts of the RNA component.

such as a nitrocellulose membrane, or in the wells of microtiter dishes), the interactions of many proteins can be studied in parallel. Traditional techniques for protein analysis and quantitation, such as western blotting and ELISA (Box 4.1) are optimized for the use of a single probe. That is, thousands of protein targets may be immobilized on the surface, but the idea is to detect interactions between a single probe and target—usually an antibody probe and a complementary antigen. These solid-phase affinity-based methods can be regarded as the forerunners of some of the higher-throughput techniques discussed later, such as phage interaction display (p. 144) and functional protein chips (p. 196).

Interactions between proteins *in vitro* and *in vivo* can be established by resonance energy transfer

When two different fluorophores are in close proximity, one of which (the donor) has an emission spectrum that overlaps the excitation spectrum of the other (the acceptor), a phenomenon known as **fluorescence resonance energy transfer** (**FRET**) can occur. When a lone donor fluorophore is excited, light is produced with a characteristic emission wavelength. However, when the donor fluorophore is excited in close proximity to the acceptor fluorophore, energy is transferred to the acceptor fluorophore with the result that the intensity of emission from the donor is reduced (quenched) while that of the acceptor is increased (enhanced). This principle can be used to investigate the interactions between two proteins *in vitro* if they are conjugated to fluorophores that undergo FRET, for example, Cy3 and Cy5. If the detector is calibrated to read the enhanced emission of the acceptor fluorophore, then a signal will be obtained only when the two proteins interact. The advantage of this method is that transient as well as stable interactions can be detected.

More recently, donor and acceptor derivatives of **green fluorescent protein** have been used to generate bioluminescent protein fusions, which can be expressed in living cells and used to detect protein interactions *in vivo* using the same principle (**bioluminescence resonance energy transfer, BRET**). For example, protein–protein interactions in living cells have been studied using bait proteins fused to cyan fluorescent protein and candidate prey fused to yellow fluorescent protein. **Luciferase** can also be used as the energy donor, and because this is an enzyme that generates light by consuming a substrate rather than in response to exogenous excitation by light (as is the case for fluorescent proteins), protein interactions can be monitored without artifacts such as photo-bleaching. One drawback of these techniques is that FRET/BRET only occur efficiently when the donor and acceptor groups are less than 10 nm apart, so fusion constructs have to be designed on a case-by-case basis to ensure the proximity is achieved.

FRET/BRET-based methods should not be confused with **protein complementation assays** involving fluorescent proteins, which exploit a different principle and are discussed on p. 149.

Surface plasmon resonance can indicate the mass of interacting proteins

Several manufacturers of **protein chips** have developed systems that exploit surface plasmon resonance (SPR) to detect protein interactions. This is an optical effect that occurs when monochromatic polarized light is reflected off thin metal films. The amount of resonance that occurs within the delocalized electrons in the film is determined by the material adsorbed to the film surface. A technique called **surface plasmon resonance spectroscopy** can measure this effect, and because there is a direct relationship between the mass of the immobilized molecules and the change in resonance energy at the metal surface, it can be used to determine the mass of interacting proteins and investigate interactions in real time. Protein chips that use SPR detection methods can be coupled directly to a mass spectrometer so that interacting proteins can be identified once captured on the chip, and this principle is discussed in more detail in Chapter 9. Other label-free methods that can be used to monitor protein interactions include **static light scattering** and **dynamic light scattering** (**photon correlation spectroscopy**), which identify protein complexes by reporting changes in the hydrodynamic radius of molecules in solution, and **isothermal titration calorimetry**, which measures changes in temperature that occur when proteins associate or dissociate in solution. Neither of these techniques is yet suitable for high-throughput applications.

7.3 LIBRARY-BASED METHODS FOR THE GLOBAL ANALYSIS OF BINARY INTERACTIONS

Each of the techniques discussed above suffers from one or more of the following three intrinsic limitations:

1. The method is suitable for the analysis of individual proteins and their interaction partners but is difficult to scale up to consider interactions at a global level.

2. The method is suitable for the analysis of proteins *in vitro* but it is unclear whether the interactions are relevant *in vivo*.

3. There is no direct link between the interacting proteins and the corresponding genes, so it is laborious to identify the interacting partners.

Affinity-based methods have now been adapted for the systematic analysis of protein complexes *in vivo*, using homologous recombination or transposon insertion to tag endogenous proteins. However, many different tagged lines are required to test all potential interactions even in the context of a simple cell.

As an example, the yeast *S. cerevisiae* produces approximately 6000 different proteins even if we ignore the additional diversity generated by post-translational modifications, which provides scope for 36,000,000 potential binary interactions. The number of genuine interactions will be much lower than this because most proteins will have a small number of specific interacting partners and because the proteome is divided both spatially and temporally into overlapping components, meaning that many proteins, even if they could interact, never exist in the same space at the same time. Mostly, however, the number of interactions is limited by the affinities of different proteins for each other. The total number of interactions in this particular organism is probably only an order of magnitude higher than the size of the proteome. But all 36,000,000 possible interactions need to be tested in order to establish the few tens of thousands that are functionally significant.

The testing of binary protein interactions on such a grand scale and under *in vivo* conditions can only be carried out using library-based methods, in which large numbers of expression clones can be tested systematically against each other. The origins of this approach lie in the standard cDNA expression library, which, although typically screened with labeled nucleic acid probes, can also be screened with labeled bait proteins to isolate clones expressing putative interaction partners. For example, labeled calmodulin has been used to screen for calmodulin-binding proteins and a probe corresponding to the phosphorylated internal domain of the epidermal growth factor receptor has been used to identify signaling proteins containing the Src homology 2 (SH2) domain. DNA and RNA probes have also been used on protein expression libraries to identify transcription factors and RNA-binding proteins (Box 7.2). Although the use of expression libraries is a good way to increase the throughput of finding prey, the screening of bait proteins using this method is still laborious. Taking the yeast example discussed above, 6000 separate library screenings would need to be carried out to test all potential binary interactions. The *in vitro* assay format also does not provide the native conditions for the folding of all proteins, so a significant number of interactions would not be detected.

An improvement in throughput can be achieved using a technique known as **phage interaction display**, where potential prey proteins are expressed on the surface of bacteriophage particles to create a phage display library (**Figure 7.10**). The wells of microtiter plates are then coated with particular bait proteins of interest. The phage display library is pipetted into each well. Phage with interacting proteins on their surface will remain bound to the surface whereas those with noninteracting proteins will be washed away. Phage display is suitable for proteome-scale interaction analysis because highly parallel screenings can be carried out and the technique is amenable to automation. In theory, 6000 bait proteins in microtiter dishes could be screened with a single phage display library and phage from each well could be eluted individually, and amplified in *E. coli* in preparation for DNA sequencing. As with standard library screening, however, the *in vitro* assay format would not allow all proteins to adopt their native structures. Furthermore, only short peptides can be displayed on the phage surface, because larger proteins disrupt replication. It is likely that some interactions, requiring more extensive contacts between the interacting partners, would not be detected for this reason. Even so, phage display libraries can be thought of as the forerunners of the current generation of protein microarrays, which present entire proteomes in a miniaturized grid format allowing the systematic evaluation of protein interactions *in vitro* (Chapter 9).

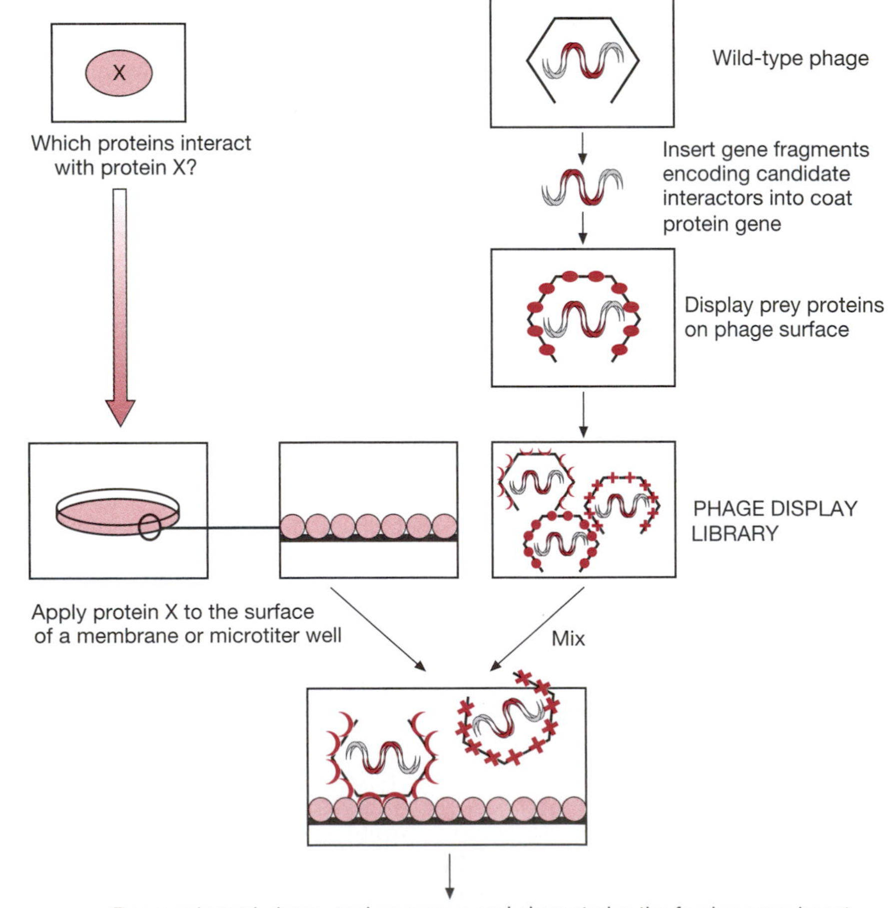

FIGURE 7.10 The principle of phage display as applied to high-throughput interaction screening. Protein X, for which interactors are sought, can be immobilized on the surface of microtiter wells or membranes. All other proteins in the proteome are then expressed on the surface of bacteriophage, by cloning within a phage coat protein gene, to create a phage display library. The wells or membranes are then flooded with the phage library. Phage-carrying interacting proteins will be retained while those displaying noninteracting proteins will be washed away. The bound phage can be eluted in a high-salt buffer and used to infect *E. coli*, producing a large number of phage particles containing the DNA sequence of the interacting protein.

7.4 TWO-HYBRID/PROTEIN COMPLEMENTATION ASSAYS

The yeast two-hybrid system works by assembling a transcription factor from two inactive fusion proteins

The yeast two-hybrid system addresses the problems of the *in vitro* assay format by testing for protein interactions within the yeast cell. The principle of the system is the assembly of an active transcription factor from two fusion proteins and the detection of this assembly by the activation of a marker gene. The general scheme is shown in **Figure 7.11**. The bait protein is expressed as a fusion with a DNA-binding domain from the transcription factor GAL4 that on its own is unable to activate the marker gene. This bait fusion is expressed in one haploid yeast strain. Another haploid yeast strain is used to create a cDNA expression library in which all the proteins in the proteome are expressed as fusions with a transactivation domain of GAL4, which is also unable to activate the marker gene on its own. The two strains of yeast are then mated to yield a diploid strain expressing both the hybrid bait protein and one candidate hybrid prey protein. In those cells where the bait and prey do not interact, the transcription factor remains unassembled and the marker gene remains silent. However, in those cells where there

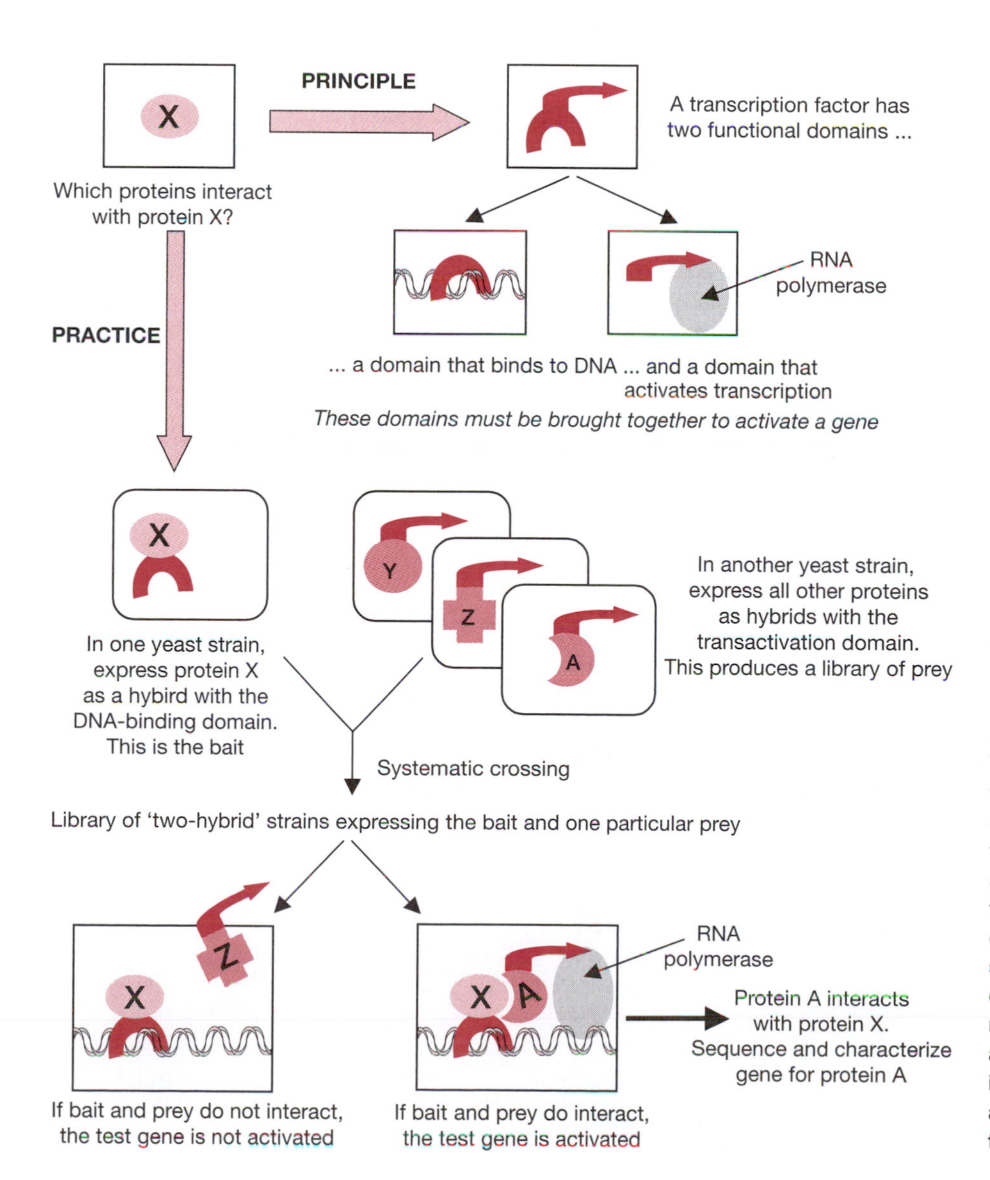

FIGURE 7.11 The principle of the yeast two-hybrid system. Transcription factors generally comprise two functionally independent domains, one for DNA binding and one for transcriptional activation. These do not have to be covalently joined together, but can be assembled to form a dimeric protein. This principle is exploited to identify protein interactions. Bait proteins are expressed in one yeast strain as a fusion with a DNA-binding domain and candidate prey are expressed in another strain as fusions with a transactivation domain. When the two strains are mated, functional transcription factors are assembled only if the bait and prey interact. This can be detected by including a reporter gene activated by the hybrid transcription factor.

is an interaction between the bait and the prey, the transcription factor is assembled and the marker gene is activated, either producing visually different colonies (visible marker gene, such as *lacZ*) or producing colonies that survive under selection (selectable marker gene, for example, conferring autotrophy). Cells with interacting proteins can therefore be identified and the corresponding cDNAs characterized.

Several large-scale interaction screens have been carried out using different yeast two-hybrid screening strategies

The yeast two-hybrid system was the first technology to facilitate global protein interaction analysis. By arraying panels of haploid yeast strains in microtiter dishes and carrying out pairwise matings in a systematic fashion, tens of thousands of interactions can be screened in a single experiment. Several comprehensive large-scale studies have been published, including complete interactome studies in bacteria and yeast, the malaria parasite *Plasmodium falciparum*, the fruit fly *D. melanogaster* and the nematode *C. elegans*, studies identifying novel disease-related signaling pathways in humans, and numerous more focused studies on individual pathways, receptors, and diseases (**Table 7.1**, **Box 7.3**).

An important aspect of the early studies was to establish the best strategies for two-hybrid screening. The **matrix screening method** is a systematic approach in which panels of defined bait and prey strains are mated in an array format (ORF × ORF, where ORF = open reading frame). Each bait and prey construct is made individually by PCR and introduced into yeast cells that are maintained in isolation. The different transformed cell lines are then arranged in microtiter dishes and crossed in all possible pairwise combinations (**Figure 7.12a**). Because individual constructs are required for every protein, whole interactome analysis is possible only in those species with fully sequenced genomes and completed gene catalogs. The advantage of the matrix approach is that it is comprehensive and can provide exhaustive interactome coverage. However, the preparation stage is laborious, and the amount of work required increases in proportion to the size of the proteome being investigated.

TABLE 7.1 KEY LARGE-SCALE INTERACTION STUDIES (YEAST TWO-HYBRID SYSTEM)

Organism	Predicted ORFs	Methods	TAD hybrids	DBD hybrids	Interactions
Bacteriophage T7	55	Matrix and library screen	ORFs or library	ORFs or library	3–22, depending on combination
Vaccinia virus	266	Matrix	ORFs	ORFs	37
Hepatitis C virus	~10	Matrix	ORFs	ORFs	0
		Library	Library	ORFs	15
Helicobacter pylori	1,590	Library	Library	ORFs	1,280
Saccharomyces cerevisiae	6,200	Matrix	ORFs	Pooled ORFs	621
		Library	Library	ORFs	2,374
Caenorhabditis elegans	19,099	Library	Library	ORFs	2,135
Drosophila melanogaster	18,000	Library	Library	ORFs	4.780
Homo sapiens	29,000	Library	Library	8,100 ORFs	2,800
		Matrix	4,456 ORFs	5,632 ORFs	911

All screens were designed to be proteomewide except the human screens, which were scaled down.
TAD, transactivation domain; DBD, DNA-binding domain.

BOX 7.3 CASE STUDY.
Large-scale interaction screens in yeast using the yeast two-hybrid approach.

Although the yeast two-hybrid method was developed by Stanley Fields and Ok-kyu Song in 1989, its true potential and scalability were not established until 10 years later when the first global interaction screen in yeast was published by Peter Uetz and colleagues (see Further Reading). Earlier investigations had confirmed that the method was suitable for systematic investigations of protein interactions, such as the 1994 study of interactions among cell cycle proteins in *D. melanogaster* by Russel Finley and Roger Brent (see Further Reading). This matrix-format study revealed 19 interactions among known cyclins and cyclin-dependent kinases.

The matrix approach was also used by Uetz *et al.* to screen a yeast proteome-wide library of prey constructs (more than 6000 open reading frames) with 192 baits, 87 of which were shown to be involved in reproducible interactions (that is, positive results in two independent screens). Each of the baits was involved in an average of three interactions, resulting in a total of 281 interactions. The same authors then used the pooled matrix approach to screen 5300 baits, reporting 692 interacting protein pairs. A pooled matrix strategy was also used by Ito and colleagues (see Further Reading), in this case with pools of baits and prey. Each pool contained 96 clones, allowing four million potential interactions to be screened in 430 combinatorial assays (equivalent to approximately 10% of all potential interactions within the yeast proteome). Nearly 850 positive colonies were identified, revealing 175 interacting protein pairs, only 12 of which were previously known. Scaling this format up to encompass the entire proteome revealed 4549 interactions among 3278 proteins, 841 of which demonstrated three or more independent interactions.

The error rate in the early screens can be estimated by the number of common interactions identified in different experiments. The matrix and pooled matrix screens carried out by Uetz and colleagues identified only 12 common interactions. The pooled matrix screens carried out by the Uetz and Ito groups revealed 692 and 841 high-confidence interacting pairs respectively, and comparison of the datasets revealed 141 interactions in common. An earlier interaction screen using the random library method and focusing on splicing proteins conducted by Fromont-Racine and colleagues revealed about 10 high-confidence interactions per bait. Some of these proteins were also included in the Uetz global screen and here it was found that there were two common interactions for two of the baits and six common interactions with the others.

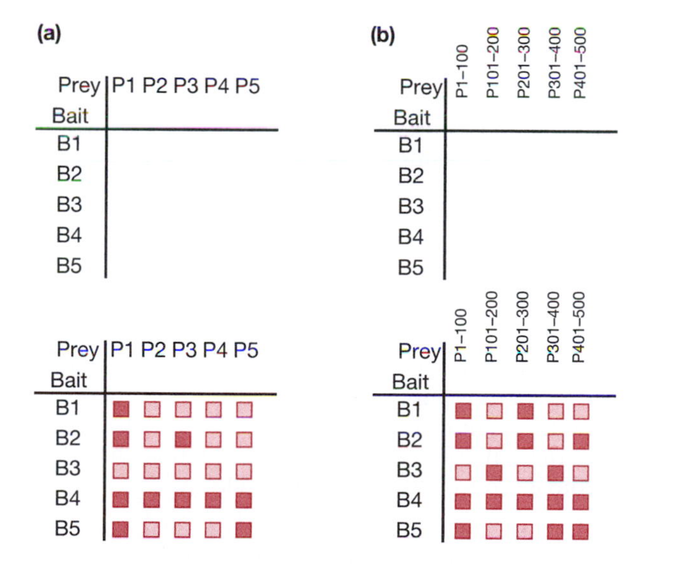

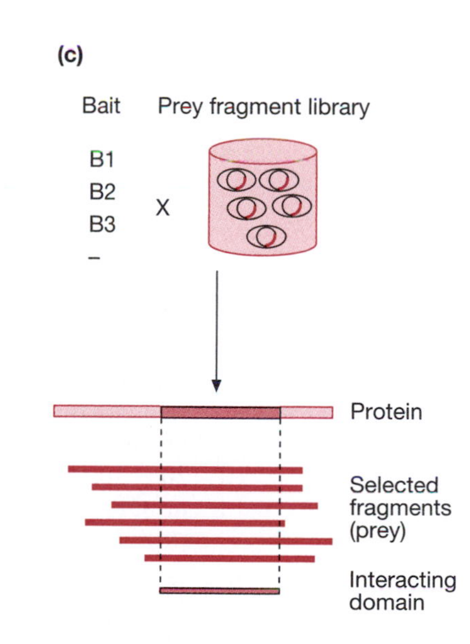

FIGURE 7.12 The matrix and library screening approaches to build large-scale protein interaction maps. (a) The matrix approach is systematic and uses the same collection of proteins (1–5) as bait (B1–B5) and prey (P1–P5). The results can be drawn in a matrix. Autoactivators (for example, B4) and "sticky" prey proteins (for example, P1 interacts with many baits) are identified and discarded. The final result is summarized as a list of interactions that can be heterodimers (for example, B2–P3) or homodimers (for example, B5–P5). (b) The pooled matrix approach is a variation on the above in which prey are pooled to allow higher-throughput screening. For example, no interactions occur between B1 and P301–500, so 200 individual screens can be omitted. Prey pool P201–300 can be deconvoluted to identify specific interactors. (c) The library screening approach is random and identifies the domain of interaction for each prey protein interacting with a given bait. Sticky prey proteins are identified as fragments of proteins that are often selected regardless of the bait protein. (From Legrain P, Wojcik J & Gauthier JM (2001) *Trends Genet.* 17, 346. With permission from Elsevier.)

In order to increase the throughput of the matrix approach, the **pooled matrix screening method** was developed to divide the screen into layers of complexity (Figure 7.12b). Like the standard matrix screening method, defined strains are produced for each bait construct but the preys are screened in pools rather than as individual strains. Cells are transformed with mixtures of prey constructs and screened en masse (ORF × pooled ORF). In this way, many more potential interactions can be screened in each mating. Where interactions are detected, the mixed strains can be deconvoluted in order to identify individual interactors.

The labor-intensive task of producing individual bait and prey constructs for every protein in the proteome can be sidestepped by the use of **random libraries** of fusion constructs. This circumvents the need to prepare expression clones for every protein, and means that redundant overlapping clones represent each protein, which provides an internal validation of interactions and allows interactions to be narrowed down to a specific protein domain (Figure 7.12b). Screens have been carried out using conventional baits against a random library of prey (ORF × library) as well as random libraries for both components (library × library).

Conventional yeast two-hybrid screens have a significant error rate

The development of different screening strategies has helped to highlight some of the potential drawbacks of the conventional yeast two-hybrid system, which manifest as **false positives** (detection of non-genuine interactions) and **false negatives** (nondetection of genuine interactions).

Where independent researchers have carried out similar large-scale studies, the degree of overlap in the reported interactions has been very low (10–15%) and a significant number of well-characterized interactions have been missed. There are many explanations for the false-negative rate, but they may reflect either failures within the assay itself (for example, a particular bait or prey clone may not be represented in a library, or it may not be expressed properly, or the protein may not fold correctly in the context of the fusion hybrid) or intrinsic properties of the protein or its requirements for interaction (for example, an interaction interface is missing, or the interaction requires a particular form of post-translational modification or the presence of a co-factor). Another important issue is that the yeast two-hybrid assay relies on the nuclear import of the fusion hybrids, a process that may be disrupted if the candidate is a membrane protein or is normally found in a different organelle. These issues appear to be more challenging when using ORFs than libraries because in the latter case there are often partial clones that may allow, for example, the nuclear import of a globular truncated protein lacking a transmembrane domain. In an extreme example, the analysis of interactions among the 10 mature polypeptides of hepatitis C virus in a 10 × 10 matrix revealed no interactions at all when each was expressed as an ORF, but all known interactions plus three novel ones when the same proteins were expressed from a random library (Table 7.1).

False positives can reflect nonspecific interactions (where one of the components, described as **sticky bait** or a **sticky prey**, interacts with many partners). This might be expected for proteins whose normal function requires diverse interactions, such as molecular chaperones or components of the protein degradation machinery, but might be induced by the non-authentic conditions created when bait and prey proteins are overexpressed. In other cases, the bait and/or prey may be capable of **autoactivation**, that is, spontaneous activation of the marker gene in the absence of an interaction, which is especially likely if either the bait or prey is a transcription factor. A further source of false positives is irrelevant interactions between proteins that would never encounter each other under normal circumstances, such as those normally found in different tissues, expressed at different stages of development, or resident in different intracellular compartments.

Several approaches have been developed to increase confidence in interaction data, including the statistical analysis of experimental reproducibility, benchmarking against independent evidence established using alternative methods, and the expectation that a single bait will interact with multiple independent clones representing the same prey in random libraries. However, the drawbacks of the conventional yeast two-hybrid system have also led to the development of many alternative platforms that strive to overcome its limitations, and we consider these platforms in more detail below.

7.5 MODIFIED TWO-HYBRID SYSTEMS FOR MEMBRANE, CYTOSOLIC, AND EXTRACELLULAR PROTEINS

The reliance of conventional yeast two-hybrid screens on the nuclear import of bait and prey has largely excluded the analysis of integral membrane proteins, which account for 30–50% of all proteins in the cell. This has been addressed by developing the **split ubiquitin system** (also known as the **ubiquitin-based split protein sensor, USPS**) in which the transcription factor that activates the reporter gene is produced indirectly following the interaction between bait and prey, allowing interactions to be monitored even if they are membrane-bound. In yeast, this approach is generally described as a **membrane-based yeast-two hybrid** (MbYTH or **MYTH**) assay.

Like the conventional two-hybrid screen, the split ubiquitin system relies on **protein fragment complementation** brought about by the interaction between bait and prey. In this case, the components are not the DNA-binding and transactivation domains of a transcription factor, but the C-terminal and N-terminal fragments of **ubiquitin**, known as Cub and Nub. One of the components (usually Cub) is fused to a complete transcription factor via a linker that is sensitive to ubiquitin-specific protease activity. Therefore, if the two components are fused to bait and prey proteins that interact, a functional pseudo-ubiquitin molecule is assembled which cleaves off the transcription factor. This can be imported into the nucleus and will activate the reporter gene (**Figure 7.13**). Variants of this approach with the transcription factor fused to Nub are useful for analyzing interactions that specifically involve the C-terminal domain of membrane proteins, such as the PDZ domain. The split ubiquitin system has also been adapted to restrict interactions to the cytosol, which is particularly useful for studying the interactions of transcription factors that might autoactivate the reporter gene in a conventional assay. This was achieved by fusing one of the ubiquitin components to Osr4p, which integrates into the endoplasmic reticulum membrane with

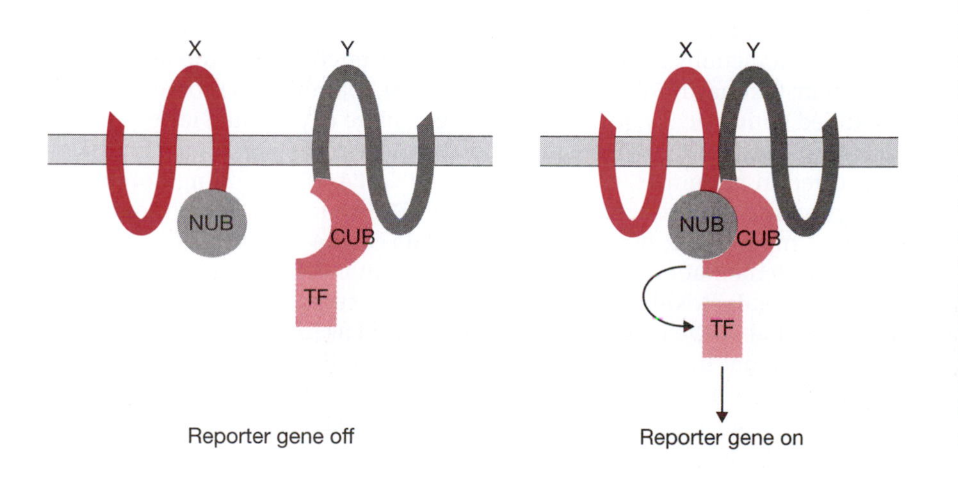

FIGURE 7.13 The split ubiquitin assay for membrane proteins. A membrane protein of interest is expressed as a fusion to the ubiquitin N-terminal component Nub and presented with a library of candidate interactors expressed as fusion proteins with the C-terminal component Cub and a transcription factor joined by a ubiquitin-cleavable linker. If there is no interaction, the transcription factor remains tethered at the membrane and the reporter gene is not activated. If there is an interaction, this permits the assembly of Nub-Cub into a functional pseudo-ubiquitin, allowing the transcription factor to be cleaved off and imported into the nucleus to activate the reporter gene.

the ubiquitin component facing the cytosol. A **split-Trp assay** has also been developed following the identification of complementary C-terminal (CTrp) and N-terminal (NTrp) fragments of the enzyme tryptophan synthase (Trp1). These fragments reconstitute the enzyme activity when brought into close proximity by prey–bait interactions, allowing the selection of interacting proteins in yeast *trp1* mutants growing on medium lacking tryptophan.

Membrane protein interactions can also be studied if cell survival is made dependent on the interaction between bait and prey. This is the principle of the **SOS-recruitment system** (**SRS**) in which the bait is expressed with a myristoylation signal that embeds it in the plasma membrane facing inwards, and prey are expressed as fusions with the SOS protein, a mammalian ortholog of the yeast protein CDC25, which is essential for survival. In yeast *csd25* mutants, cells can only survive if SOS is recruited to the membrane via the interaction between bait and prey, automatically selecting for interacting proteins. A similar concept is used in the **RAS recruitment system** (**RRS**) in yeast cells deficient for RAS.

Another variant approach known as **screening for interactions between extracellular proteins** (**SCINEX-P**) is used for the analysis of protein–protein interactions in the oxidizing environment of the endoplasmic reticulum, by using the yeast unfolded protein response (UPR) to induce the dimerization of the transmembrane protein Ire1p when unfolded proteins accumulate, in turn inducing the transcription factor Hac1p, which is required for the production of molecular chaperones. The SCINEX-P method involves the expression of bait proteins as fusions with a mutated version of Ire1p lacking the N-terminal oligomerization domain that normally projects into the lumen of the endoplasmic reticulum. Oligomerization therefore becomes dependent on interactions between the bait and prey, resulting in the activation of UPR signaling. The inclusion of a Hac1p response element in the promoter of the reporter gene completes the circuit and allows interacting protein pairs to be identified.

7.6 BACTERIAL AND MAMMALIAN TWO-HYBRID SYSTEMS

The development of two-hybrid systems in bacteria is also a convenient way to overcome the limitations imposed by nuclear import in yeast. Additional advantages include the higher transformation efficiency and faster growth rate of *E. coli*, which allows more complex libraries to be used, and the ability to express proteins that are toxic in eukaryotic cells. Prey clones expressed in bacteria can also be prepared in phagemid constructs, which allows them to be isolated either as plasmids or bacteriophage. Systems have been developed that work exclusively in bacteria and that function in both bacteria and yeast, allowing the benefits of both systems to be exploited.

One of the drawbacks of bacterial two-hybrid screens is that eukaryotic protein interactions may not be reported accurately due to the absence of post-translational modifications. Even in yeast, mammalian proteins are not modified authentically, providing another source of potential false-positive and false-negative data. The conventional yeast two-hybrid system has been replicated in mammals, typically using the GAL4 DNA-binding domain to recognize the reporter construct and the VP16 transactivator protein from herpes simplex virus, but more sensitive data can be acquired by using protein complementation assays that generate a quantifiable signal directly. In the **split β-galactosidase assay**, the reconstituted protein itself acts as the reporter; that is, the bait and prey are expressed as fusions with N-terminal and C-terminal fragments of the *E. coli* enzyme β-galactosidase, which is widely used as a reporter in mammalian systems. This is a straightforward approach because the enzyme has been widely used for complementation, and subdomains that can be separated and that reconstitute enzymatic activity when brought together are well-characterized. However, the fact that these subdomains can fold spontaneously means that weak affinity generates a background signal. A more refined approach is to use protein fragments that cannot fold correctly unless they are in proximity, and this has been used with ubiquitin in the mammalian equivalent of the split ubiquitin assay, with light-emitting reporter proteins such as luciferase and green fluorescent protein derivatives, and also with enzymes such as β-lactamase and dihydrofolate reductase.

Assays that rely on the reconstitution of split fluorescent proteins (**bimolecular fluorescence complementation, BiFC**) are particularly valuable because they can be used to monitor aspects of the interaction in real time, such as protein translocation events following interaction and the impact of external factors (such as the activation of particular signaling pathways). It is also possible to monitor different interactions simultaneously using fluorescent proteins with different spectral qualities. As discussed above, this technique should not be confused with FRET and BRET, which involve the co-localization of intact fluorescent proteins and the transfer of energy between them to quench the signal. In contrast, BiFC involves a single fluorescent protein per interaction and the signal is dependent on the interaction between split components of that protein. Although real-time analysis can be achieved with BiFC, fluorescent proteins tend to fold slowly and the interactions are irreversible, precluding the analysis of rapid and dynamic interactions. In contrast, BiFC with **split luciferase** is suitable for such applications because the enzyme folds rapidly and reversibly. This system has been used to study signaling pathways where the association and dissociation of interacting components can be followed in near real time.

The mammalian systems described above reconstitute the reporter protein directly, but, like the split ubiquitin assay in yeast, it is often beneficial to generate the reporter indirectly, particularly in the case of mammalian cells, because this achieves signal amplification and increases the sensitivity of the assay, adjusting for the smaller scale of the experiments. The **split TEV assay** achieves this goal by reconstituting the protease from tobacco etch virus, and using it to cleave off the reporter molecule, which may be a transcription factor as in the split ubiquitin assay or a reporter enzyme. The irreversible nature of the protease activity, the ability to respond to transient interactions, and the inherent amplification step mean that this assay is highly sensitive, although it cannot follow dynamic interactions. The split luciferase assay and the split TEV assay therefore show how different assay components can be recruited to focus on different aspects of protein–protein interactions.

7.7 LUMIER AND MAPPIT HIGH-THROUGHPUT TWO-HYBRID PLATFORMS

The advantages of direct two-hybrid approaches in mammalian cells are somewhat limited by their low scalability compared with microbial systems. However, two recent advances have demonstrated how the throughput of such experiments can be increased. In the **LUMIER approach (luminescence-based mammalian interactome mapping)** the two-hybrid proteins are augmented with luciferase to enable the detection of interactions and also a specific epitope to facilitate the affinity capture of interacting proteins. Importantly, this assay can be automated using 96-well microtiter plates. In the **MAPPIT (mammalian protein–protein interaction trap) approach**, the two-hybrid system incorporates components of the JAK–STAT signaling pathway so that signaling capacity is only restored when the components are reconstituted and the normal ligand is supplied, thereby helping to suppress false positives. This assay involves the reverse transfection of cells expressing bait proteins, which can be seeded into plates containing panels of prey clones. This has been carried out using 384-well microtiter plates, but the same principle could be extended to cells seeded on glass slides arrayed with prey clones. This further downscaling would effectively convert the assay into a cell array, a concept we discuss in more detail in Chapter 9.

7.8 ADAPTED HYBRID ASSAYS FOR DIFFERENT TYPES OF INTERACTIONS

The original yeast two-hybrid system was developed to test or screen for interactions between pairs of proteins. Once it was established, however, investigators turned their attention toward improvements and enhancements that allowed the detection of different types of interactions. One of the earliest adaptations made it possible to detect interactions between proteins and small peptides, which can be useful to define minimal sets of conserved sequences in interaction partners. Other derivatives allowed higher-order complexes to be studied, by expressing the bait with a known interaction partner in the hope of attracting further complex components. The inability of yeast cells to carry out many of the post-translational modifications that occur in mammals was one of the driving forces behind the development of mammalian two-hybrid approaches (see above), but this challenge has also been addressed by carrying out two-hybrid screening in yeast cells expressing mammalian kinases, ensuring that the phosphorylation target sites on mammalian proteins are occupied. The one-hybrid system and its derivatives for the detection of DNA-binding proteins, and the bait-and-hook/three-hybrid systems for the detection of RNA-binding proteins and protein–ligand interactions are described in Box 7.2.

Another interesting variant is the **reverse two-hybrid system**, which uses counterselectable markers to screen for the loss of protein interactions. In the conventional (forward) system, the gene driven by the reassembled transcription factor is either essential for survival (selecting for interactions) or encodes a reporter protein (allowing cells containing interacting bait–prey pairs to be selected visually). In the reverse system, cells can only survive if the gene driven by the reassembled transcription factor is inactive. This can be used to identify mutations that disrupt specific interaction events (**Figure 7.14a**) and to find drugs that disrupt interactions between disease-causing proteins (Figure 7.14b). This approach should not be confused with the **reverse transactivator system**, in which protein interactions are detected by the loss of reporter gene expression. Here, the goal is the same as in the classic yeast two-hybrid approach (to screen for positive interactions) but the successful interaction between bait and prey results in the repression of the reporter gene (usually achieved by fusing the prey constructs to a transcriptional repressor). This system can also be used to screen for drugs that

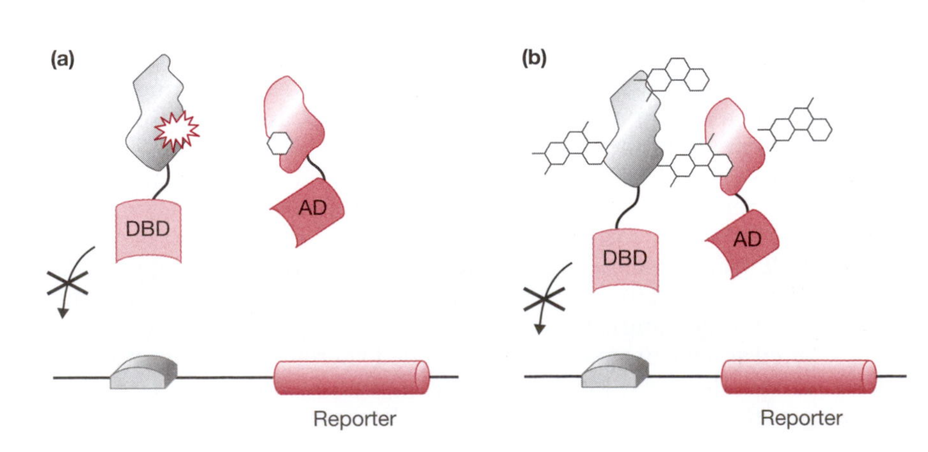

FIGURE 7.14 In the reverse yeast two-hybrid system, reconstitution of the transcription factor by bait–prey interaction drives a counterselectable reporter that generates a toxic metabolite and leads to cell death. (a) The reverse yeast two-hybrid system can be used to find interaction domains on protein surfaces and to show which residues are required for the interactions. Viability is restored only by mutations that abolish the interaction. (b) The reverse yeast two-hybrid system can be used for drug screening. Viability is restored only by compounds that interfere with the bait–prey interaction and prevent the reconstitution of the transcription factor.

inhibit protein interactions, with the advantage that where such inhibition occurs, the reporter gene is activated, allowing positive clones to be selected in the conventional manner. Reverse variants of the RAS recruitment assay have also been developed, although they have not become well established. Various systems using **dual baits** have also been described, and these can be used to find mutations that block specific interactions between a given prey protein and one of two distinct baits.

7.9 SYSTEMATIC COMPLEX ANALYSIS BY TANDEM AFFINITY PURIFICATION–MASS SPECTROMETRY

Affinity-based methods for the analysis of individual protein–protein interactions (see p.138) have evolved into efficient strategies for the global characterization of protein complexes. The major initial bottleneck, that of identifying the proteins in each complex, has been eliminated by advances in mass spectrometry that allow the characterization of very-low-abundance protein samples (in the low femtomole range). These developments have culminated in the use of affinity purification and mass spectrometry for the systematic analysis of all protein complexes in the cell (sometimes described as the **complexome**).

The first reports of complex analysis by mass spectrometry involved the antibody-based affinity purification of known components, and thus relied on the availability of suitable antibodies. A more versatile approach is required to prepare baits for whole-complexome analysis, and affinity tags are ideal because they can be attached to any protein and used to capture that protein on a suitable affinity matrix. Two large-scale studies of the yeast complexome were carried out in 2002, one using the transient expression of proteins containing a FLAG epitope for single-step immunocapture with an anti-FLAG antibody, and the other using a technique known as **tandem affinity purification** (**TAP**), which uses two different affinity tags (**Box 7.4**). The prototype TAP cassette included a calmodulin-binding peptide and staphylococcal protein A, with the two elements separated by a cleavage site for TEV protease (**Figure 7.15**). Instead of expressing these constructs transiently, the investigators used gene targeting to replace nearly 2000 yeast genes with a TAP fusion cassette. Yeast cells expressing each bait–TAP fusion

BOX 7.4 CASE STUDIES.
Large-scale interaction screens in yeast using affinity purification.

The first large-scale studies of the yeast complexome were carried out in 2002 using two different methods. In the first investigation (see Further Reading, Ho et al.), 725 bait proteins selected to represent multiple functional classes were transiently expressed as fusions with the FLAG epitope, allowing affinity capture using an anti-FLAG antibody. More than 1500 captured complexes were isolated and characterized by MS/MS, revealing 3617 interactions among 1578 proteins. In the other investigation (see Further Reading, Gavin et al.), 2000 yeast genes were replaced, by homologous recombination, with a tandem affinity purification cassette comprising the calmodulin-binding peptide and staphylococcal protein A. After cultivation, cell lysates from each strain were passed through an immunoglobulin affinity column to capture the protein A component of the tag, and the bound protein complexes were released by protease treatment after washing to remove nonspecific binders. Highly selective binding was then carried out in a second round of affinity chromatography using calmodulin as the affinity matrix in the presence of calcium ions. The proteins retained in this step were eluted by adding the chelating agent ethylene glycol tetraacetic acid (EGTA), and were examined by mass spectrometry, revealing 4111 interactions involving 1440 proteins. Only 10% of the identified complexes were already fully characterized, whereas 30% contained previously unknown components and 60% were entirely novel. The largest complex contained 83 proteins. However, approximately 60% of known interactions from the literature were not identified (possibly because the method favors the recovery of the most stable complexes rather than transient ones). There was a low degree of overlap between the two studies discussed above, perhaps reflecting the differences in abundance of the bait. Transient expression would lead to nonphysiological high levels of bait protein whereas the TAP method allowed each gene to be expressed under the control of its native promoter.

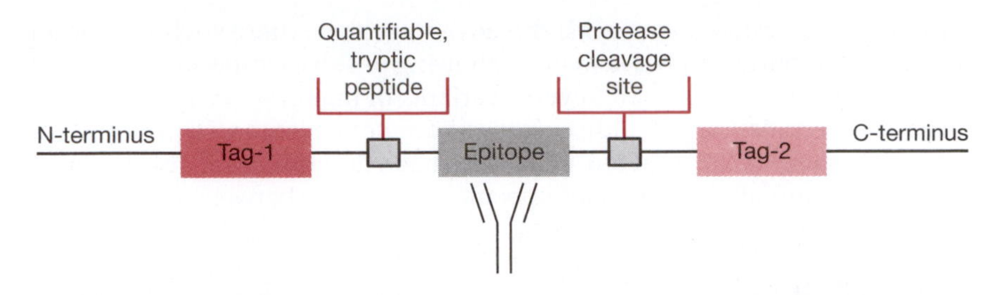

Tag name, MW	Tag-1	Tag-2	Enzyme cleavage via (cleavage site)	Organism/ comments
AC-TAP, 20 kDa	Protein A	CBP	TEV (ENLYFQ*G)	Prokaryotes, yeast/ most widely used tag to date
GS-TAP, 19 kDa	Protein G	SBP	TEV	Higher eukaryotes
LAP, 36 kDa	EGFP	S-peptide HIS_6	1. TEV 2. PreScission (LEVLFQ*GP)	Higher eukaryotes/ allows protein localization via GFP
SH-TAP, 5 kDa	SBP	Hemagglutinin	–	Higher eukaryotes/ small tag, lower risk of steric interference
SPA, 8 kDa	3 x FLAG	CBP	TEV	Prokaryotes, yeast/ small tag, lower risk of steric interference
Flag-HA, 3 kDa	FLAG	Hemagglutinin	–	Higher eukaryotes/ small tag, lower risk of steric interference

FIGURE 7.15 Tandem affinity tags for complexome analysis—generic construct structure and variants. CBP = calmodulin-binding peptide; SBP = streptavidin-binding peptide; EGFP = epidermal growth factor peptide; FLAG = epitope containing the amino acids F, L, A, and G; TEV = tobacco etch virus protease site.

cassette were lysed and the cell lysate was passed through an immunoglobulin affinity column to capture the protein A component of the bait fusion. After washing to remove nonspecific binding, the bound complexes were selectively eluted by the addition of the protease. Highly selective binding was then carried out in a second round of affinity chromatography using calmodulin as the affinity matrix in the presence of calcium ions. The proteins retained in this step were eluted by adding the calcium-chelating agent ethylene glycol tetraacetic acid (EGTA), and were examined by mass spectrometry. The advantage of the TAP method was the increased sensitivity and specificity because of the two-step purification, and the use of gene targeting rather than transient expression meant that bait proteins were expressed at physiological levels (under the control of endogenous promoters). The TAP method has been used for complexome studies in a wide range of species (**Table 7.2**). Single-tag methods using FLAG or the influenzavirus hemagglutinin protein continue to be used in some studies because they are better for the analysis of transient complexes.

As discussed above for two-hybrid systems, the TAP-MS approach also has a significant error rate, failing to detect about 60% of known interactions but generating fewer false positives in benchmarking tests. In part, this may reflect the fact that affinity-based methods favor the recovery of stable complexes rather than transient ones. In contrast, the yeast two-hybrid system can detect transient interactions because even short-lived interactions will cause some activation of the reporter gene. The sensitivity of TAP-MS is highest in yeast and bacteria, where homologous recombination can be used, whereas adaptations of the procedure for higher organisms generally require the use of plasmids or artificial transgenes with the TAP constructs under the control of constitutive or inducible promoters. Site-specific

TABLE 7.2 KEY LARGE-SCALE INTERACTION STUDIES (AFFINITY PURIFICATION MASS SPECTROMETRY)

Organism	Scope	Tag-type	Expression system
Eukaryotes			
Saccharomyces cerevisiae	Kinases and phosphatases	Flag and HA	Plasmid vectors (inducible GAL1 promoter)
		AC-TAP	Homologous recombination (endogenous promoter)
	Genome-wide	AC-TAP	Homologous recombination (endogenous promoter)
	Genome-wide	AC-TAP	Homologous recombination (endogenous promoter)
Homo sapiens	Autophagy system	Flag-HA	Retroviral and lentiviral vectors (LTR-driven constitutive expression or transient transfections)
	Mitosis	LAP	Bacterial artificial chromosome (endogenous promoter)
	Chromatin remodeling	Flag	Site-specific recombination via flippase/flippase recognition target (Flp/FRT)
	Deubiquitinating enzymes	Flag-HA	Retroviral vectors (LTR-driven constitutive expression or inducible cytomegalovirus promoter)
	Disease candidate genes	Flag	Plasmid vectors (transient transfections)
Drosophila melanogaster	Notch signaling pathway	AC-TAP	Plasmid vectors (inducible metallothionein or Hsp70Bb promoter)
Oryza sativa	Kinases	AC-TAP	Transgenic plant (constitutive maize ubiquitin promoter)
Arabidopsis thaliana	Cell cycle	GS-TAP	Plasmid vector (constitutive Cauliflower mosaic virus 35S promoter)
Prokaryotes			
Mycoplasma pneumoniae	Genome-wide	AC-TAP	Transposon (endogenous clpB promoter)
Escherichia coli	Uncharacterized genes	SPA	Homologous recombination (endogenous promoter)

Names of tags are explained in Figure 7.15 opposite. From Gavin AC, Maeda K & Kühner S (2011) *Curr. Opin. Biotechnol.* 22, 42. With permission from Elsevier.

recombination has recently been used in human cell lines to generate isogenic clones expressing TAP cassettes, and is also common in *D. melanogaster* and chicken cells. The increasing popularity of TAP-MS has resulted in the diversification of protocols and reagents to meet the demands of different species. The calmodulin-binding tag is not always suitable in mammalian cells, because these can express high levels of calmodulin, so a number of alternatives have been introduced, including streptavidin-binding peptides, the FLAG epitope, hemagglutinin, and synthetic peptides (Figure 7.15). To avoid interference with protein folding, some of these tags are very small (for example, FLAG-hemagglutinin is 3 kDa, compared with the 20 kDa tag on the original TAP cassette).

7.10 ANALYSIS OF PROTEIN INTERACTION DATA

The protein–protein interaction community uses a quality standard for data submission called MIMIx, the minimum information required for reporting a molecular interaction experiment (p. 21). Both of the main approaches for gathering large-scale interaction data are subject to errors, so careful quality control is required to validate the data. Putative interactions are often evaluated on the basis of reproducibility in independent experiments, by reference to literature reports on interactions that have been independently confirmed using different methods, and by filtering out promiscuous proteins (sticky baits and prey) that feature in many binary interactions and complexes. Frequency filtering must be applied carefully to avoid excluding proteins that genuinely have many interaction partners, and this was initially achieved by examining the topology of known interaction networks using **socioaffinity scoring methods**. Such methods include direct reciprocal

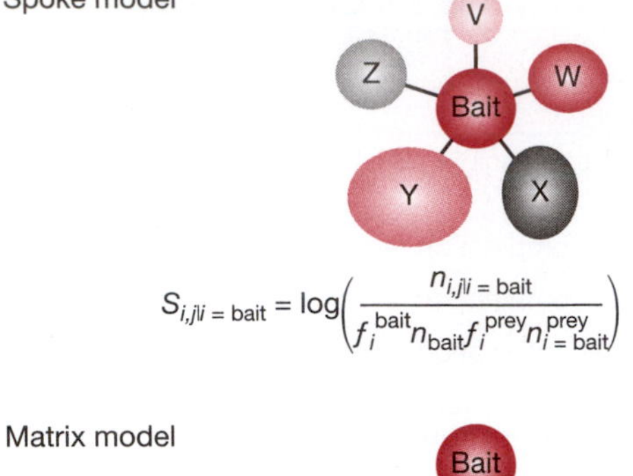

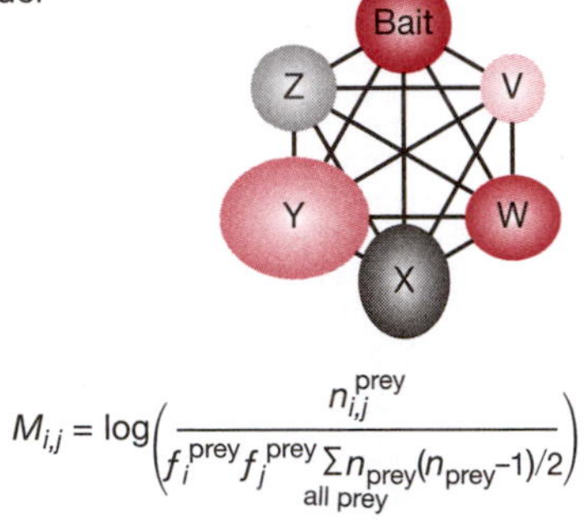

FIGURE 7.16 Socioaffinity scoring. The spoke model quantifies the tendency for proteins to identify each other when tagged, and the matrix model to co-purify when other proteins are tagged. These models were developed for tandem affinity purification, but the spoke model is applicable to binary interactions indicated by two-hybrid/protein complementation assays. (From Gavin AC, Maeda K & Kühner S (2011) *Curr. Opin. Biotechnol.* 22, 42. With permission from Elsevier.)

interactions between proteins (**spoke interactions**) as well as the direct and indirect interactions within a complex (**matrix interactions**) using data derived from two-hybrid screens, affinity purifications, and mass spectrometry scores such as normalized spectral abundance factors (**Figure 7.16**).

The original socioaffinity scoring method was enhanced by making full use of repetitive purifications and focusing on spoke interactions to generate so-called **improved socioaffinity scores (ISAs)**. Alternative methods have been proposed, including **purification enrichment scores**, which use a sophisticated statistical approach to score individual observed interactions separately, and **Hart scores**, which focus instead on combining observations from different screens. **IDBOS (interaction detection based on shuffling) scores** look specifically for direct physical interactions and assume that purified complexes can be randomly permutated, generating more accurate predictions although requiring more computer resources. Finally, **significance analysis of interactome (SAINT) scores** have been introduced more recently to incorporate peptide counts from mass spectrometry data, and use Poisson distributions for the heuristic computation of posterior probabilities relating to specific interactions between proteins. The requirement for experimental peptide count data means that SAINT scores cannot be applied to historical screens, unlike the other methods. A number of software platforms that process spectral data for interaction mapping have been developed, including SAINT and CompPASS.

7.11 PROTEIN INTERACTION MAPS

Interaction data are usually presented as graphs or networks with the interacting partners shown as nodes and the interactions shown as interconnecting lines or edges. The presentation of interaction data is challenging because of the complexity of the data in large interaction datasets. Each interaction must be annotated to show how it was identified, binary interactions must if necessary be distinguished from interactions within complexes, transient interactions must be distinguished from permanent associations, there needs to be some representation of interaction stoichiometry, and there should also be a system to assign confidence to interactions identified in different ways. All this must be integrated with existing sequence, structure, metabolic pathway, and ontology databases and must be presented in such a way that the interested researcher can switch between simplified views encompassing the entire cell or subcellular compartment and detailed views of particular interaction networks. Other cellular components, such as DNA, RNA, and small molecules, will also have to be built in. Finally, the purpose of protein interaction maps is not only to simplify visualization but also to enable computer analysis.

With these issues in mind, a number of interaction databases have been established that can be accessed over the Internet (**Table 7.3**). Most of them originated from the large-scale interaction screens listed in Tables 7.1 and 7.2, and although the early examples largely focus on the yeast proteome (for example, the Biomolecular Interaction Network Database, the Database of Interacting Proteins, the Comprehensive Yeast Genome Database, and the Saccharomyces Genome Database) others have since been developed to focus on more diverse organisms. Several tens of thousands of interactions are listed, many of which await further functional validation. These databases have been augmented with additional data from other sources. Importantly, a potentially very large amount of data concerning individual protein interactions is "hidden" in the scientific literature going back many years. It will be a challenge to extract this information and integrate it with that obtained from recent high-throughput experiments. Several bioinformatics tools have been developed to trawl through the literature databases

TABLE 7.3 PROTEIN–PROTEIN INTERACTION (PPI) DATABASES

Acronym	Database full name and URL	PPI sources	Type of molecular interaction	Species	*n* proteins (Dec. 2009)	*n* interactions (Dec. 2009)
Primary databases: PPI experimental data (curated from specific small-scale and large-scale (Ssc, Lsc) published studies)						
BIND	Biomolecular Interaction Network Database, http://bond.unleashedinformatics.com/	Ssc and Lsc published studies (literature-curated)	PPIs and others	All	[31,972]	[58,266]
BioGRID	Biological General Repository for Interaction Datasets, http://www.thebiogrid.org/	Ssc and Lsc published studies (literature-curated)	PPIs and others	All	[28,717]	[108,691]
DIP	Database of Interacting Proteins, http://dip.doe-mbi.ucla.edu/dip/	Ssc and Lsc published studies (literature-curated)	Only PPIs	All	20,728	57,683
HPRD	Human Protein Reference Database, http://www.hprd.org/	Ssc and Lsc published studies (literature-curated)	Only PPIs	Human	27,081	38,806
IntAct	IntAct Molecular Interaction Database, http://www.ebi.ac.uk/intact/	Ssc and Lsc published studies (literature-curated)	PPIs and others	All	[60,504]	[202,826]
MINT	Molecular INTeraction database, http://mint.bio.uniroma2.it/mint/	Ssc and Lsc published studies (literature-curated)	Only PPIs	All	30,089	83,744
MIPS-MPact	MIPS protein interaction resource on yeast, http://mips.gsf.de/genre/proj/mpact/	Derived from CYGD	Only PPIs	Yeast	1,500	4,300
MIPS-MPPI	MIPS Mammalian Protein–Protein Interaction Database, http://mips.gsf.de/proj/ppi	Ssc published studies (literature-curated)	Only PPIs	Mammalian	982	937
Meta-databases: PPI experimental data (integrated and unified from different public repositories)						
APID	Agile Protein Interaction DataAnalyzer, http://bioinfow.dep.usal.es/apid/	BIND, BioGRID, DIP, HPRD, IntAct, MINT	Only PPIs	All	56,460	322,579
MPIDB	The Microbial Protein Interaction Database, http://www.jcvi.org/mpidb/	BIND, DIP, IntAct, MINT, other sets (experimental and literature-curated)	Only PPIs	Microbial	7,810	24,295
PINA	Protein Interaction Network Analysis platform, http://cbg.garvan.unsw.edu.au/pina/	BioGRID, DIP, HPRD, IntAct, MINT, MPact	Only PPIs	All	[?]	188,823
Prediction databases: PPI experimental and predicted data ("functional interactions," that is, interactions sensu lato *derived from different types of data)*						
MiMI	Michigan Molecular Interactions, http://mimi.ncibi.org/MimiWeb/	BIND, BioGRID, DIP, HPRD, IntAct, and non-PPI data	PPIs and others	All	[45,452]	[391,386]
PIPs	Human PPI Prediction database, http://www.compbio.dundee.ac.uk/www-pips/	BIND, DIP, HPRD, OPHID, and non-PPI data	PPIs and others	Human	[?]	[37,606]
OPHID	Online Predicted Human Interaction Database, http://ophid.utoronto.ca/	BIND, BioGRID, HPRD, IntAct, MINT, MPact, and non-PPI data	PPIs and others	Human	[?]	[424,066]
STRING	Known and Predicted Protein–Protein Interactions, http://string.embl.de/	BIND, BioGRID, DIP, HPRD, IntAct, MINT, and non-PPI data	PPIs and others	All	[2,590,259]	[88,633,860]
UniHI	Unified Human Interactome, http://www.mdc-berlin.de/unihi/	BIND, BioGRID, DIP, HPRD, IntAct, MINT, and non-PPI data	PPIs and others	Human	[22,307]	[200,473]

Numbers in square brackets include PPIs and other types of interactions (for example, protein–ligand interactions or, for the case of prediction databases, non-PPI data). From De Las Rivas J & Fontanillo C (2010) *PLoS Comput. Biol.* 6, e1000807. With permission from Public Library of Science.

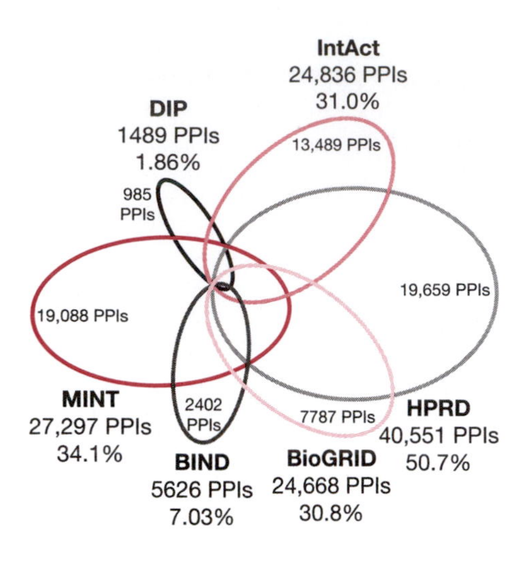

FIGURE 7.17 Overlapping coverage of human protein–protein interactions in major interaction databases surveyed in 2009. (From De Las Rivas J & Fontanillo C (2010) *PLoS Comput. Biol.* 6, e1000807. With permission from Public Library of Science.)

and identify keywords that indicate protein interactions so that such references can be scrutinized by the human curators of interaction databases. Even so, there is a significant variation in coverage among the databases, as illustrated in **Figure 7.17** with respect to known human protein–protein interactions.

The complexity of interaction networks is shown in **Figure 7.18**. Figure 7.18a shows a binary interaction network representing 1548 yeast proteins (25% of the proteome) and a total of 2358 interactions that had been identified at the time of publication (2000). As might be expected, proteins with a similar general function (for example, membrane transport) tend to interact with each other rather more than with functionally unrelated proteins, such as those involved in the maintenance of chromatin structure. Figure 7.18b shows a simplified version in which proteins have been clustered according to their function. This provides a good overview of the whole interaction map, which can be probed for more detail if required. Although the map is not topological (that is, it does not reflect the architecture of the cell), proteins that are co-localized also tend to interact more often than those that are located primarily in different compartments, and tracing interaction routes from protein to protein does allow signaling, regulatory, and metabolic pathways to be identified. **Figure 7.19** (see also color plates) shows the complex interaction map resulting from the yeast protein complex screen published in 2002. This has been simplified by omitting any protein that is found in more than nine complexes. As shown in the insert, each complex can be inspected for individual proteins, again providing the researcher with multiple levels of detail. As with the binary map, complexes with similar functions tend to share components and interactions, while there are fewer interactions between functionally unrelated complexes. One of the most popular software packages for visualizing protein interaction data is Cytoscape because of the large range of plug-ins available for different purposes. Others are listed in **Table 7.4**.

7.12 PROTEIN INTERACTIONS WITH SMALL MOLECULES

We touch briefly on the subject of protein interactions with small molecules to finish this chapter, but we shall return to the subject in more detail in Chapter 9, which considers activity-based proteomics (p. 196), and Chapter 10, which looks at some of the applications of proteomics in drug development (p. 219). Small molecules can act as co-factors, enzyme substrates, ligands for receptors, or allosteric modulators, and the function of many proteins is to transport or store particular molecules. On a proteomic scale, screening methods can be employed to isolate proteins that interact with particular small molecules, for example, through the use of labeled substrates as probes or the immobilization of those substrates on chips. Protein interactions with small molecules can be studied at atomic resolution using X-ray crystallography, and occasionally a protein is accidentally purified along with its ligand, providing valuable information about its biochemical function.

Large-scale screens for protein interactions with small molecules are often carried out to identify lead compounds that can be developed into drugs. This process can be simplified considerably by choosing compounds based on a known ligand or screening for potential interacting compounds *in silico* using a chemical library. Where the structure of the target protein is available at high resolution, docking algorithms can be used in an attempt to fit small molecules into binding sites using information on steric constraints and bond energies. Some of this software is available for free on the Internet, and is listed in **Table 7.5**. One of the most widely-used docking algorithms is AutoDock, which establishes ligand coordinates, bonds that have axes of

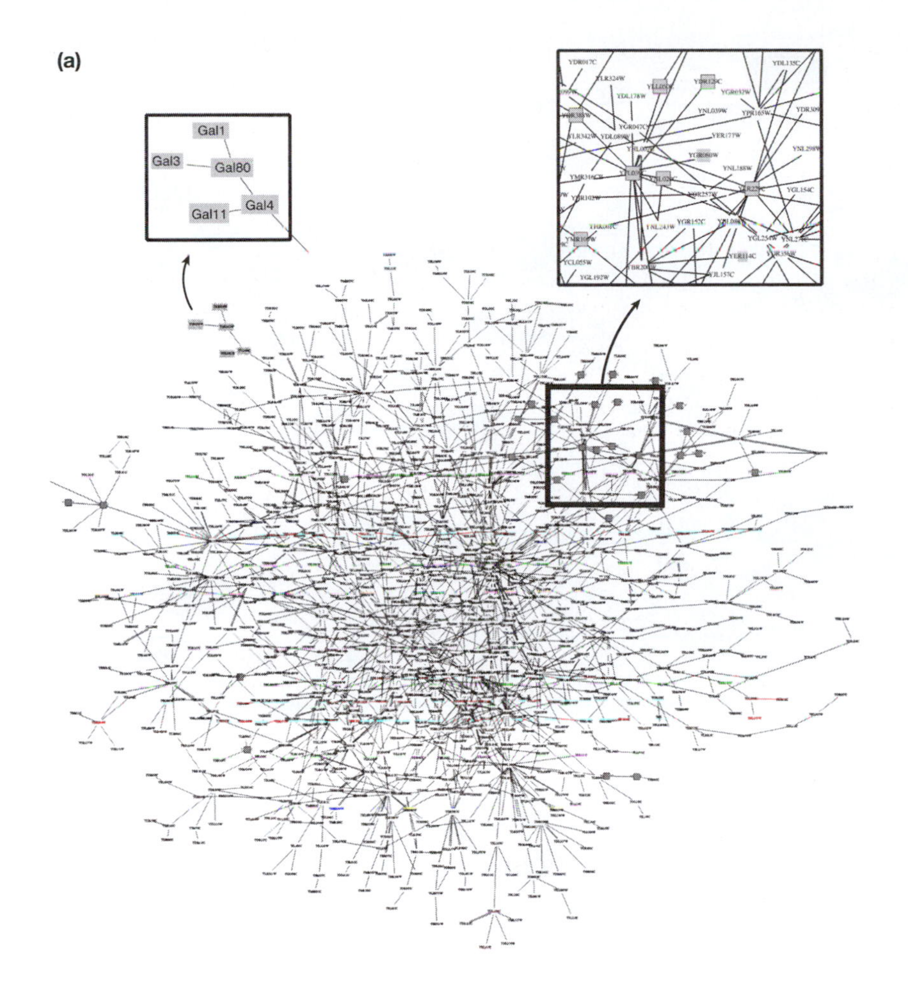

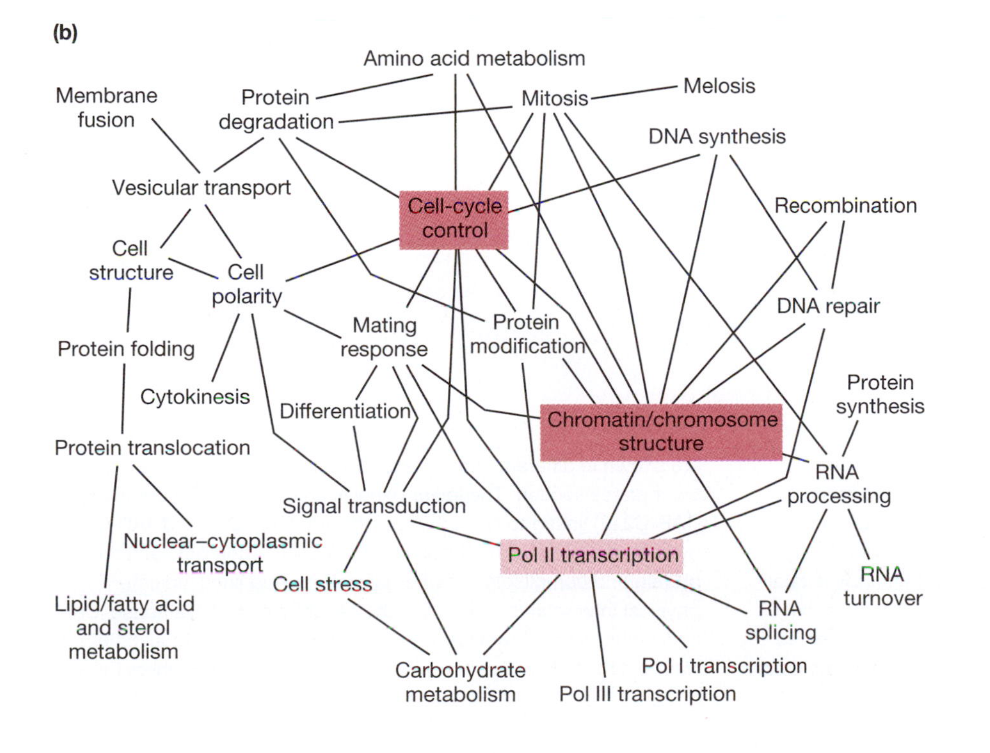

FIGURE 7.18 Visualization of protein interaction networks. (a) Binary interaction map including 1200 interacting proteins based on published interactions. The inset shows close-up of region highlighted in box. Highlighted as *dark gray* boxes are cell structure proteins (a single functional class). Proteins in this category can be observed to cluster primarily in one region. Although interacting proteins are not depicted in a way that is consistent with their known cellular location (that is, those proteins known to be present in the nucleus in the center of the interaction map and those present in plasma membranes in the periphery), signal-transduction pathways (or at least protein contact paths) can be inferred from this diagram. (b) Functional group interaction map derived from the detailed map. Each line indicates that there are 15 or more interactions between proteins of the connected groups. Connections with fewer than 15 interactions are not shown, because one or a few interactions occur between almost all groups and often tend to be spurious, that is, based on false positives in two-hybrid screens or other assays. Note that only proteins with known function are included and that about one-third of all yeast proteins belong to several classes. (From Tucker CL, Gera JF & Uetz P (2001) *Trends Cell Biol.* 11, 102. With permission from Elsevier.)

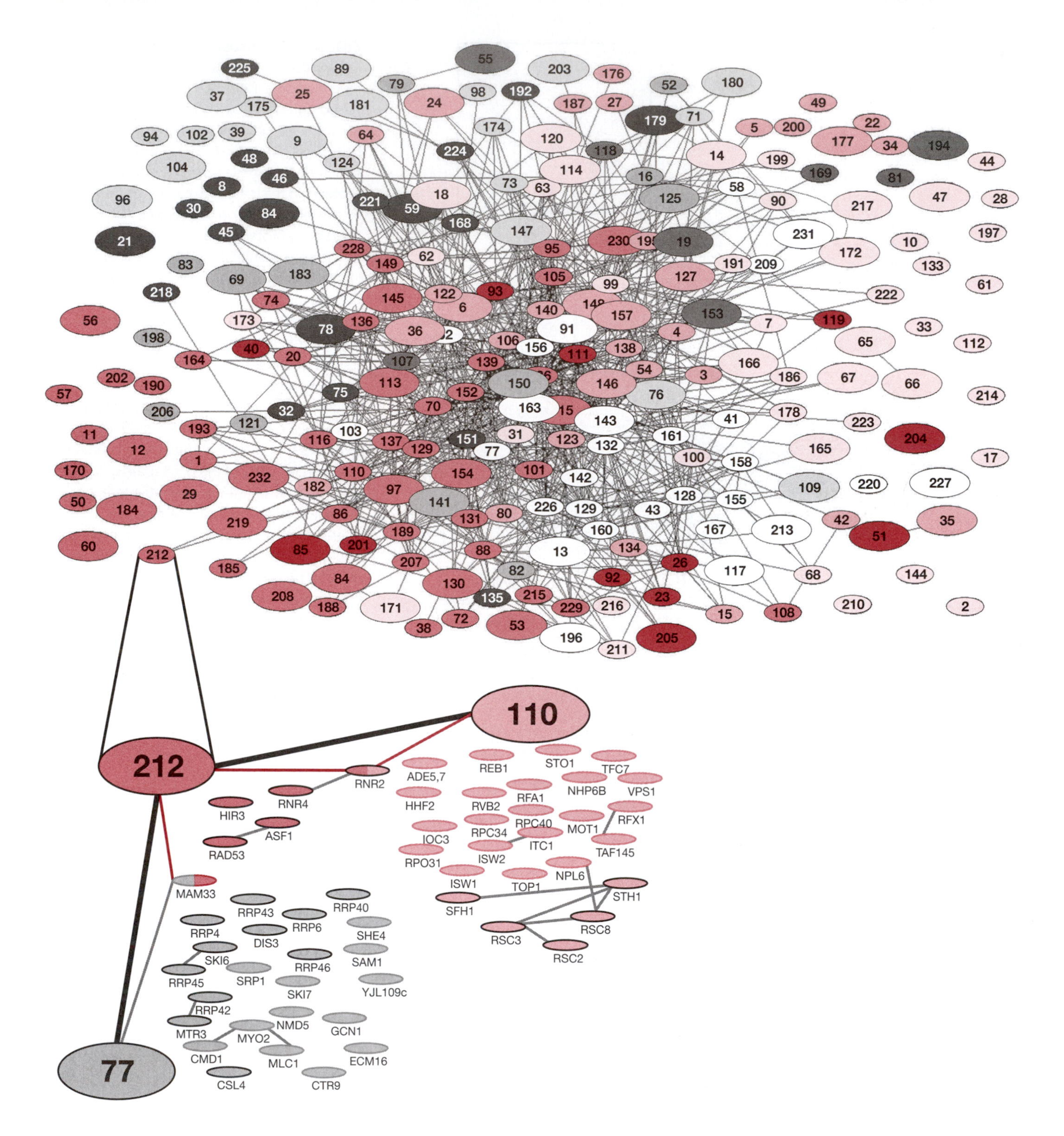

FIGURE 7.19 The protein complex network, and grouping of connected complexes. Links were established between complexes sharing at least one protein. For clarity, proteins found in more than nine complexes were omitted. The graphs were generated automatically by a relaxation algorithm that finds a local minimum in the distribution of nodes by minimizing the distance of connected nodes and maximizing the distance of unconnected nodes. In the upper panel, cellular roles of the individual complexes are shown in different shades and identified by color codes in the color plates section. The lower panel is an example of a complex (TAP-C212) linked to two other complexes (TAP-C77 and TAP-C110) by shared components. It illustrates the connection between the protein and complex levels of organization. *Red* lines indicate physical interactions as listed in the Yeast Proteome Database. See also color plates. (From Gavin AC, Bösche M, Krause et al. (2002) *Nature* 415, 141. With permission from Macmillan Publishers Ltd.)

TABLE 7.4 VISUALIZATION TOOLS FOR INTERACTION NETWORKS

Name	Cost	OS	Description	URL
Stand-alone				
Arena 3D	Free	Win, Mac, Linux	Visualization of biological multi-layer networks in 3D	http://www.arena3d.org/
BiNA	Free	Win, Mac, Linux	Exploration and interactive visualization of pathways	http://www.bina.unipax.info/
BioLayout Express 3D	Free	Win, Mac, Linux	Generation and cluster analysis of networks with 2D/3D visualization	http://www.biolayout.org/
BiologicalNetworks	Free	Win, Mac, Linux	Analysis suite; visualizes networks and heat map; abundance data	http://www.biologicalnetworks.org/
Cytoscape	Free	Win, Mac, Linux	Network analysis; extensive list of plug-ins for advanced visualization	http://www.cytoscape.org/
GENeVis	Free	Win, Mac, Linux	Network and pathway visualization; abundance data	http://tinyurl.com/genevis/
Medusa	Free	Win, Mac, Linux	Basic network visualization tool	http://coot.embl.de/medusa/
N-Browse	Free	Win, Mac, Linux	Network visualization software for heterogeneous interaction data	http://www.gnetbrowse.org/
NAViGaTOR	Free	Win, Mac, Linux	Visualization of large protein-interaction data sets; abundance data	http://tinyurl.com/navigator1/
Ondex	Free	Win, Mac, Linux	Integrative workbench: large network visualizations; abundance data	http://www.ondex.org/
Osprey	Free	Win, Mac, Linux	Tool for visualization of interaction networks	http://tinyurl.com/osprey1/
Pajek	Free	Win	Generic network visualization and analysis tool	http://pajek.imfm.si/
ProViz	Free	Win, Mac, Linux	Software for visualization and exploration of interaction networks	http://www.cb.uu.se/research/proviz/
SpectralNET	Free	Win	Network visualizations; scatter plots for dimensionality reduction methods	http://broadinstitute.org/software/spectralnet
Tulip	Free	Win, Mac, Linux	Generic visualization tool; extremely large networks; 3D support	http://tulip.labri.fr/TulipDrupal/
VANTED	Free	Win, Mac, Linux	Combined visualization of abundance data, networks, and pathways	http://tinyurl.com/vanted/
yEd	Free	Win, Mac, Linux	Generic network visualization software; offers many layout algorithms	http://tinyurl.com/yEdGraph/
Cytoscape plug-in				
BiNoM	Free	Win, Mac, Linux	Extensive support for common systems biology network formats	https://binom.curie.fr/
Cerebral	Free	Win, Mac, Linux	Biologically motivated layout algorithm; maps abundance data; clustering	http://tinyurl.com/cerebral1/
MCODE	Free	Win, Mac, Linux	Network clustering algorithm; support for manual cluster refinement	http://tinyurl.com/MCODE123/
VistaClara	Free	Win, Mac, Linux	Mapping of abundance data to nodes and 'heat strips'; provides heat map	http://apps.cytoscape.org/apps/vistaclaraplugin
Web-based				
Graphle	Free		Distributed client/server network exploration and visualization tool	http://tinyurl.com/graphle/
Lichen	Free		Library for web-based visualization of network and abundance matrix data	http://tinyurl.com/Lichen1/
MAGGIE Data Viewer	Free		Visualization of networks; abundance data in heat maps and profile plots	http://maggie.systemsbiology.net/
STITCH	Free		Construction and visualization of networks from a wide range of sources	http://stitch.embl.de/
VisANT	Free	Win, Mac, Linux	Analysis, mining, and visualization of pathways and integrated omics data	http://visant.bu.edu/

From Gehlenborg N, O'Donoghue SI, Baliga NS, et al. (2010) *Nat. Methods* 7 (Suppl.), S56. With permission from Macmillan Publishers Ltd.

TABLE 7.5 CHEMICAL DOCKING SOFTWARE AVAILABLE OVER THE INTERNET

URL	R/F	Description	Availability
http://autodock.scripps.edu/	F	AutoDock, discussed in main text	Download for Unix/Linux
http://sgedg.weizmann.ac.il/ligin/	R	LIGIN, a robust ligand–protein interaction predictor limited to small ligands	Download for Unix or as a part of the WHATIF package
http://www.sbg.bio.ic.ac.uk/docking/ftdock.html	R	FTDock and associated programs RPScore and MultiDock. Can deal with protein–protein interactions. Relies on a Fourier transform library.	Download for Unix/Linux
http://vakser.bioinformatics.ku.edu/resources/gramm/grammx/	R	GRAMM (Global Range Molecular Matching), an empirical method based on tables of inter-bond angles. GRAMM has the merit of coping with low-quality structures.	Download for Unix or Windows.
http://www.biosolveit.de/flexx/	F	FlexX, which calculates energetically favorable molecular complexes consisting of the ligand bound to the active site of the protein, and ranks the output.	Apply on-line for FlexX workspace on the server.

R/F means rigid or flexible, and indicates whether the program regards the ligand as a rigid or flexible molecule.

free rotation, and interaction energies before carrying out the docking simulation. Another widely used program is DOCK, in which the arrangement of atoms at the binding site is converted into a set of spheres called site points. The distances between the spheres are then used to calculate the exact dimensions of the binding site, and these are compared with a database of chemical compounds. Matches between the binding site and a potential ligand are given a confidence score, and ligands are then ranked according to their total scores. This modeling approach has the disadvantage that the binding site and every potential ligand are considered to be stiff and inflexible. Other algorithms can incorporate flexibility into the structures. A more recent development called CombiDOCK considers each potential ligand as a scaffold decorated with functional groups. Only spheres on the scaffold are initially used in the docking prediction and then individual functional groups are tested using a variety of bond torsions. Finally, the structure is bumped (checked to make sure none of the positions predicted for individual functional groups overlap) before a final score is presented.

Chemical databases can be screened not only with a binding site (searching for complementary molecular interactions) but also with another ligand (searching for identical molecular interactions). Several available algorithms can compare two-dimensional or three-dimensional structures and build a profile of similar molecules. This approach is important, for example, in a drug development if a ligand has been shown to interact with a protein but has negative side effects, or if a structurally distinct ligand is required to avoid intellectual property issues. In each case a molecule of similar shape with similar chemical properties is required, but with a different structure.

FURTHER READING

Charbonnier S, Gallego O & Gavin AC (2008) The social network of a cell: recent advances in interactome mapping. *Biotechnol. Annu. Rev.* 14, 1-28.

Choi H (2012) Computational detection of protein complexes in AP-MS experiments. *Proteomics* 12, 1663–1668.

Choi H, Larsen B, Lin ZY et al. (2011) SAINT: probabilistic scoring of affinity purification–mass spectrometry data. *Nat. Methods* 8, 70–73.

De Las Rivas J & Fontanillo C (2010) Protein–protein interactions essentials: key concepts to building and analyzing interactome networks. *PLoS Comp. Biol.* 6, e1000807.

Drewes G & Bouwmeester T (2003) Global approaches to protein–protein interactions. *Curr. Opin. Cell Biol.* 15, 1–7.

Finley RL Jr & Brent R (1994) Interaction mating reveals binary and ternary connections between Drosophila cell cycle regulators. *Proc. Natl Acad. Sci.* 91, 12980-12984.

Flajolet M, Rotondo G, Daviet L et al. (2000) A genomic approach of the hepatitis C virus generates a protein interaction map. *Gene* 241, 369–379.

Fukao Y (2012) Protein–protein interactions in plants. *Plant Cell Physiol.* 53, 617–625.

Gavin AC, Aloy P, Grandi P et al. (2006) Proteome survey reveals modularity of the yeast cell machinery. *Nature* 440, 631–636.

Gavin AC, Bosch M, Krause R et al. (2002) Functional organization of the yeast proteome by systematic analysis of protein complexes. *Nature* 415, 141–147.

Gavin AC, Maeda K & Kühner S (2011) Recent advances in charting protein–protein interaction: mass spectrometry-based approaches. *Curr. Opin. Biotechnol.* 22, 42–49.

Gehlenborg N, O'Donoghue SI, Baliga NS et al. (2010) Visualization of omics data for systems biology. *Nat. Methods* 7 (Suppl), S56–S68.

Häuser R, Blasche S, Dokland T et al. (2012) Bacteriophage protein–protein interactions. *Adv. Virus Res.* 83, 219–298.

Ho Y, Gruhler A, Heilbut A et al. (2002) Systematic identification of protein complexes in *Saccharomyces cerevisiae* by mass spectrometry. *Nature* 415, 180–183.

Ito T, Chiba T, Ozawa R et al. (2001) A comprehensive two-hybrid analysis to explore the yeast protein interactome. *Proc. Natl Acad. Sci. USA* 98, 4569–4574.

Ito T, Tashiro K, Muta S et al. (2000) Toward a protein–protein interaction map of the budding yeast: a comprehensive system to examine two-hybrid interactions in all possible combinations between the yeast proteins. *Proc. Natl Acad. Sci. USA* 97, 1143–1147.

Koegl M & Uetz P (2007) Improving yeast two-hybrid screening systems. *Brief. Funct. Genomic. Proteomic.* 6, 302–312.

Koh GC, Porras P, Aranda B et al. (2012) Analyzing protein–protein interaction networks. *J. Proteome Res.* 11, 2014–2031.

Krogan NJ, Cagney G, Yu H et al. (2006) Global landscape of protein complexes in the yeast *Saccharomyces cerevisiae*. *Nature* 440, 637–643.

Lee TI, Rinaldi NJ, Robert F et al. (2002) Transcriptional regulatory networks in *Saccharomyces cerevisiae*. *Science* 298, 799–804.

Lievens S, Eyckerman S, Lemmens I & Tavernier J (2010) Large-scale protein interactome mapping: strategies and opportunities. *Expert Rev. Proteomics.* 7, 679–690.

Lievens S, Lemmens I & Tavernier J (2009) Mammalian two-hybrids come of age. *Trends Biochem. Sci.* 34, 579–588.

McCraith S, Hotzam T, Moss B & Fields S (2000) Genome-wide analysis of vaccinia virus protein–protein interactions. *Proc. Natl Acad. Sci. USA* 97, 4879–4884.

Orchard S (2012) Molecular interaction databases. *Proteomics* 12, 1656–1662.

Orchard S, Salwinski L, Kerrien S et al. (2007) The minimum information required for reporting a molecular interaction experiment (MIMIx). *Nat. Biotechnol.* 25, 894–898.

Phizicky E, Bastiaens PIH, Zhu H et al. (2003) Protein analysis on a proteomic scale. *Nature* 422, 208–215.

Phizicky EM & Fields S (1995) Protein–protein interactions: methods for detection and analysis. *Microbiol. Rev.* 59, 94–123.

Ritchie DW (2008) Recent progress and future directions in protein–protein docking. *Curr. Protein Pept. Sci.* 9, 1–15.

Salwinski L, Licata L, Winter A et al. (2009) Recurated protein interaction datasets. *Nat. Methods* 6, 860–861.

Sardiu ME & Washburn MP (2011) Building protein–protein interaction networks with proteomics and informatics tools. *J. Biol. Chem.* 286, 23645–23651.

Schwikowski B, Uetz P & Fields S (2000) A network of protein–protein interactions in yeast. *Nat. Biotechnol.* 18, 1257–1261.

Suter B, Kittanakom S & Stagljar I (2008) Two-hybrid technologies in proteomics research. *Curr. Opin. Biotechnol.* 19, 316–323.

Tong AH, Evangelista M, Parsons AB et al. (2001) Systematic genetic analysis with arrays of yeast deletion mutants. *Science* 294, 2364–2368.

Tong AHY, Lesage G, Bader GD et al. (2004) Global mapping of the yeast genetic interaction network. *Science* 303, 808–813.

Tucker CL, Gera JF & Uetz P (2001) Towards an understanding of complex protein networks. *Trends Cell Biol.* 11, 102–106.

Uetz P (2001) Two-hybrid arrays. *Curr. Opin. Chem. Biol.* 6, 57–62.

Uetz P, Giot L & Cagney G et al. (2000) A comprehensive analysis of protein–protein interactions in *Saccharomyces cerevisiae*. *Nature* 403, 623–627.

Walhout AJM & Vidal M (2001) Protein interaction maps for model organisms. *Nat. Rev. Mol. Cell Biol.* 2, 55–62.

Werner T (2003) Promoters can contribute to the elucidation of protein function. *Trends Biotechnol.* 21, 9–13.

Yu H, Braun P, Yildirim MA et al. (2008) High-quality binary protein interaction map of the yeast interactome network. *Science* 322, 104–110.

CHAPTER 8

Protein modification in proteomics

8

8.1 INTRODUCTION

Almost all proteins are modified in some way during or after synthesis, either by cleavage of the polypeptide backbone or covalent chemical modification of specific amino acid side chains. This phenomenon, which is known as **post-translational modification (PTM)**, provides a direct mechanism for the regulation of protein activity and greatly enhances the structural diversity and functionality of proteins by providing a larger repertoire of physical and chemical properties than is possible using the 20 standard amino acids specified by the genetic code. Several hundred different forms of chemical modification have been documented, some of which influence protein structure, some are required for proteins to interact with ligands or each other, some have a direct impact on biochemical activity, and some help sort proteins into different subcellular compartments (**Table 8.1**). Some proteins are modified in particular compartments of the cell but not in others.

Modifications are often permanent, but some, such as phosphorylation, are reversible and can be used to switch protein activity on and off in response to intracellular and extracellular signals. Post-translational modification is therefore a dynamic phenomenon with a central role in many biological processes. Importantly, inappropriate post-translational modification is often associated with disease, allowing particular post-translational variants to be used as disease biomarkers or therapeutic targets (Chapter 10). Whereas the above types of modification are typical of normal physiological processes or pathological states, others are associated with damage or ageing, and still others occur as artifacts when proteins are extracted and exposed to unnatural environments.

The complexity of the proteome is increased significantly by post-translational modification, particularly in eukaryotes, where many proteins exist as a heterogeneous mixture of alternative modified forms. Ideally, it would be possible to catalog the proteome systematically and quantitatively in terms of the types of post-translational modifications that are present, and specify the modified sites in each case. However, such attempts are frustrated by the sheer diversity involved. Every protein could potentially be modified in hundreds of different ways, and many contain multiple modification target sites allowing different forms of modification to take place either singly or in combination. Many post-translational modifications are still discovered accidentally when individual proteins, complexes, or pathways are studied. Modifications cannot be predicted accurately from the genome sequence, since even the presence of a known modification motif does not necessarily confirm that modification takes place. Indeed, the reverse process is usually more common, that is, the discovery of a modified protein by mass spectrometry can be used to annotate the genome sequence by adding a modification target site, an approach known as proteogenomics

TABLE 8.1 A SUMMARY OF PROGRAMMED ENZYMATIC PROTEIN MODIFICATIONS WITH ROLES IN PROTEIN STRUCTURE AND FUNCTION, PROTEIN TARGETING OR PROCESSING, AND THE FLOW OF GENETIC INFORMATION

Covalent modification	Examples
Substitutions (minor side-chain modifications)	
Minor side chain modification—permanent and associated with protein function	Hydroxylation of proline residues in collagen stabilizes triple-helical coiled-coil tertiary structure Sulfation of tyrosine residues in certain hormones Iodination of thyroglobulin γ-carboxylation of glutamine residues in prothrombin
Formation of intra- and intermolecular bonds	Formation of disulfide bonds in many extracellular proteins, e.g. insulin, immunoglobulins
Minor side-chain modification—reversible and associated with regulation of activity	Phosphorylation of tyrosine, serine, and threonine residues regulates enzyme activity, e.g., receptor tyrosine kinases, cyclin-dependent kinases Many side chains are also methylated, although the function of this modification is unknown Acetylation of lysyl residues of histones regulates their ability to form higher-order chromatin structure and has an important role in the establishment of chromatin domains
Augmentations (major side or main chain modifications)	
Addition of chemical groups to side chains—associated with protein function	Addition of nucleotides required for enzyme activity, e.g., adenyl groups added to glutamine synthase in *E. coli* Addition of *N*-acetylglucosamine to serine or threonine residues of some eukaryotic cytoplasmic proteins Addition of cholesterol to Hedgehog family signaling proteins controls their diffusion Addition of prosthetic groups to conjugated proteins, e.g., heme group to cytochrome *c* or globins
Addition of chemical groups to side chains—associated with protein targeting or trafficking	Acylation of cysteine residue targets protein to cell membrane Addition of GPI membrane anchor targets protein to cell membrane *N*-glycosylation of asparagine residues in the sequence Asn-Xaa-Ser/Thr is a common modification in proteins entering the secretory pathway *O*-glycosylation of Ser/Thr occurs in the Golgi apparatus Ubiquitinylation of proteins targeted for degradation
End-group modification	Acetylation of N-terminal amino acid of many cytoplasmic proteins appears to relate to rate of protein turnover Acylation of N-terminal residue targets proteins to cell membrane, e.g., myristoylation of Ras
Cleavage (removal of residues)	
Cleavage of peptide bonds	Co- or post-translational cleavage of initiator methionine occurs in most cytoplasmic proteins
	Co-translational cleavage of signal peptide occurs during translocation across endoplasmic reticulum membrane for secreted proteins
	Maturation of immature proteins (proproteins) by cleavage, e.g., activation of zymogens (inactive enzyme precursors) by proteolysis, removal of internal C-peptide of proinsulin, cleavage of Hedgehog proteins into N-terminal and C-terminal fragments
	Processing of genetic information, e.g., cleavage of polyproteins synthesized from poliovirus genome and mammalian tachykinin genes, splicing out of inteins

(p. 90). Some proteins are modified in certain individuals but not others, a key example being the differences in protein glycosylation that account for the ABO blood groups in humans.

8.2 METHODS FOR THE DETECTION OF POST-TRANSLATIONAL MODIFICATIONS

With the increasing recognition that post-translational modification is more a rule than an exception and that it often plays an important role in protein structure and function, a number of approaches have been developed to detect and characterize modified proteins and peptides on a global scale. Initially, modified proteins were identified through the replication *in vitro* of biochemical reactions that take place *in vivo*, for example, using radiolabeled substrates to confirm the addition of novel chemical adducts. Modified proteins tend to differ from their "parent" protein in mass and therefore migrate at different rates during one-dimensional SDS-PAGE, allowing their detection as novel bands in stained gels or, if an antibody is available, on western blots. More recently, antibodies have been developed that recognize particular types of modification, for example tyrosine phosphorylation, an example we consider later in the chapter. Western blot detection methods have also been developed using the enzymes, such as protein kinases and methyltransferases, that are responsible for post-translational modification, but all the above methods are hampered by their low throughput.

More progress has been made by adapting the larger-scale techniques for protein fractionation, identification, and quantitation discussed in earlier chapters. For example, 2DGE can resolve post-translational variants that differ from the parent protein in terms of mass and/or charge, and gels can be stained with reagents that recognize particular types of modified proteins to provide a visual overview of the modified proteome. If the modified group can be removed by chemical or enzymatic treatment, then "before and after" two-dimensional gels can identify the positions of modified proteins. However, the most promising methods are based on adaptations of shotgun proteomics that focus on the analysis of modified peptides. The typical workflow is shown in **Figure 8.1** and generally involves protein isolation and proteolytic digestion followed by an enrichment procedure that selects for modified peptides, followed by the analysis of this enriched population in a mass spectrometer. Such enrichment procedures are not generic but are targeted to select specific forms of modification. For this reason, researchers often talk of analyzing **sub-proteomes** with common forms of modification. Because of their prevalence, we focus on the **phosphoproteome** (all proteins modified by phosphorylation) and the **glycoproteome** (all proteins with glycan chains) in more detail later in this chapter but other types of modifications are discussed by Chen et al., Hoofnagle & Heinecke, and Mischerikow & Heck (see Further Reading).

Once an enriched protein or peptide population is available, downstream analysis by mass spectrometry must take into account the fact that modifications have a predictable impact on peptide masses in the first spectrum, as well as influencing the nature of the fragmentation products in MS/MS and MS^n analysis. These **mass deviations** and the nature of diagnostic **marker ions** can be incorporated into the algorithms for peptide mass fingerprinting and ion fragment analysis (see Chapter 3). Some typical fragmentation ions associated with different forms of modification are listed in **Table 8.2**, and their relevance in the fields of phosphoproteomics and glycoproteomics is discussed in more detail later.

Enrichment is important not only to simplify the analysis of modified proteins, but also because the sensitivity of detection is a key issue. Reversible modifications such as phosphorylation are often used to control the activities

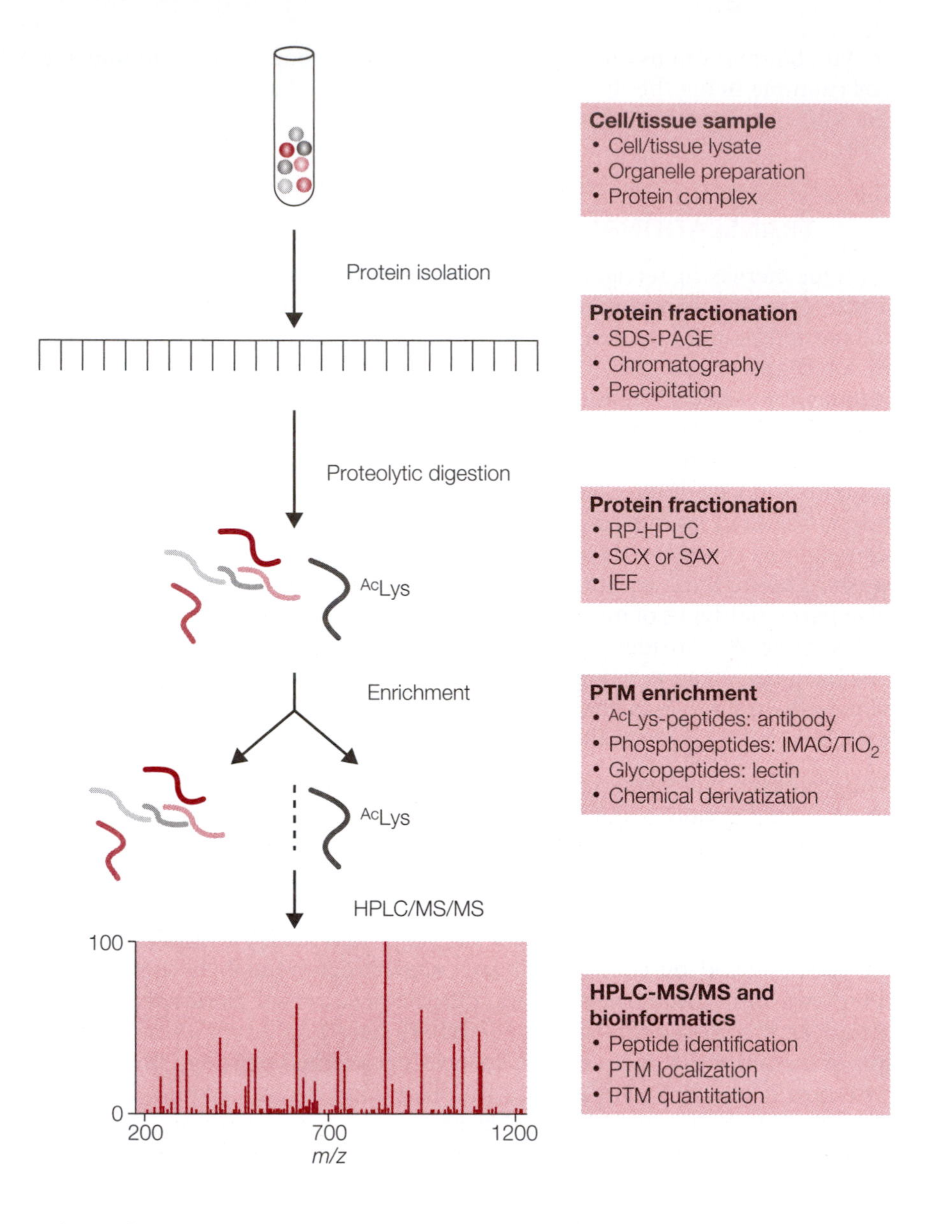

FIGURE 8.1 General scheme for the analysis of modified proteins in an adaptation of shotgun proteomics that enriches for specific types of modification. Picture shows acetylated lysyl peptides as an example, but the methods, shown in the boxes, represent a number of different modifications. (From Zhao Y & Jensen ON (2009) *Proteomics* 9, 4632. With permission from John Wiley & Sons, Inc.)

of signaling proteins and regulatory molecules such as transcription factors, which are the least abundant proteins in the cell. Furthermore, the stoichiometry of phosphorylation is usually low, that is, only a small proportion of the total intracellular pool of a given protein is likely to be modified at a particular time, and therefore the modified target protein may be present in limiting amounts and may be difficult to detect and quantify. Even when the protein is abundant, the heterogeneity of modification can mean that the quantity of each protein with a defined, single modification state is very low, as is often the case with glycoproteins.

8.3 ENRICHMENT STRATEGIES FOR MODIFIED PROTEINS AND PEPTIDES

There are four main strategies for the enrichment of modified proteins and peptides, all of which depend on some form of affinity purification. Therefore, adapted shotgun strategies for modified proteins often incorporate affinity chromatography or another bind-and-elute chromatography step.

The first approach is the use of antibodies to isolate specific modified variants (**immunoaffinity enrichment**). This approach works because many antibodies recognize a short linear epitope even if it is embedded in a larger

TABLE 8.2 MASS CHANGES AND FREQUENTLY DETECTED NEUTRAL LOSSES AND DIAGNOSTIC IONS FOR COMMON POST-TRANSLATIONAL MODIFICATIONS

Modification (site)	Mass change (MS)	Neutral loss (MS/MS)	Diagnostic ion (MS/MS)
Phosphorylation			
(Ser, Thr)	79.966	97.977	–78.959
(Tyr)	79.966	79.966	–78.959
Glycosylation, N-*linked*			
(*N*-acetylglucosamine)	203.079	203.079	204.087
(Hexose)	162.053	162.053	163.061
(Hexoyl-*N*-acetylhexosamine)	365.148	365.148	366.156
(*N*-acetylneuraminic acid)	291.095	291.095	292.103
Sulfation (Tyr)	79.956	79.956	
Acetylation (Lys)	42.011		126.091, 143.118
Methylation			
(Lys, mono)	14.016		98.097
(Lys, di)	28.031		
(Lys, tri)	42.047		
Ubiquitin (Lys)	114.043 (with Gly-Gly tag)		
Palmitoylation (Cys)	238.230	272.217	
Farnesylation (Cys)	204.188	204.188	
Myristoylation (Lys, Arg)	210.198	210.198	

protein, and many forms of post-translational modification occur within the context of a conserved sequence because this is required for the substrate to be recognized by the modification enzyme. The immunoisolation of modified peptides is more efficient than that of whole proteins because the epitope is more exposed. This means a greater range of modification-specific antibodies may be suitable for shotgun proteomics than, for example, identifying modified proteins on a western blot. **Modification-specific antibodies** have therefore been successful for the proteomic analysis of arginine methylation, lysine acetylation, and tyrosine nitration as well as the much more widely cited example of tyrosine phosphorylation, which is discussed in more detail later in the chapter. The analysis of lysine acetylation is particularly interesting because this can be used to investigate the functional modification of histones and transcription factors. This approach could be expanded in the future through the use of more diverse affinity reagents, such as aptamers (oligonucleotides with specific binding properties that can be isolated from randomized libraries). As discussed later, the affinity-based enrichment of glycoproteins and glycopeptides can be accomplished by binding to lectins, and different types of lectins can be used for the more detailed analysis and classification of glycoproteins because they are specific for different sugars and oligosaccharides.

The second enrichment method can in theory be used with any type of modification because it involves specific **chemical derivatization**, which converts an intractable site into one suitable for the attachment of a more robust affinity ligand such as biotin. This approach is usually performed once proteins have been isolated, either before or after digestion into peptides. In some cases, a simple chemical modification may be involved, such as the β-elimination of *O*-linked phosphate and *O*-linked acetylated glycans, whereas in other cases a more complex chemical adduct is involved. For example, peptides with *N*-glycan chains can be isolated by oxidation followed by conjugation to an immobilized hydrazine substrate. Once other proteins have been washed away, the glycoproteins can be released using a glycan-specific enzyme such as PNGase F and analyzed by mass spectrometry. The main drawback of derivatization methods is that the identification and particularly the accurate quantitation of modified proteins rely on the efficiency and specificity of the derivatization reaction. If the reaction is not efficient, or if it has off-target effects, it can produce spurious false-positive and false-negative data.

Although derivatization is usually carried out *in vitro*, it is also possible to achieve some forms of chemical modification *in vivo*, a variant approach known as **metabolic tagging**. For example, *in vivo* exposure to azide allows the subsequent tagging and affinity purification of proteins modified by myristoylation, palmitoylation, and farnesylation. A specialized form of metabolic tagging suitable for the isolation of ubiquitinylated proteins involves the introduction of a purification tag such as an affinity epitope or a string of consecutive histidine residues into the ubiquitin gene by genetic engineering.

The third enrichment method exploits more general principles, such as ionic interactions. The key example here is the isolation of phosphoproteins and phosphopeptides, which have a strong negative charge, by IMAC and/or TiO_2 chromatography using strong cationic resins. This example is discussed in more detail on page 175.

The final method involves the use of modification-specific enzymes to facilitate the purification of particular types of modified proteins. This can vary from the simple concept of using a substrate recognition site as an affinity reagent, for example a methyltransferase to isolate proteins that are potentially methylated, to the sophisticated use of enzymes as targeted release agents. The use of enzymes as probes can be unreliable as this tends to generate a significant number of false positives (proteins that are recognized by a methyltransferase, for example, are not necessarily methylated *in vivo*). However, targeted release agents can be valuable for identifying particular classes of modified proteins. A good example is the use of phospholipases that are highly specific for the cleavage of phosphatidylinositol to release proteins tethered to the plasma membrane by glycosylphosphatidylinositol bridges (GPI anchors).

8.4 PHOSPHOPROTEOMICS

Protein phosphorylation is a key regulatory mechanism

We choose phosphoproteomics as a prime case study of protein modification proteomics because phosphorylation is a ubiquitous form of post-translational modification and is probably the most important form of regulatory modification in both prokaryotic and eukaryotic cells. Phosphorylation lies at the heart of many biological processes, including signal transduction, gene expression, and the regulation of cell division. In humans, the aberrant phosphorylation of proteins is often associated with disease, particularly cancer.

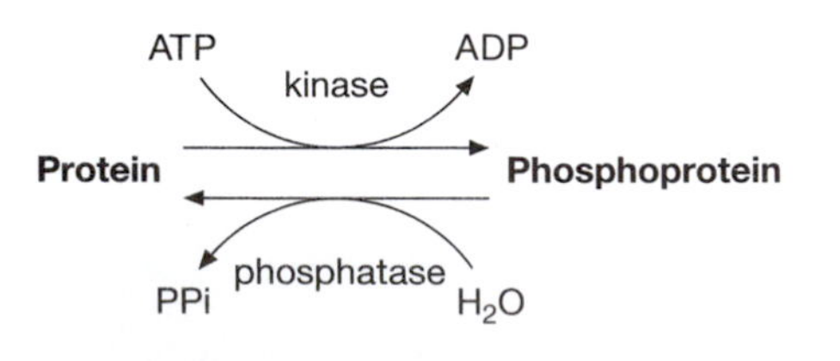

FIGURE 8.2 Protein phosphorylation is a reversible modification catalyzed by kinases and phosphatases.

The esterification of an amino acid side chain through the addition of a phosphate group introduces a strong negative charge, which can change the conformation of the protein and alter its stability, activity, and potential for interaction with other molecules. The enzymes that phosphorylate proteins are termed protein kinases and those that remove phosphate groups are termed **protein phosphatases** (**Figure 8.2**). The substrates for these enzymes, that is, the proteins that are subject to phosphorylation and dephosphorylation, are termed **phosphoproteins**. In bacteria, proteins are phosphorylated predominantly on aspartic acid, glutamic acid, and histidine residues, but this is rare in eukaryotes, where serine, threonine, and tyrosine are the major targets. Some proteins in both prokaryotes and eukaryotes are also phosphorylated on arginine, lysine, or cysteine residues (**Figure 8.3**). Genes encoding protein kinases and phosphatases account for 2–4% of the eukaryotic genome (there are about 120 kinase and 40 phosphatase genes in the yeast genome and about 500 kinase and 100 phosphatase genes in the human genome). It is thought that up to one-third of all the proteins in a eukaryotic cell are phosphorylated at any one time and that there may be as many as 100,000 potential phosphorylation sites in the human proteome, the majority of which are presently unknown. Many phosphoproteins have more than one phosphorylation site and exist as a mixture of alternative **phosphoforms**.

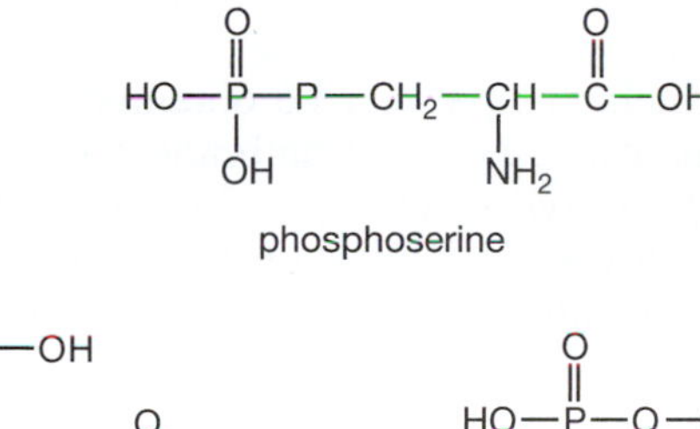

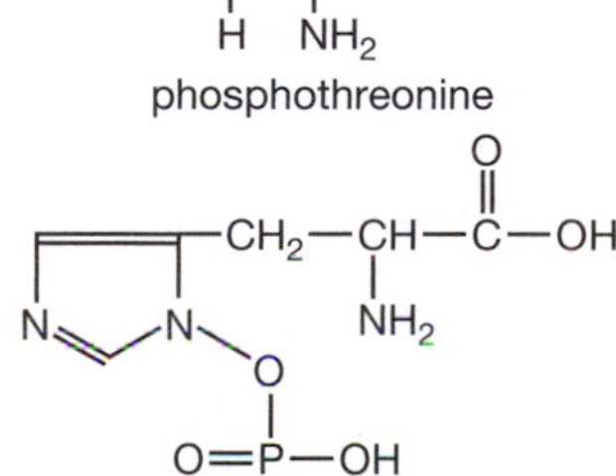

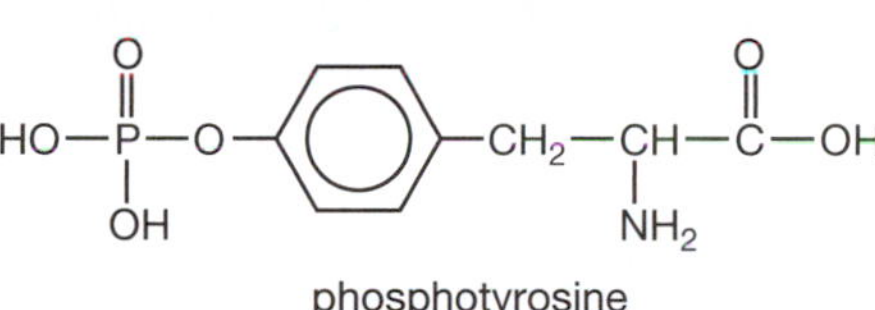

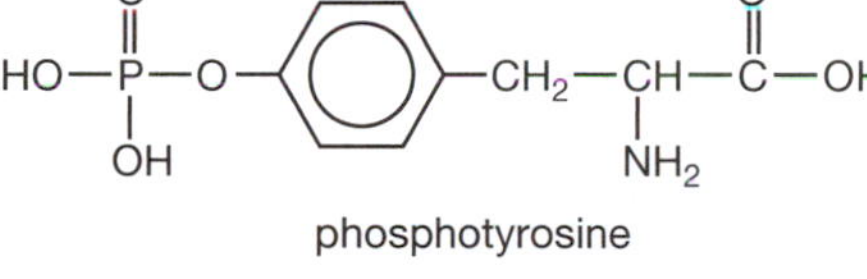

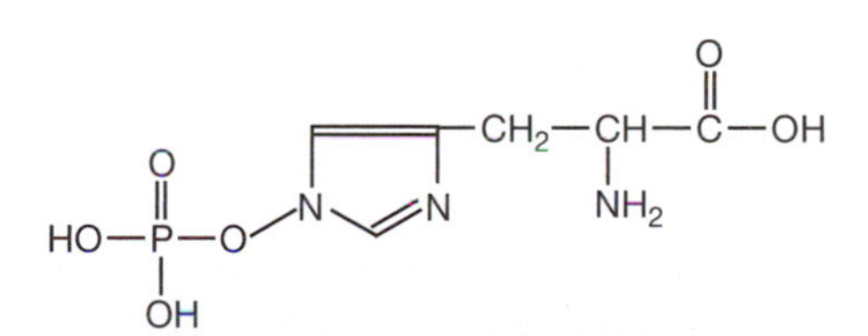

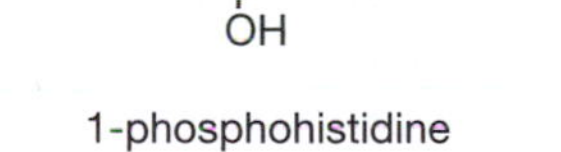

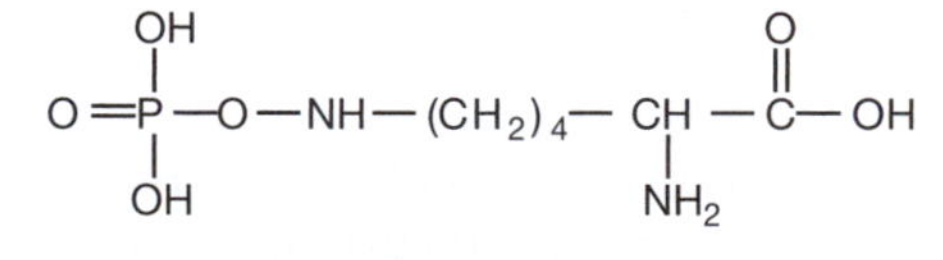

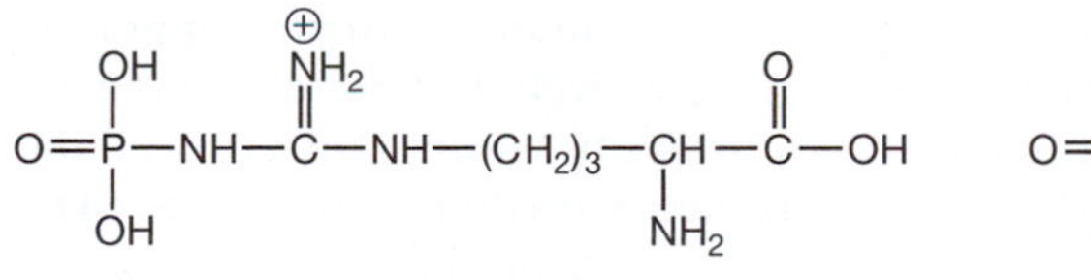

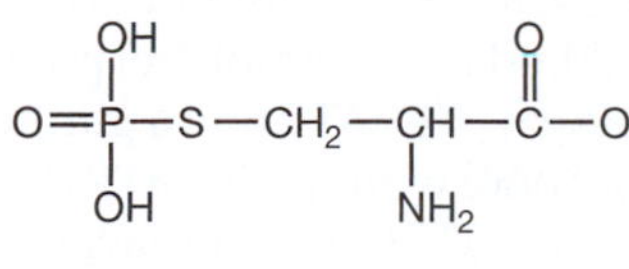

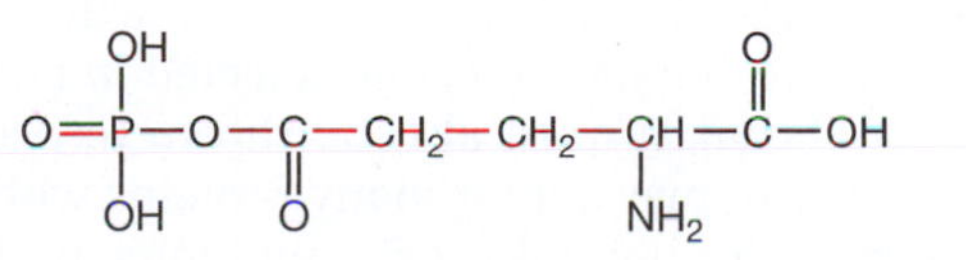

FIGURE 8.3 Ten different amino acids are known to be phosphorylated in a natural biological context. Whereas pSer, pThr, and pTyr are common in eukaryotes, only pHis, pAsp, and pGlu are common in prokaryotes.

Since phosphorylation plays such an important role in the regulation of cellular activities, our understanding of the functioning cell would be incomplete without a comprehensive inventory of the phosphoproteome, that is, a catalog of all the phosphoproteins in the cell, showing the distribution of phosphorylation sites and the abundance of alternative phosphoforms under different conditions. The quantitative aspect of this inventory is important because the phosphoproteome is not only complex but also extremely dynamic. Such analysis has become possible with the development of phosphoprotein enrichment methods and improved techniques for the analysis of phosphopeptides by mass spectrometry. However, similar methods could in principle be applied to any form of modification that involve the addition of a small chemical group, for example, sulfation, methylation, and hydroxylation.

Separated phosphoproteins can be detected with specific staining reagents

If a relatively pure protein sample can be obtained, for example, a spot excised from a two-dimensional gel or membrane, partial hydrolysis under alkaline conditions or via enzyme cleavage can release individual phosphoamino acids, which can be used to confirm the presence of a phosphoprotein and identify the phosphorylated residue. If related samples are available, for example, healthy and diseased tissue, this method also allows the abundance of phosphoamino acids in each sample to be compared. The method used for separation of the phosphoamino acids depends on the type of sample and how it has been treated. Proteins taken directly from an *in vitro* kinase reaction, in which [γ-^{32}P]-ATP has been used to incorporate a radiolabel, are generally separated by gel electrophoresis or thin-layer chromatography after digestion, because phosphoserine, phosphothreonine, and phosphotyrosine are readily identified by this method. For proteins labeled *in vivo* with $^{32}PO_4$ (orthophosphate), there may be contaminating labeled compounds and two-dimensional separation is required to give better resolution. If proteins or released phosphoamino acids can be derivatized with a fluorogenic reagent, then HPLC can be used in combination with UV detection to identify and quantify the phosphorylated residues. Mass spectrometry can also be used to identify phosphoamino acids, although improved techniques for the analysis of phosphopeptides (see below) make this approach obsolete.

A general picture of the phosphoproteome can be gained by selectively labeling or staining phosphoproteins, allowing them to be identified in gels or on membranes. Gel or membrane staining is not a particularly sensitive technique and only allows the most abundant phosphoproteins to be detected, but it helps to identify groups of phosphoforms (these tend to migrate in gels at slightly different rates, forming chains of spots) and can show on/off differences and overt quantitative differences between samples.

The classical technique, which is applicable to all downstream separation methods, is to radiolabel phosphoproteins selectively with ^{32}P, either *in vitro* using [γ-^{32}P]-ATP and a purified kinase or *in vivo* by equilibrating the cellular ATP pool with [γ-^{32}P]-ATP or $^{32}PO_4$. The proteins are then separated by SDS-PAGE, 2DGE, or thin-layer chromatography, and the labeled proteins are detected by autoradiography or phosphorimaging. Individual proteins can then be excised from the gel or membrane for further analysis. Alternatively, radiolabeled fractions can be collected as they elute from a chromatography column. Before a phosphorylation event detected *in vitro* can be accepted as biologically significant, it must also be shown to occur *in vivo*. This is because kinases *in vitro* may act on many proteins with which they never come into contact under physiological conditions, perhaps because they are expressed in different cells or located in different subcellular compartments. Unfortunately, *in vivo* incorporation also depends on the metabolic phosphorylation rate and the equilibrium between phosphorylated and nonphosphorylated forms of the target protein. If the pool of a given protein is already saturated with phosphate groups, then no further incorporation will occur regardless of the activity of the kinase, and the phosphoprotein will not be detected.

To avoid such problems, a number of generic phosphoprotein stains have been developed that work in one- and two-dimensional gels, on membranes, and in chromatography fractions (for example, Pro-Q Diamond marketed by Invitrogen). These can detect as little as 1 ng of phosphoprotein and can be used in combination with other fluorescent stains that identify all proteins (Chapter 3). Western blotting with anti-phosphotyrosine antibodies and the chemical modification of phosphate residues with fluorescent labels are examples of general detection methods based on the enrichment procedures discussed below. Although anti-phosphoserine and anti-phosphothreonine antibodies have insufficient selectivity to be used for the enrichment of peptides, they are adequate for the detection of whole proteins on membranes.

Sample preparation for phosphoprotein analysis typically involves enrichment using antibodies or strongly cationic chromatography resins

Phosphoprotein analysis begins with a mixture of phosphorylated and nonphosphorylated proteins, for example, a cell lysate or serum sample. The overall aim is to recognize the phosphoproteins, identify them, determine the phosphorylated sites, and, if possible, carry out a quantitative analysis of phosphorylation under different conditions. There are various different experimental methods that can be used to achieve these aims, and these are summarized in **Figure 8.4**. However, the low abundance of many phosphoproteins means that enrichment of the sample prior to analysis is essential. As discussed below, enrichment is also necessary to overcome some of the limitations of phosphoprotein analysis by mass spectrometry. Enrichment can be achieved by affinity purification either with or without chemical modification of the phosphate groups.

Affinity purification without chemical modification is advantageous because only small amounts of starting material are required. The simplest method involves antibodies that bind specifically to phosphorylated proteins or peptides. If one particular phosphoprotein is sought, it may be possible to use an antibody that binds specifically to that protein. However, for large-scale analysis, a more general approach is required. A number of companies market **anti-phosphotyrosine antibodies** that can be used as generic reagents to isolate or enrich phosphotyrosine-containing proteins and peptides. Antibodies that bind other phosphorylated residues, for example serine and threonine, have also been produced, but they are less specific and less sensitive for the isolation of peptides, and are not widely used. This means that antibody enrichment is generally confined to the analysis of proteins phosphorylated on tyrosine residues (**Box 8.1**).

An alternative affinity-based enrichment strategy that is widely used to isolate phosphopeptides from pre-digested samples is **immobilized metal-affinity chromatography** (**IMAC**). This exploits the attraction between negatively charged phosphate groups and positively charged metal ions, particularly Fe^{3+} and Ga^{3+}, but more recently also Al^{3+}. The method is advantageous because it is relatively easy to combine with mass spectrometry. Thus, several research groups have carried out off-line analysis of phosphopeptides isolated by IMAC, and others have coupled IMAC to on-line ESI-mass spectrometry either with or without an intervening fractionation step. It has even been possible to analyze phosphopeptides by MALDI-TOF mass spectrometry while still bound to the IMAC column.

One drawback of this method is that IMAC columns also bind other negatively charged amino acid residues, such as aspartic acid and glutamic acid. This becomes a significant problem when the phosphopeptides are scarce but there are abundant anionic peptides that compete for binding,

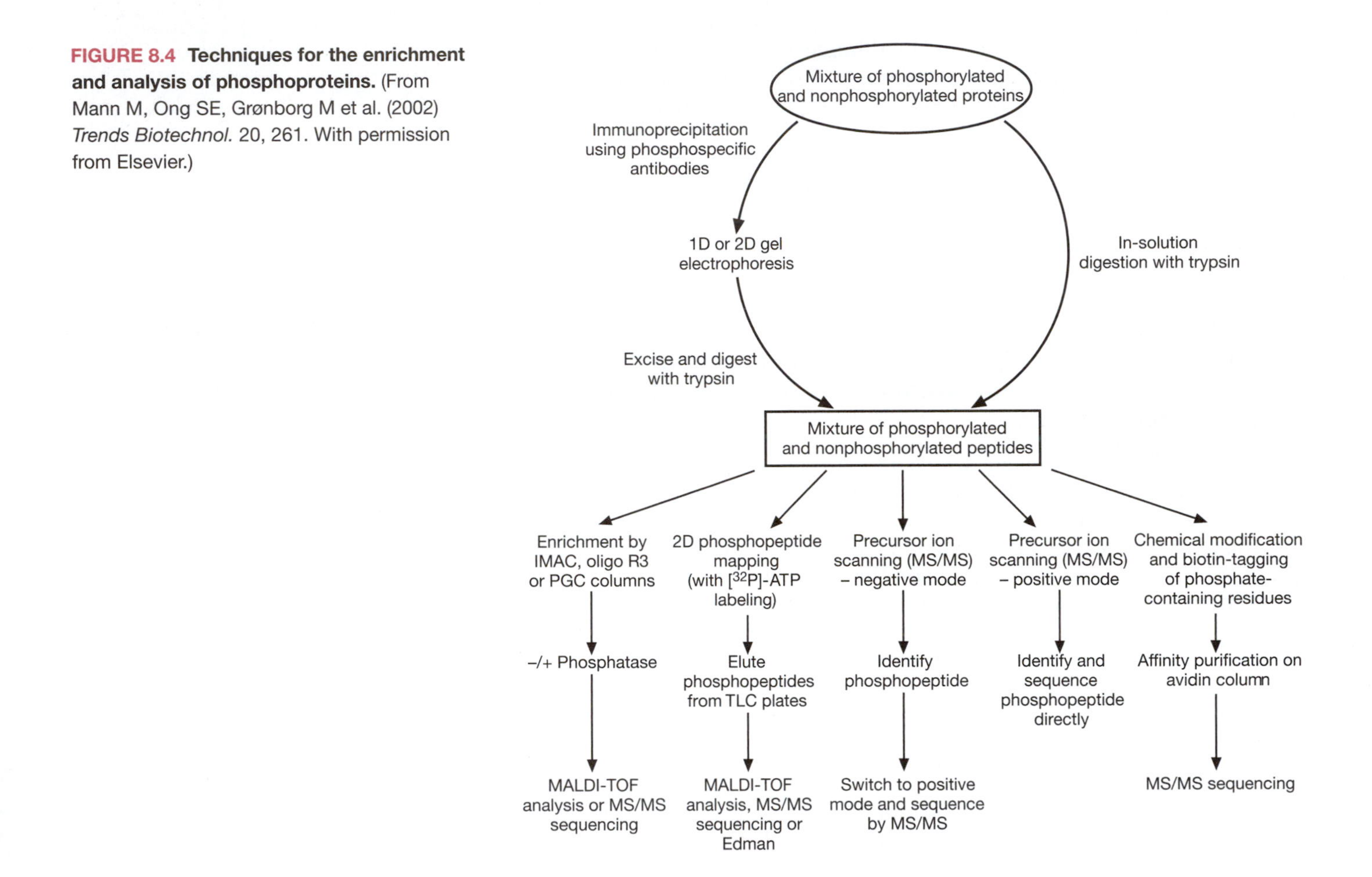

FIGURE 8.4 Techniques for the enrichment and analysis of phosphoproteins. (From Mann M, Ong SE, Grønborg M et al. (2002) *Trends Biotechnol.* 20, 261. With permission from Elsevier.)

BOX 8.1 CASE STUDY.
Probing signaling pathways with anti-phosphotyrosine antibodies.

Protein phosphorylation plays an important role in the regulation of signaling pathways in both prokaryotes and eukaryotes. Two studies published by Pandey et al. in 2000 (see Further Reading) neatly demonstrated the power of phosphotyrosine-specific antibodies for the analysis of the epidermal growth factor (EGF) pathway, which is regulated by tyrosine phosphorylation. The EGF receptor (EGFR) oligomerizes in response to EGF, inducing a latent kinase activity and resulting in the phosphorylation of each subunit of the oligomeric receptor by the tyrosine kinase activity of another subunit (autotransphosphorylation). Cytosolic proteins are then recruited to the receptor if they contain an SRC homology 2 domain (SH2) or phosphotyrosine interaction domain (PID), which bind specifically to phosphotyrosine residues. These cytosolic proteins may also be phosphorylated.

To analyze the entire EGF signaling pathway in a single step, a combination of two anti-phosphotyrosine antibodies was used to immunoprecipitate phosphotyrosine-containing proteins from HeLa cells that had or had not been stimulated with EGF. Side-by-side comparison of the recovered proteins identified nine proteins that were tyrosine-phosphorylated specifically in response to EGF. Seven of these proteins were known components of the EGF pathway but the eighth, a ubiquitous protein called VAV2, was not previously known to be a substrate of EGFR. A ninth protein, named STAM2, was novel, but was found to be related to a signaling protein induced by interleukin-2.

In 2002, Steen et al. (see Further Reading) used a novel mass spectrometry technique known as phosphotyrosine-specific immonium ion scanning (or PSI scanning, see main text) to look at the EGF pathway. This identified 10 pathway components, two of which were novel, but more importantly it found five novel phosphorylation sites on the proteins SHIP-2, Hrs, Cbl, STAM, and STAM2.

as is the case in human serum. The binding of nonspecific anionic peptides can be reduced by using low-pH buffers or by the methyl esterification of all carboxylate groups in the protein sample prior to chromatography. More sophisticated multidimensional chromatography approaches can also be used, such as IMAC combined with HILIC, strong anion exchange, and/or low-pH strong cation exchange steps. Other chromatography methods that can be used for phosphoprotein or phosphopeptide enrichment include elution from reversed-phase or porous graphitic carbon beads, and most recently the use of solid-phase matrices based on titanium dioxide (TiO_2). The combination of IMAC and TiO_2 resins (sequential elution from IMAC, or **SIMAC**) has proven useful for the sensitive separation of mono-phosphoproteins from proteins with two or more phosphorylated residues.

Affinity purification with prior chemical modification of phosphates is limited by the requirement for larger amounts of starting material, but two methods have been described that may be useful for the isolation of more abundant phosphoproteins and phosphopeptides. In the first method, ethanedithiol is used to replace the phosphate group of phosphoserine and phosphothreonine residues by β-elimination under strongly alkaline conditions, leaving a thiol group that can be used to attach a biotin affinity tag (**Figure 8.5**). The phosphoproteins or phosphopeptides can then be isolated from complex mixtures using streptavidin-coated beads. Cysteine and methionine residues are also derivatized by this method, so they must be oxidized with performic acid prior to the reaction. Ethanedithiol treatment does not work with phosphotyrosine residues, so this is a good way to select for serine/threonine phosphorylation. However, serine and threonine residues that are *O*-glycosylated (see later in the chapter) are also derivatized,

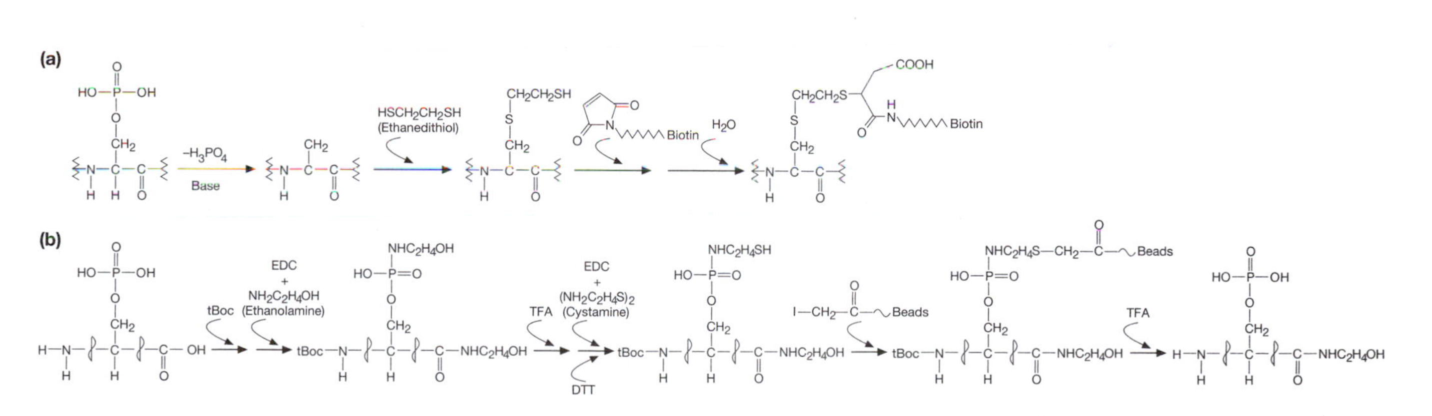

FIGURE 8.5 Chemical derivatization for the isolation of phosphoproteins and phosphopeptides. (a) Chemical modification based on β-elimination: samples containing phosphoproteins are first treated with a strong base, leading to β-elimination reactions in the case of phosphoserine (pictured here) and phosphothreonine residues. A reactive species containing an α,β unsaturated bond is formed. This serves as a Michael acceptor for the nucleophile (in this case, ethanedithiol or an isotopic variant may be substituted for quantitation purposes). The biotinylated reagent reacts with sulfhydryl (–SH) groups at acidic to neutral pH. Biotinylated phosphoprotein is now tagged for enrichment on avidin columns in later steps. (b) Chemical modification based on carbodiimide condensation reaction. The N-termini of peptides are first protected with *tert*-butyloxycarbonyl (tBoc) chemistry. A condensation reaction then occurs between the carboxyl groups as well as the phosphate moiety in the presence of excess amine (ethanolamine) in a reaction catalyzed by *N*-ethyl-*N′*-(3-dimethylaminopropyl)carbodiimide hydrochloride (EDC). The condensation reaction results in the formation of an amide bond and a phosphoamidate bond from carboxyl and phosphate bonds, respectively. The phosphate group is regenerated by rapid hydrolysis with trifluoroacetic acid (TFA) and the sample is desalted on reversed phase material (this intermediate step is not shown here). A second condensation reaction (also catalyzed by EDC) is performed next with excess cystamine. The sample is reduced with dithiothreitol (DTT), converting the disulfide bond of cystamine to a sulfhydryl group and thereby tagging the phosphate moiety. The sample is again desalted using reversed phase material. The tagged peptides are captured on glass beads containing bound iodoacetyl groups that will react with sulfhydryl groups. The recovery of phosphopeptides is performed by strong acid hydrolysis that cleaves both the phosphoamidate bond and the tBoc protective group, thus regenerating the phosphate moiety and the N-terminus, respectively. (From Mann M, Ong SE, Grønborg M et al. (2002) *Trends Biotechnol.* 20, 261. With permission from Elsevier.)

so further experiments are necessary to confirm that the protein is phosphorylated rather than glycosylated. In contrast, all three phosphorylated residues are modified in the second method, in which cystamine is added to phosphate groups via a carbodiimide condensation reaction (Figure 8.5). This allows affinity-based isolation using iodoacetylated beads.

8.5 ANALYSIS OF PHOSPHOPROTEINS BY MASS SPECTROMETRY

A combination of Edman degradation and mass spectrometry can be used to map phosphorylation sites

Until the late 1990s, the standard approach for phosphorylation site analysis was to label an isolated phosphoprotein with ^{32}P, digest it into peptides, and then separate the peptides in the first dimension by thin-layer electrophoresis and in the second dimension by thin-layer chromatography (an approach known as **two-dimensional peptide mapping**). Phosphopeptides could then be identified by autoradiography, and these were excised and sequenced using Edman chemistry. The phosphorylated site was determined by the release of a radiolabeled amino acid in the corresponding sequencing cycle. Variations on this theme included the use of chemically derivatized phosphopeptides, which released amino acids that could be recognized by their particular retention times or because they carried a fluorescent label. Although two-dimensional peptide mapping and Edman sequencing continues to be used today for phosphorylation site analysis, it is too laborious to apply on a large scale.

As in other areas of proteomics, the analysis of phosphoproteins has been revolutionized by mass spectrometry. Two main principles are exploited, namely, that peptides containing a single phosphate modification will show a mass shift of about +80 (actually +79.983) compared with the nonphosphorylated peptide, and that phosphopeptides will yield diagnostic marker ions and neutral loss products that are not produced when unmodified peptides are fragmented. In practice, MS analysis is more challenging because phosphopeptides are often far less abundant than their unmodified counterparts. Furthermore, due to factors than are not completely understood, an equimolar mixture of peptides will generate signals of varying intensity and some will be lost altogether, resulting in incomplete coverage of the protein. This phenomenon affects phosphopeptides more strongly than unmodified peptides because they are more difficult to ionize; the signal from the unmodified peptide is said to suppress that of the modified peptide. These problems can be addressed by enriching the phosphopeptide pool using one of the methods discussed earlier in the chapter and fractionating the peptides prior to mass spectrometry. Another difficulty is that phosphate groups tend to inhibit proteolytic cleavage by trypsin at nearby peptide bonds, making full coverage of the protein even more unlikely, although this can be anticipated by the algorithms used to correlate theoretical and experimental peptide masses.

Intact phosphopeptide ions can be identified by MALDI-TOF mass spectrometry

As discussed in Chapter 3, MALDI-TOF mass spectrometry is most often used to analyze intact peptides, and correlative database searching (peptide mass fingerprinting) allows the derived masses to be matched against the theoretical peptides of known proteins. Therefore, if the identity of the protein is known or can be deduced from the peptide masses, phosphopeptides can be identified simply by examining the mass spectrum for mass shifts of 79.983 or multiples thereof compared with predicted masses. Parallel analysis in which the sample has been treated with alkaline phosphatase can also be helpful, because peaks corresponding to phosphopeptides in the untreated sample should be absent from the treated sample (**Figure 8.6**). This method does not identify phosphorylated residues directly, but if the peptide sequence contains only one possible phosphorylation site, if a consensus kinase target site is present (**Table 8.3**), or if the phosphorylated residue has been chemically identified using one of the methods described above, then it is more likely that the site can be identified accurately.

Phosphopeptides yield diagnostic marker ions and neutral loss products

The analysis of fragment ions (see Chapter 3 for a general overview) serves two purposes in phosphoproteomics. First, phosphopeptides preferentially yield diagnostic, phosphate-specific marker ions such as $H_2PO_4^-$, PO_3^-, and PO_2^-, which have masses of approximately 97, 79, and 63, respectively. Phosphoserine and phosphothreonine are more labile than phosphotyrosine and therefore yield marker ions at lower collision energies. Second, fragmentation along the polypeptide backbone can yield peptide fragments that allow a sequence to be built up *de novo* (p. 61). This sequence will include the phosphoamino acid, so if the difference in mass between two consecutive fragments in, for example, the *b*-series of ions is equivalent to that of phophoserine residue, this provides a definitive location.

Phosphate-specific fragment ions can be obtained by MS/MS using an ESI or MALDI ion source and a standard triple quadrupole or hybrid quadrupole-TOF mass spectrometer, with fragmentation achieved by conventional **collision induced dissociation** (**CID**) or **in-source CID**, the latter requiring excess energy during ionization to induce multiple collisions and produce the phosphate reporter ions in the emerging ion stream (**Figure 8.7**). This method also fragments the peptide backbone to a lesser extent and can therefore provide peptide sequence information (see below). Normal ionization energy levels are used in **precursor ion scanning**. In this mode, the first quadrupole (Q_1) is used to scan the ion stream, CID occurs in q_2 (running in RF mode), and the third analyzer (Q_3 or TOF) is set to detect phosphate reporter ions, such as PO_3^-, induced by collision. Phosphopeptides are thus identified when a precursor ion scanned in Q_1 yields a phosphate fragment that is detected in Q_3 (**Figure 8.8**). In **neutral loss scan mode**, both Q_1 and Q_3 are set to scan the ion stream. Q_1 scans the full mass range, q_2 is used as the collision cell, and

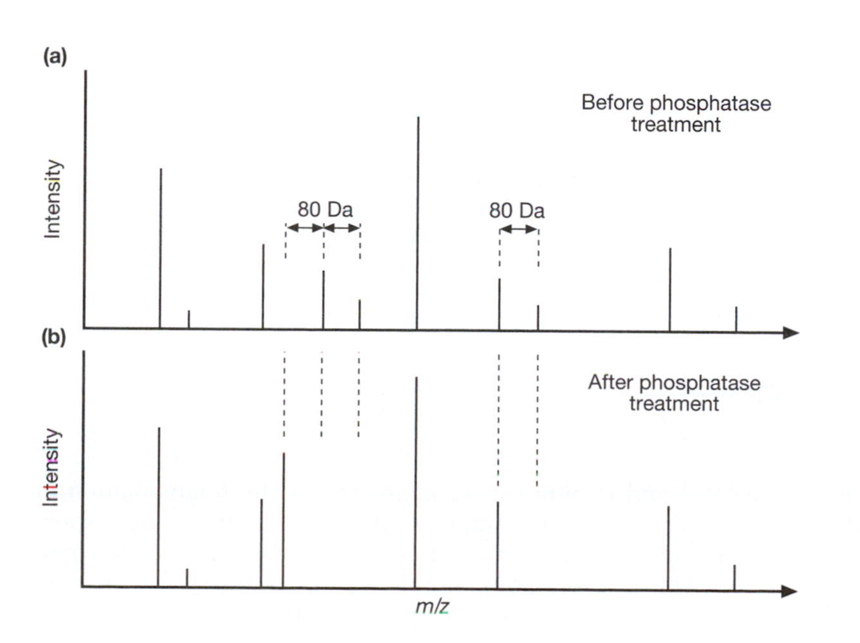

FIGURE 8.6 Phosphopeptide identification by MALDI-TOF MS mapping combined with alkaline phosphatase treatment. (a) The MALDI-TOF MS spectrum of a proteolytic digest. Phosphopeptides are indicated by peaks shifted by multiples of 80 Da (HPO_3 = 80 Da) relative to predicted unphosphorylated peptide masses. (b) The disappearance of such peaks upon treatment with a phosphatase confirms their identity as phosphopeptides. (From McLachlin DT & Chait BT (2001) *Curr. Opin. Chem. Biol.* 5, 591. With permission from Elsevier.)

TABLE 8.3 SUMMARY OF PHOSPHORYLATION SEQUENCE MOTIFS RECOGNIZED BY VARIOUS KINASES

Sequence motif	Enzyme	Protein (substrate)
Ser/Thr phosphorylation		
Ser-Ser-Xaa-Ser(P)	Bone morphogenetic proteins receptor kinase	TGF-β family mediator Smad 1
Arg-Xaa-Arg-Yaa-Zaa-Ser(P)/Thr(P)-Hyd	Protein kinase B	Synthetic peptide
Ser(P)-Xaa-His	Protein kinase C	Alpha 1 Na, K-ATPase
Ser(P)-Leu-Gln-Xaa-Ala	cGMP-dependent kinase	Lentivirus Vif proteins
Glu-Val-Glu-Ser(P)	c-Myb kinase	Vertebrate c-Myb proteins
Arg-Xaa-Xaa-Ser(P)	Phosphotransferase	Serum response factor, c-Fos, Nur77, and the 40S ribosomal protein S6
Ser(P)/Thr(P)-Pro-Xaa (basic), Pro-Xaa-Thr(P)-Pro-Xaa (basic)	Cyclin-dependent kinase 2	Cyclin A or E
Thr(P)-Leu-Pro	Ceramide-activated protein kinase	Raf protein
Arg-Xaa-Ser(P)	cAMP-dependent protein serine kinase	PII protein (*glnB* gene product)
Ser-Xaa-Xaa-Xaa-Ser(P)	Glycogen synthase kinase 3	cAMP responsive element binding protein
Arg-Xaa-(Xaa)-Ser(P)/Thr(P)-Xaa-Ser/Thr	Autophosphorylation-dependent protein kinase	Myelin basic protein
Hyd-Xaa-Arg-Xaa-Xaa-Ser(P)/ Thr(P)-Xaa-Xaa-Xaa-Hyd	Ca^{2+}/calmodulin-dependent protein kinase Ia	Peptide analogs
Ser-Xaa-Glu-Ser(P)	Casein kinase	Bovine osteopontin, vitamin K-dependent matrix Gla protein from shark, lamb, rat, cow, and human
Ser-Xaa-Xaa-Glu-Ser(P)	Casein kinase II	Bovine osteopontin
Xaa-Ser(P)/Thr(P)-Pro-Xaa	Proline-directed protein kinase	Tau protein
Ser-Pro-Arg-Lys-Ser(P)-Pro-Arg-Lys Ser(P)-Pro-Lys/Arg-Lys/Arg	Histone H1 kinase	Sea-urchin, sperm-specific histones H1 and H2B
Lys-Ser(P)-Pro	Serine kinase	Murine neurofilament protein
Tyr phosphorylation		
Tyr(P)-Met-Asn-Met, Tyr(P)-Xaa-Xaa-Met, Tyr(P)-Met-Xaa-Met	Phosphatidylinositol 3-kinase, in cytoplasmic tail	CD28 T cell co-stimulatory receptor
Asn-Pro-Xaa-Tyr(P)	Focal adhesion kinase, in cytoplasmic domain	Integrin β_3
Tyr(P)-Xaa-Xaa-Leu	Protein Tyr kinase, in cytoplasmic tail	Mast cell function-associated antigen
Glu-Asp-Ala-Ile-Tyr(P)	Protein Tyr kinase	Synthetic peptides
Dual Thr and Tyr phosphorylation		
Thr(P)-Xaa-Tyr(P)	Mitogen-activated protein kinase	p38 mitogen-activated protein
Thr(P)-Glu-Tyr(P)	Mitogen-activated protein kinase kinase	Mitogen-activated protein kinase

Xaa is any amino acid; Yaa and Zaa are small residues other than Gly; Hyd is a hydrophobic amino acid residue Phe or Leu. (Reprinted from Yan A et al. (1998) *J. Chromatogr.* 808, 23. With permission from Elsevier.)

Q_3 scans a parallel range to Q_1 but at an m/z ratio that is $98/z$ lower, with the intention of detecting the neutral loss of H_3PO_4 (**Figure 8.9**).

Analogous methods can be used with the most recent generation of high-sensitivity FT-ICR and Orbitrap mass analyzers for the fragmentation of individual phosphoproteins. Ion traps are the most common mass spectrometers used for phosphoproteomics in CID mode, particularly when combined with MS^3 (three rounds of collision) or multistage activation, which induce additional backbone cleavages that allow peptide sequencing.

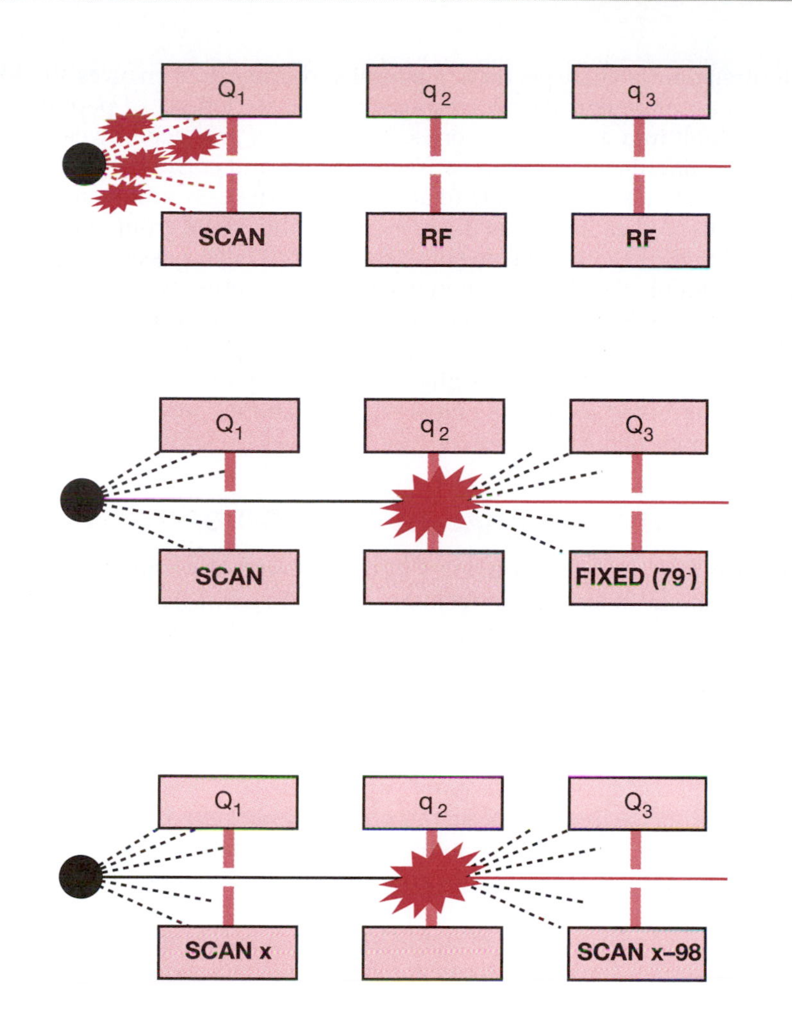

FIGURE 8.7 Detection of phosphopeptides using in-source CID in a triple quadrupole mass spectrometer. Excess energy used during ionization causes fragmentation to occur at the ion source. The ion stream is scanned in Q_1 for phosphate reporter ions such as PO_3^- (m/z = 79$^-$). These pass through the other quadrupoles (running in RF mode) to the detector.

FIGURE 8.8 Detection of phosphopeptides using precursor ion scan mode in a triple quadrupole mass spectrometer. The entire ion stream is scanned in Q_1, allowing selected ions through to the collision chamber in q_2. The fragmented ions then pass through Q_3, which is fixed to detect phosphate reporter ions such as PO_3^- (m/z = 79$^-$). Only if intact phosphopeptide ions pass through Q_1 will reporter ions pass through Q_3 to the detector.

FIGURE 8.9 Detection of phosphopeptides using neutral loss scan mode in a triple quadrupole mass spectrometer. The entire ion stream is scanned in Q_1, allowing selected ions (x) through to the collision chamber in q_2. Q_3 is set to scan the fragmented ions in parallel to Q_1, but at a lower mass range (for example, x–98, where 98 is the mass of H_3PO_4). Only phosphopeptide ions that lose H_3PO_4 during CID will pass through Q to the detector.

Electron transfer dissociation (**ETD**) and **electron capture dissociation** (**ECD**) can be used instead of CID or post-source decay by bombarding the ion stream with subthermal electrons because the phosphate group remains attached during and after activation. This method is more sensitive and offers more comprehensive coverage, resulting in extended ion series that are much easier to interpret than CID spectra. Phosphorylation sites can also be identified in small proteins without proteolytic digestion (**top-down phosphoproteomics**).

More recently, instruments have been developed that can easily detect the immonium ion (mass 216.043) generated by double fragmentation of the polypeptide backbone on either side of a phosphotyrosine residue (**phosphotyrosine-specific immonium ion scanning**, **PSI scanning**). This method is not applicable to other phosphorylated residues but can be carried out in positive-ion mode, which means it can be used in combination with strategies to identify the phosphorylation site. PSI scanning has been used to identify components of the EGF signaling pathway, as discussed in Box 8.1.

Generally, sequencing and the determination of phosphorylation sites are carried out by MS/MS using the full **product scan mode**, where specific precursor ions are gated at Q_1, fragmented in q_2, and scanned for product ions in Q_3 (Chapter 3). The uninterpreted CID spectra can be used to search for matching proteins in the databases as long as mass shifts caused by the phosphate group are taken into account, or the mass spectra can be interpreted to produce an amino acid series. As discussed in Chapter 3, the interpretation of CID spectra is a complex process that involves the identification of fragment ions corresponding to a nested set of peptides that can be built into a complete series. The series is produced by calculating the mass differences between consecutive ions and correlating those mass differences to a standard table of amino acids. The only adjustment necessary in this application is that the mass of the phosphate group must also be allowed in the calculations. In some cases, it has been possible to facilitate the recognition of phosphorylated residues in ion series by chemical derivation. For example, β-elimination can be used to convert phosphoserine into *S*-ethylcysteine and phosphothreonine into β-methyl-*S*-ethylcysteine. This also removes these labile phosphate groups so that fragmentation is more evenly distributed along the polypeptide backbone, providing more complete coverage of the peptide.

8.6 QUANTITATIVE ANALYSIS OF PHOSPHOPROTEINS

We have discussed the principles of quantitative proteomics in detail in Chapter 4, and many of the techniques discussed in that chapter for the analysis of proteins in general are equally applicable to the analysis of protein modifications. Rough estimates of the stoichiometry of phosphorylation in a single sample, or differences in phosphorylation levels between two samples, can be made by separating the phosphorylated and corresponding unmodified peptides by HPLC, carrying out quantitative amino acid analysis, and comparing the peaks. Rough quantitative data can also be obtained by comparing the intensity of signals on two-dimensional gels. As discussed above, however, it is not possible simply to compare the intensity of signals obtained from the phosphorylated and unmodified peptides in a mass spectrometer due to the occurrence of suppression. Instead, quantitation is based on the incorporation of stable isotopes or mass tags, which produce two very similar molecules that can be detected with equal efficiency, but which can be distinguished due to a mass shift when the samples are combined and analyzed together.

For the analysis of phosphoproteins, the general labeling methods discussed in Chapter 4 are not always suitable because they are not selective for phosphopeptides. For example, the standard ICAT method (p. 79) labels cysteine-containing peptides, and its use in phosphoprotein analysis would depend on whether the phosphorylated residue resided on the same peptide as the labeled cysteine. Metabolic labeling *in vivo* with ^{15}N or using heavy amino acids and the SILAC procedure (p. 83) would be more useful, since these are nonselective and all peptides would be labeled in the same way. When comparing two related samples, each phosphopeptide should be represented by four peaks, one phosphorylated and labeled, one unmodified and labeled, one phosphorylated and unlabeled, and one unmodified and unlabeled. The labeled/unlabeled pairs should occur as doublets separated by the mass value of the chosen label, whereas the mass difference between each unmodified peptide and its phosphorylated counterpart should be about 80 (**Figure 8.10**). A variation of the ICAT method, which results in the specific labeling of phosphoserine and phosphothreonine residues, has been described. In this case, only phosphopeptides would be labeled and each would be represented by just two peaks, one labeled and one unlabeled, allowing direct comparison between samples.

As an alternative to labeling, quantitative differences in the phosphorylation of a given peptide between samples can be determined by including a chemically synthesized heavy derivative of a peptide in each sample as an internal standard. This approach cannot be applied on a proteomic scale but it is useful for the analysis of known phosphoproteins.

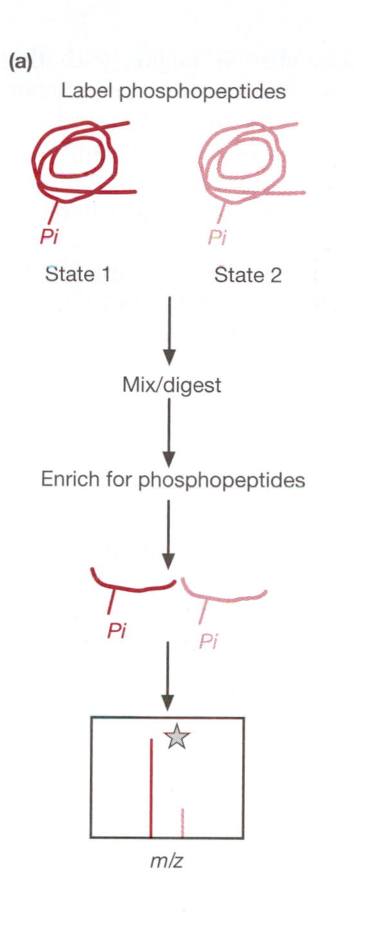

FIGURE 8.10 Quantitation of phosphoproteins by mass spectrometry. (a) If phosphoproteins can be labeled with mass tags and isolated by affinity capture, direct quantitative comparison across samples is possible by comparing peak intensities. (b) If peptides are labeled uniformly, peaks corresponding to the phosphorylated and unmodified versions of the same peptide will appear about 80 mass units apart. If protein abundance is the same in each sample, only phosphopeptides should show any quantitative variation. (From Sechi S & Oda Y (2003) *Curr. Opin. Chem. Biol.* 7, 70. With permission from Elsevier.)

8.7 GLYCOPROTEOMICS

Glycoproteins represent more than half of the eukaryotic proteome

Glycosylation involves the addition of short-chain carbohydrate residues (**oligosaccharides** or **glycans**) to proteins during or after synthesis. This type of modification is very common in eukaryotes (more than 50% of all proteins are glycosylated), but it also occurs to a lesser extent in prokaryotes. In eukaryotes, the vast majority of glycosylated proteins, or **glycoproteins**, pass through the secretory pathway. However, not all of them are actually secreted. Some are retained within the endoplasmic reticulum (ER) or Golgi apparatus, some are targeted to lysosomes, and many are inserted into the plasma membrane. The three major types of glycosylation that occur in the secretory pathway of mammalian cells are *N*-linked glycosylation, *O*-linked glycosylation, and the addition of GPI anchors. *N*-linked glycans are attached to asparagine residues in the context Asn-Xaa-Ser/Thr, where Xaa can be any amino acid except proline. *O*-linked glycans are linked to the hydroxyl group of serine or threonine residues. GPI anchors are attached to the C-terminus of the protein following the removal of a C-terminal signal sequence. More rarely, proteins in the cytosol or those targeted to the nuclear pore complex may be modified by the addition of a single *O*-linked GlcNAc residue. There are additional forms of glycosylation such as *C*-linked glycosylation of tryptophan and *S*-linked glycosylation of cysteine that are not well characterized.

***N*-linked glycosylation** occurs only in eukaryotes. It begins with the attachment of a branched, 14-residue oligosaccharide—the **core glycan** $GlcNAc_2Man_9Glc_3$. This modification occurs only in the ER because the enzyme responsible for the reaction is localized in the ER membrane. The core glycan is then trimmed by **glycosidases** to remove some of the residues, and the partially glycosylated protein moves to the Golgi apparatus. In this compartment, further modifications take place, involving the substitution of certain core glycan residues and the elaboration of the glycan chains. More than 30 different types of sugar molecule can be added, and the structure and architecture of chains can vary significantly. This process of elaboration produces three major types of glycan structure, known as **high-mannose**, hybrid, and **complex** types (**Figure 8.11**). The modifications that take place are different in plants, mammals, yeast, and insects, resulting in glycan structures that are distinct to each species in terms of complexity, composition, branching structure, and the linkages between residues. Since *N*-linked glycosylation does not take place in bacteria, recombinant mammalian proteins produced in bacteria lack the glycan chains. For some proteins, this appears to have no effect on biological activity (for example, growth hormone) whereas in others the aglycosylated version is less active or completely nonfunctional (for example, interleukin-2, thyroperoxidase).

There is generally a degree of heterogeneity in *N*-glycan modification, so that each protein is produced not as a single, defined molecule but a collection of **glycoforms**. Proteins may have more than one acceptor site for *N*-linked glycosylation, so there may be different glycan chains at the same site on different molecules (**microheterogeneity**) and different site occupancy (**macroheterogeneity**). This provides immense scope for structural diversity, although it is apparent that glycoproteins do not necessarily exist as all possible glycoforms and not all potential acceptor sites are modified.

Many proteins with *N*-linked glycans also undergo ***O*-linked glycosylation** in the Golgi apparatus, which involves the addition of sugars to exposed hydroxyl groups on serine and threonine side chains (and occasionally hydroxylysine and hydroxyproline). There does not appear to be a consensus sequence for the addition of mucin-type glycans, the most common form of *O*-linked glycosylation, indicating that modification may depend on secondary and tertiary structure. However, it is not uncommon to find *O*-linked glycosylation in proline-rich domains. Mucin-type glycans are structurally

FIGURE 8.11 The three major types of *N*-linked glycans—high-mannose, hybrid, and complex—are all built from a common pentasaccharide. This pentasaccharide is generated by trimming the 14-residue core glycan (see main text) and is then elaborated into the forms shown.

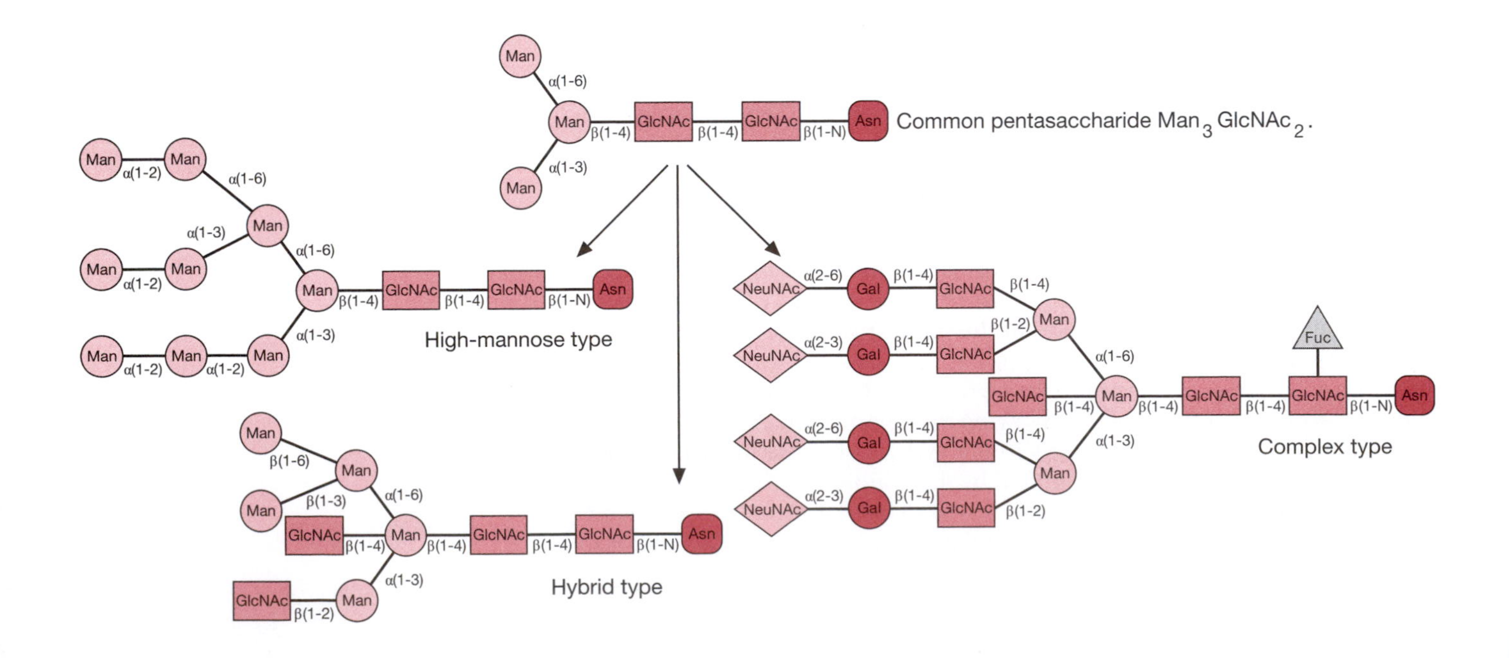

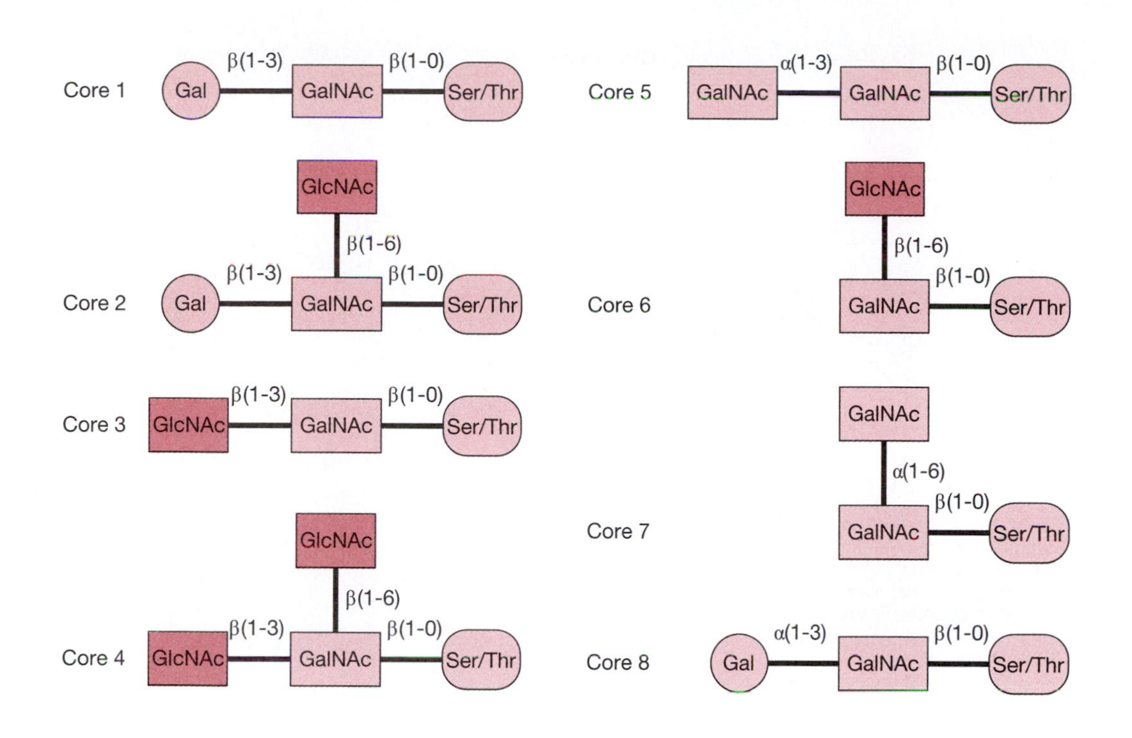

FIGURE 8.12 ***O*-linked glycans are very diverse and are classified according to the core residues.**

very heterogeneous and are usually classified according to their core structure (**Figure 8.12**). Other *O*-linked glycans include single glucosamine, fucose, galactose, mannose, or xylose residues, and these require specific consensus sequences.

A third major form of glycosylation that occurs in the secretory pathway is the addition of GPI moieties to proteins that need to be anchored in the plasma membrane. GPI-anchored proteins are attached at the C-terminus to a trimannosyl-non-acetylated glucosamine (Man_3-GlcN) core through a phosphodiester linkage involving phosphoethanolamine. This may be further modified in some cell types. The reducing end of the GlcN residue is linked through another phosphodiester bond to phosphatidylinositol (PI), which is anchored in the membrane (**Figure 8.13**).

Glycans play important roles in protein stability, activity, and localization, and are important indicators of disease

Although most secreted proteins in eukaryotes are glycosylated, the purpose of the glycan chains is not always clear. It is apparent that glycans are required for some proteins to fold properly, and in other cases the carbohydrate residues act as address labels that facilitate protein sorting to the appropriate subcellular compartments. This is the case for proteins targeted to the lysosome (which have exposed mannose 6-phosphate residues) and membrane proteins containing GPI anchors. The stability of some proteins appears to be improved by the presence of glycan chains, possibly because they prevent proteases gaining access to the protein surface. One of the most important functions, however, is the control of protein interactions with each other and with other ligands. In this context, glycan chains are important for cell signaling or recognition during fertilization, in development, and in the modulation of the immune response. Deficiencies or alterations to the glycan component of several proteins have been linked to disease (see **Table 8.4**) and a number of glycoproteins are already approved by the US FDA (Food and Drug Administration) as cancer biomarkers (see **Table 8.5**).

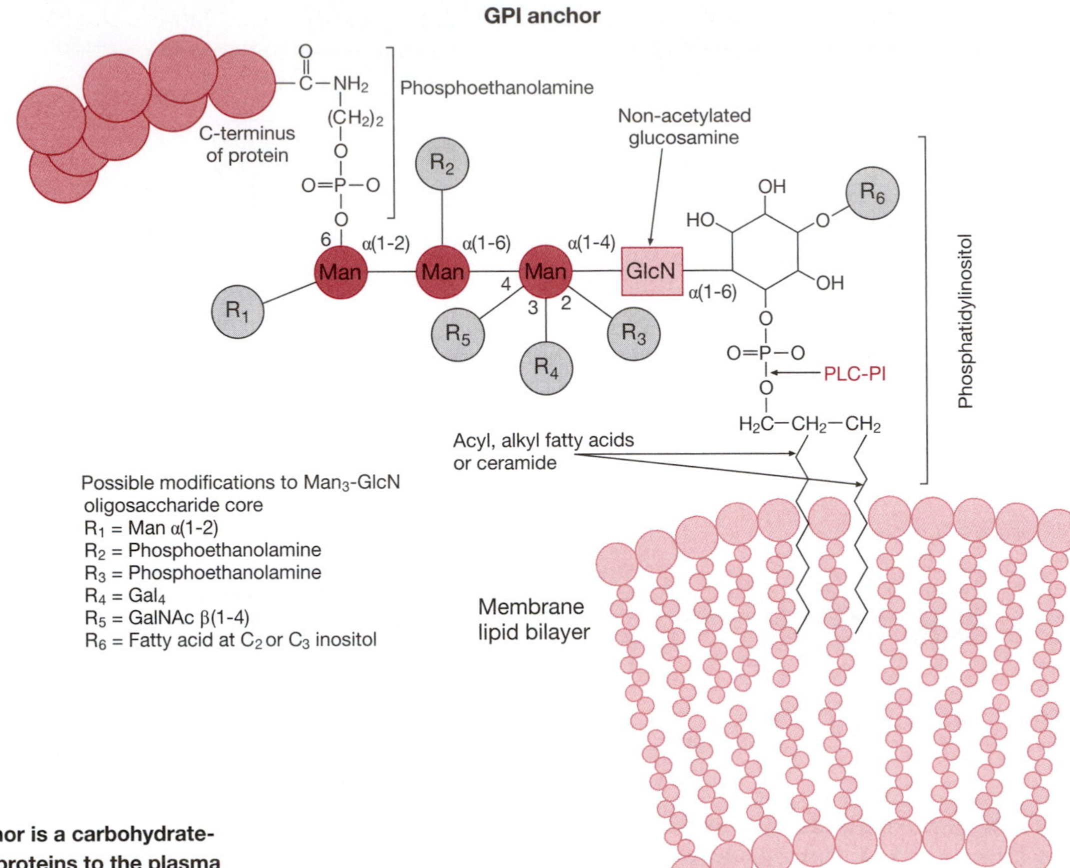

FIGURE 8.13 The GPI anchor is a carbohydrate-rich structure that tethers proteins to the plasma membrane.

TABLE 8.4 SOME DISEASES ASSOCIATED WITH ALTERED GLYCAN CHAINS ON GLYCOPROTEINS

Disease	Glycoprotein	Alteration
Hepatic cancer	α-Fetoprotein	Different *N*-glycan structures
Immune disorders	CD43	Different *O*-glycan structures
Rheumatoid arthritis	Immunoglobulin G	*N*-glycans, reduction in terminal galactosylation
Choriocarcinoma	Human chorionic gonadotropin	*N*- and *O*-linked glycans, hyperbranching
Alcohol abuse	Transferrin	*N*-glycans, desialylation

Conventional glycoanalysis involves the use of enzymes that remove specific glycan groups and the separation of glycoproteins by electrophoresis

Conventional glycoanalysis techniques are laborious and time-consuming because multiple steps are required, and they do not provide exhaustive data (**Figure 8.14**). Initially, the glycan or glycans must be removed from the isolated parent protein. With *N*-glycans, this can be achieved in a single step using the enzyme **peptide-*N*-glycosidase F (PNGase F)**. Despite its name, PNGase F is a deamidase rather than an endoglycosidase, which means that the Asn residue to which the glycan chain is attached is deamidated to Asp. This modification provides a useful signature when mass spectrometry is used to identify potential glycopeptides, since a database

TABLE 8.5 CANCER BIOMARKERS APPROVED BY THE US FOOD AND DRUGS ADMINISTRATION

Protein	Glycosylation	Detection	Source	Disease	Biomarker purpose
α-Fetoprotein	Yes	Glycoprotein	Serum	Nonseminomatous testicular cancer	Diagnosis
Human chorionic gonadotropin β	Yes	Glycoprotein	Serum	Testicular cancer	Diagnosis
CA19–9	Yes	Carbohydrate	Serum	Pancreatic cancer	Monitoring
CA125	Yes	Glycoprotein	Serum	Ovarian cancer	Monitoring
CEA: carcinoembryonic antigen	Yes	Protein	Serum	Colon cancer	Monitoring
Epidermal growth factor receptor	Yes	Protein	Tissue	Colon cancer	Therapy selection
KIT	Yes	Protein	Tissue	Gastrointestinal (GIST) cancer	Diagnosis/ therapy selection
Thyroglobulin	Yes	Protein	Serum	Thyroid cancer	Monitoring
PSA: prostate-specific antigen (Kallikrein 3)	Yes	Protein	Serum	Prostate cancer	Screening/ monitoring/ diagnosis
CA15–3	Yes	Glycoprotein	Serum	Breast cancer	Monitoring
CA27–29	Yes	Glycoprotein	Serum	Breast cancer	Monitoring
HER2/NEU	Yes	Protein	Tissue, serum	Breast cancer	Prognosis/ therapy selection/ monitoring
Fibrin/FDP (fibrin degradation protein)	Yes	Protein	Urine	Bladder cancer	Monitoring
BTA: bladder tumor-associated antigen (complement factor H-related protein)	Yes	Protein	Urine	Bladder cancer	Monitoring
CEA and mucin (high-molecular-weight)	Yes	Protein (immunofluorescence)	Urine	Bladder cancer	Monitoring

From Pan S, Chen R, Aebersold R & Brentnall TA. (2011) *Mol. Cell. Proteomics* 10, R110. 003251.

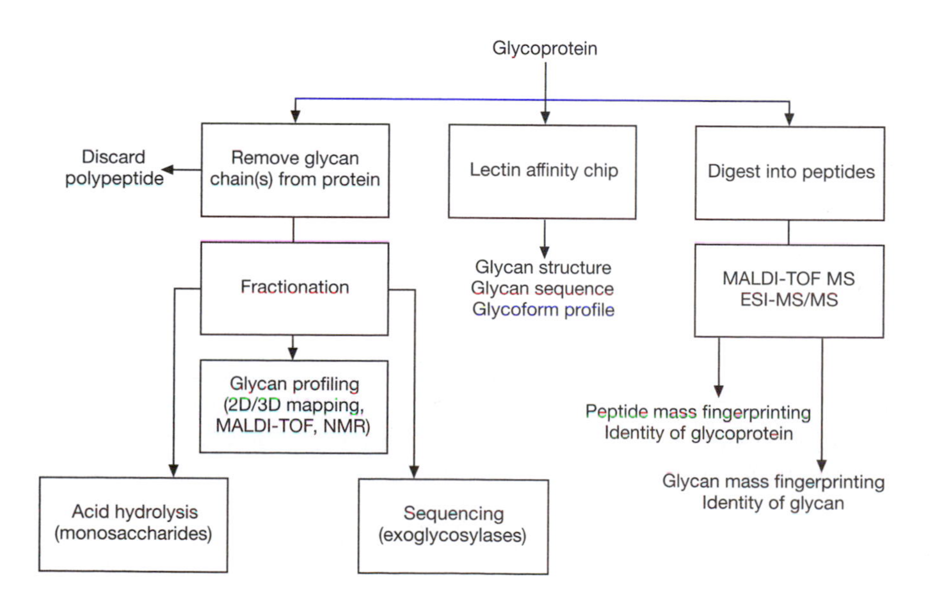

FIGURE 8.14 The full analysis of glycoproteins must involve characterization of both the polypeptide and glycan components. At the current time, this is usually achieved by parallel analysis of unglycosylated peptides and released glycans. Lectin chips, a new innovation, now allow the characterization of intact glycoproteins.

error of +1 is obtained compared with the predicted mass of the unmodified protein. However, this enzyme does not work on glycans containing α(1-3)-fucose (a residue that is found in plants and some worms, but not in mammals) and the alternative PNGase A must be used. There is no equivalent enzyme for *O*-linked glycans, although mild alkaline treatment resulting in β-elimination is a useful and successful approach.

If the protein has multiple glycosylation sites or is present as multiple glycoforms, the collection of glycan chains must be fractionated and individual glycan species must be isolated. Several different types of analysis can then be carried out to determine the glycan structure. Each glycan can be broken up into its constituent monosaccharides by acid hydrolysis and the individual sugars can be identified. This provides the **monosaccharide composition** of each glycan, but does not reveal any sequence or structural information. Nevertheless, monosaccharide composition is still a useful approach since all hexoses have the same mass (as do GlcNAc and GalNAc glycans) and cannot therefore be distinguished by mass spectrometry. One traditional way to obtain sequences and structures is to carry out two- or three-dimensional mapping, where glycans are resolved into their sugars by multiple rounds of chromatography, and the structures are worked out by comparing the elution positions to those of standards. Further structural information can be derived from mass spectrometry and NMR spectroscopy experiments. Each glycan can also be labeled and digested with specific **exoglycosylases** to determine the sequence. By combining these techniques, the sequence and branching pattern of each glycan and its relative abundance in the original protein sample can be deduced. This sort of approach has been optimized for the analysis of individual glycoproteins produced in recombinant expression systems and is the method of choice for monitoring batch-to-batch differences in recombinant protein drugs.

What this type of analysis does not reveal is the distribution of glycan chains within a protein sample. For example, if a protein has two glycosylation sites (A and B) and there are two types of glycan chain (1 and 2), what does it mean if conventional glycoanalysis shows that each glycan is equally abundant? It could mean that all the proteins in the sample have glycan 1 at site A and glycan 2 at site B, or the reverse could be true, or 50% of the proteins might have two glycan 1 chains while the other 50% have two glycan 2 chains (**Figure 8.15**). Novel lectin chip technologies have been developed to address such problems as discussed in Chapter 9.

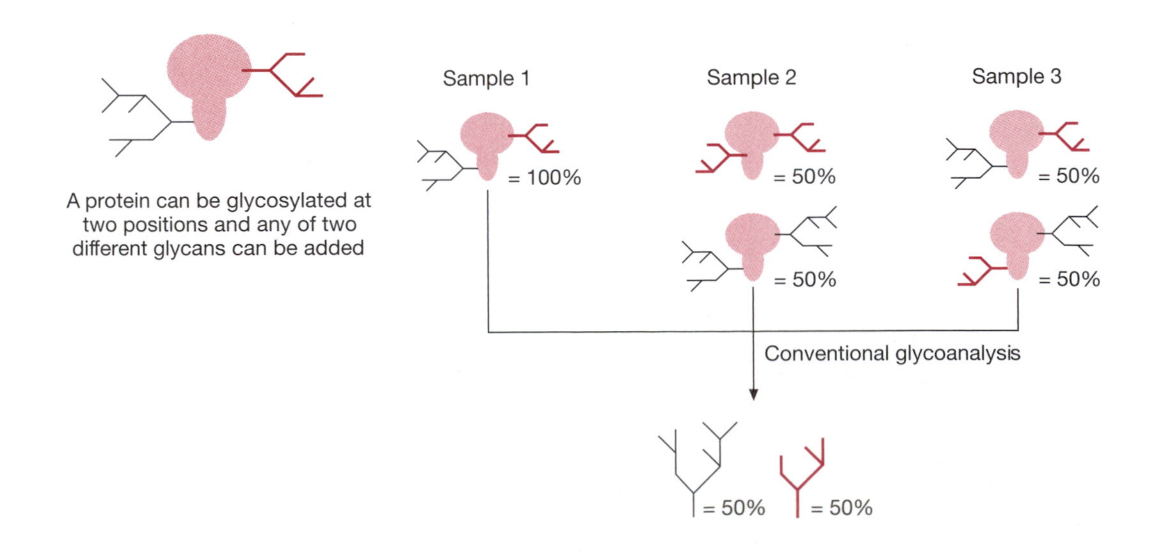

FIGURE 8.15 Conventional glycoanalysis, which involves removal of the glycan chains, can reveal glycan sequence and structure, and the relative proportions of each type of glycan in a mixture, but cannot resolve a mixture of different glycoforms. The three samples shown in the figure are not the same, but they cannot be distinguished by conventional methods.

Glycoprotein-specific staining allows the glycoprotein to be studied by 2DGE

For glycoanalysis at the whole-proteome level, higher throughput and more information-rich analysis techniques are required. As discussed above for phosphoproteins, glycoproteomic analysis requires methods for the selective detection and isolation of glycoproteins and the identification of the glycosylated sites. In this case, however, structural and compositional analysis of the glycan chains has to be included in the investigation, making comprehensive glycoprotein analysis an order of magnitude more complex than the analysis of phosphoproteins.

All of the separation techniques discussed in Chapter 2 are applicable to glycoproteins, and 2DGE is the method of choice for visualizing the glycoproteome. Standard techniques for in-gel or on-membrane protein staining appear to work poorly with glycoproteins and special methods are required. One of the earliest to be described was the **periodic acid/Schiff (PAS)** reaction, which relies on the oxidation of sugar moieties for the detection of glycoproteins in gels or on membranes. This can be combined with a more sensitive detection assay if the glycans are modified, for example with biotin hydrazide (detected with streptavidin–peroxidase and appropriate colorimetric or chemiluminescent substrates) or with fluorescein semicarbazide (for sensitive fluorometric detection). More recently, several companies have developed universal stains that work well with glycoproteins, such as the SYPRO reagents (see p. 72), as well as stains that are specific for glycoproteins (for example, Pro-Q Emerald, which is 50 times more sensitive than the PAS method and can be combined with SYPRO staining).

A problem with 2DGE is that glycan heterogeneity causes each glycoprotein to appear as a series of discrete spots, representing glycoforms with different molecular masses or different pI values (**Figure 8.16**). The same protein can therefore appear in several or many different spots, often in a chain, and the concentration of that protein in each individual spot is reduced. The analysis of individual spots by MS (see below) often reveals heterogeneity in terms of glycan acceptor site, glycan size, and charge, suggesting that the relationship between glycoprotein structure and the migration on two-dimensional gels is complex. Higher-resolution glycoform separation is possible by combining methods such as reversed-phase HPLC, strong anion exchange chromatography and capillary electrophoresis.

More recently, robust methods have been described that allow the analysis of *N*- and *O*-linked oligosaccharides released from glycoproteins separated by two-dimensional PAGE and then electroblotted onto PVDF (polyvinylidene difluoride) membranes. Nicole Packer and colleagues (see Wilson et al., Further Reading) have described a technique in which *N*-linked

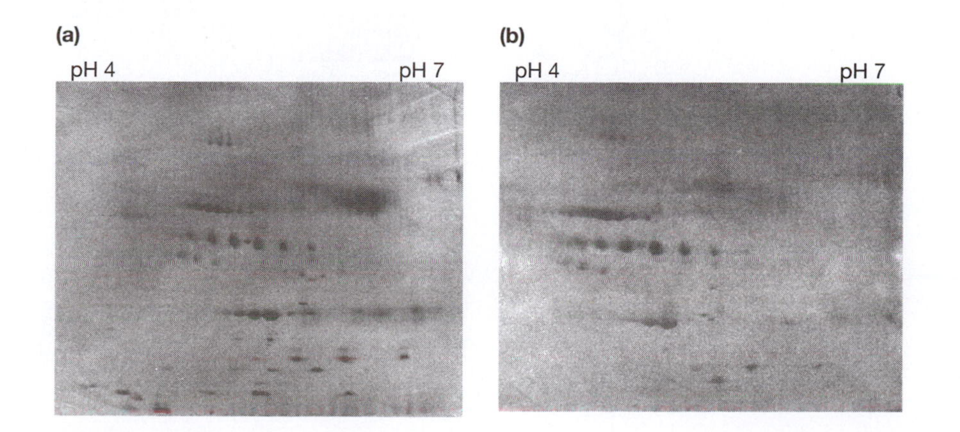

FIGURE 8.16 Fluorescent staining of glycoproteins on a PVDF membrane. (a) Coomassie Brilliant Blue stain. (b) Fluorescein carbohydrate stain. One microliter of serum was loaded by rehydration, separated by 2DGE and electroblotted to a PVDF membrane. The fluorescent stain used periodate oxidation of the sugars as contained in the Bio-Rad glycoprotein detection kit, but used a solution of 0.05 mg/ml fluorescein semicarbazide (Molecular Probes) in 50% (v/v) methanol in 1 mM sodium acetate pH 5.5, instead of sandwich antibody color development. Fluorescence was visualized with a Fluor S imager (Bio-Rad). (Courtesy of Nicolle Packer.)

oligosaccharides are released by treatment with PNGase F followed by the chemical release of *O*-linked oligosaccharides using β-elimination. The released species are then separated and characterized by LC-ESI mass spectrometry. *N*-linked site-specific information was obtained by tryptic digestion of the remaining proteins and analysis by MALDI-MS.

There are two principal methods for glycoprotein enrichment that have complementary uses

Current glycoproteomic analysis follows the general workflow of shotgun proteomics with additional components specific to the glycans, and therefore involves an enrichment step specific for glycoproteins (**Figure 8.17**). There are two principal methods. The first is **lectin-affinity chromatography**, which involves the use of carbohydrate-specific binding proteins known as **lectins**. Many lectins are available that have specific ligands, including concanavalin A (Con A), which binds specifically to mannosyl and glucosyl residues; jacalin, which recognizes galactosyl (β1-3) *N*-acetylgalactosamine; wheat germ agglutinin, which binds specifically to di-*N*-acetylglucosamine and *N*-acetylneuraminic acid; and *m*-aminophenylboronic acid, which binds to 1,2-*cis*-diol groups. Other lectins, such as *Riccinus communis* lectin, have broader specificities. Several lectin-affinity procedures may be used one after the other to select different classes of glycan chains progressively, as most recently applied in the filter-aided sample preparation method (FASP). Alternatively, fractionation within a class of glycoforms can be achieved by gradient elution (using stepped increases in a competitive binding agent). Lectin-affinity chromatography is also useful for the purification of glycopeptides following proteolytic digestion, and the purification of glycan chains derived from glycoproteins in preparation for enzymatic or mass-spectrometry-based sequencing.

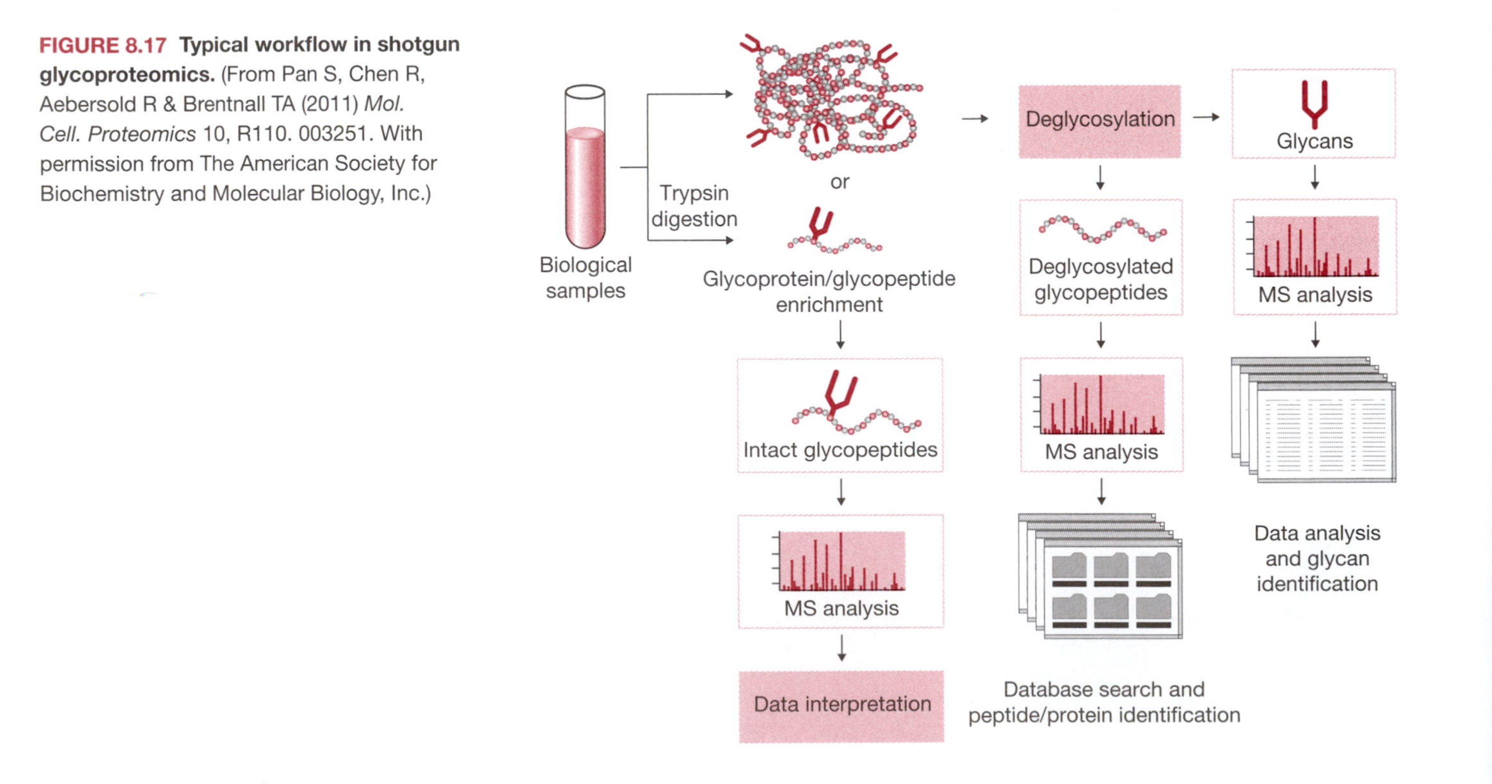

FIGURE 8.17 Typical workflow in shotgun glycoproteomics. (From Pan S, Chen R, Aebersold R & Brentnall TA (2011) *Mol. Cell. Proteomics* 10, R110. 003251. With permission from The American Society for Biochemistry and Molecular Biology, Inc.)

The second major method is **hydrazine-based solid-phase extraction**, in which glycoproteins are oxidized to their aldehyde forms, allowing them to react with immobilized hydrazine resins. This is an efficient method for capturing both *N*-linked and *O*-linked glycoproteins, although it relies on the enzymatic release of the proteins after capture and washing using an enzyme such as PNGaseF. Therefore, only the protein component and not the glycan component is available for downstream analysis. Furthermore, PNGase F can only release *N*-linked glycans, so *O*-linked glycoproteins remain bound to the resin and are excluded from the analysis. Lectin-based enrichment and hydrazine capture are therefore considered complementary techniques, with the former more flexible and useful for the isolation of specific glycans and the latter more comprehensive and efficient for capturing all glycoproteins. There have been occasional reports of glycoprotein enrichment using HILIC and size exclusion chromatography, retention on graphite powder microbeads, and selective glycopeptide enrichment with boronic acid, although these methods are not as yet widely applied.

Mass spectrometry is used for the high-throughput identification and characterization of glycoproteins

Mass spectrometry can be used to identify glycoproteins using the now classic bottom-up approach (analysis of enriched glycopeptides) or the more recent top-down approach, which involves fragmenting whole glycoprotein ions. Both approaches can also provide information about glycosylation sites and (if the glycans are intact) the oligosaccharide structures. The bottom-up approach is more widely used because it is not constrained by an upper mass limit. The direct analysis of intact glycopeptides with attached glycans is challenging because the fragmentation yields ions that represent the peptide backbone, the carbohydrate group, and the combinations of both, thus it can be more straightforward to analyze the deglycosylated peptides and isolated glycans separately (Figure 8.17). The analysis of deglycosylated peptides (generated by enzymatic removal of the glycans) provides the same data as standard MS and MS/MS analysis as discussed in Chapters 3 and 4 but no specific information about the glycans (although the position of an *N*-glycan acceptor site can be deduced from the presence of particular sequence motifs once the peptide has been identified by correlative database searching).

The introduction of sophisticated mass analyzers that allow multiple rounds of fragmentation means that complementary information about the peptide and glycan structures can be obtained from the fragmentation of intact glycoproteins and glycopeptides. MS/MS spectra generated by standard CID are generally dominated by *b*-series and *y*-series peptide fragments and glycosidic cleavage ions, but the nature of the products can be controlled to a certain extent by modifying the collision energy. Thus, low-energy CID favors the production of glycosidic cleavage ions whereas higher energy CID favors the production of peptide ions, and MALDI sources, which tend to produce precursor ions with a single charge, are more likely to produce an even mixture of peptide and glycosidic cleavage ions with either CID or PSD fragmentation methods. More recent development such as ECD and infrared multiphoton dissociation (IRMPD), which are used with high-resolution FT-ICR and Orbitrap mass analyzers, tend to cleave the peptide backbone with the undamaged glycan attached, which can provide highly specific data about the position of a glycan acceptor site. When combined with the ability to select particular fragment ions and subject them to additional rounds of fragmentation (MS^n analysis), this can not only provide rich data for peptide identification, but also go a long way to mapping the precise structure of the glycans too.

FURTHER READING

Beltran L & Cutillas PR (2012) Advances in phosphopeptide enrichment techniques for phosphoproteomics. *Amino Acids* 43, 1009–1024.

Chen PC, Na CH & Peng J (2012) Quantitative proteomics to decipher ubiquitin signaling. *Amino Acids* 43, 1049–1060.

Ficarro SB, McCleland ML, Stukenberg PT et al. (2002) Phosphoproteome analysis by mass spectrometry and its application to *Saccharomyces cerevisiae. Nat. Biotechnol.* 20, 301–305.

Hoofnagle AN & Heinecke JW (2009) Lipoproteomics: using mass spectrometry-based proteomics to explore the assembly, structure, and function of lipoproteins. *Lipid Res.* 50, 1967–1975.

Mann M & Jensen ON (2003) Proteomic analysis of post-translational modifications. *Nat. Biotechnol.* 21, 255–261.

Merbl Y & Kirschner MW (2011) Protein microarrays for genome-wide posttranslational modification analysis. *Wiley Interdiscip. Rev. Syst. Biol. Med.* 3, 347–356.

Mischerikow N & Heck AJ (2011) Targeted large-scale analysis of protein acetylation. *Proteomics* 11, 571–589.

Morelle W, Faid V, Chirat F & Michalski JC (2009) Analysis of *N*- and *O*-linked glycans from glycoproteins using MALDI-TOF mass spectrometry. *Methods Mol. Biol.* 534, 5–21.

Nilsson CL (2012) Advances in quantitative phosphoproteomics. *Anal. Chem.* 84, 735–746.

Pan S, Chen R, Aebersold R & Brentnall TA (2011) Mass spectrometry based glycoproteomics—from a proteomics perspective. *Mol. Cell. Proteomics* 10, R110.003251.

Pandey A, Podtelejnikov AV, Blagoev B et al. (2000) Analysis of receptor signaling pathways by mass spectrometry: identification of Vav-2 as a substrate of the epidermal and platelet-derived growth factor receptors. *Proc. Natl Acad. Sci. USA* 97, 179–184.

Pandey A, Fernandez MM, Steen H et al. (2000) Identification of a novel immunoreceptor tyrosine-based activation motif-containing molecule, STAM2, by mass spectrometry and its involvement in growth factor and cytokine receptor signaling pathways. *J. Biol. Chem.* 275, 38633–38639.

Pasing Y, Sickmann A & Lewandrowski U (2012) *N*-glycoproteomics: mass spectrometry-based glycosylation site annotation. *Biol. Chem.* 393, 249–258.

Roth AF, Wan J, Green WN et al. (2006) Proteomic identification of palmitoylated proteins. *Methods* 40, 135–142.

Steen H, Kuster B, Fernandez M et al. (2002) Tyrosine phosphorylation mapping of the epidermal growth factor receptor signaling pathway. *J. Biol. Chem.* 277, 1031–1039.

Tichy A, Salovska B, Rehulka P et al. (2011) Phosphoproteomics: searching for a needle in a haystack. *J. Proteomics* 74, 2786–2797.

Wilson NL, Schulz BL, Karlsson NG & Packer NH (2002) Sequential analysis of *N*- and *O*-linked glycosylation of 2D-PAGE separated glycoproteins. *J. Proteome Res.* 1, 521–529.

Witze ES, Old WM, Resing KA & Ahn NG (2007) Mapping protein post-translational modifications with mass spectrometry. *Nat. Methods* 4, 798–806.

Ytterberg AJ & Jensen ON (2010) Modification-specific proteomics in plant biology. *J. Proteomics* 73, 2249–2266.

Zhang Y, Yin H & Lu H (2012) Recent progress in quantitative glycoproteomics. *Glycoconj. J.* 29, 249–258.

Zhao Y & Jensen ON (2009) Modification-specific proteomics: strategies for characterization of post-translational modifications using enrichment techniques. *Proteomics* 9, 4632–4641.

CHAPTER 9

Protein microarrays

9

9.1 INTRODUCTION

All of the key proteomics technologies described in this book share one common feature, namely, that they have been developed for the parallel analysis of large numbers of proteins in a single experiment. As in other areas of biological research, the trend toward higher throughput has been matched by a trend toward miniaturization and automation. Almost any conceivable proteomics method can be miniaturized and automated, including protein separation by chromatography or electrophoresis, the processing of samples for mass spectrometry, the determination of protein sequences, structures, and interactions, and the large-scale analysis of fluids, cells, or tissue samples for protein abundance and localization. Typically, the assay format is reduced to the size of a microscope slide or smaller and the resulting device is described as a **chip**, in reference to the miniaturization of circuit boards to produce microchips.

We can distinguish between two major types of device. The first is conceptually similar to a **DNA microarray** (Chapter 1) but contains proteins instead of nucleic acids. This is known as a **protein microarray** and is the subject of the remainder of this chapter. The second type is more variable in design. Some devices are based on microfluidics and/or nanotechnology and are used to separate molecules such as proteins by controlling the movement of small volumes of liquids and gases. Several companies are developing such **lab-on-a-chip** devices. Others are based on different surface chemistries and are used to capture broad classes of chemically similar proteins. These are chip-based equivalents of bind-and-elute chromatography. Some examples of these devices are discussed in Box 9.1.

9.2 THE EVOLUTION OF PROTEIN MICROARRAYS

As discussed in Chapter 1, **DNA microarrays** are miniature appliances upon which many cDNA sequences, genomic DNA fragments, or oligonucleotides can be arrayed in the form of a grid. The resulting device is about the size of a microscope slide, but it allows thousands of genes to be analyzed in parallel using only a few tens of microliters of analyte. The use of small sample volumes means that the assays have a greater sensitivity and specificity than equivalent larger assay formats (hybridization to nylon or nitrocellulose filters) and the compatibility with automation reduces operator errors and increases reproducibility and reliability. Microarray-based assays are therefore suitable for the repetitive quantitative analysis of large numbers of samples. The establishment of in-house facilities for custom DNA

BOX 9.1 RELATED TECHNOLOGIES.
Protein chips.

Several companies have developed miniature devices that allow proteomics separations technology to be applied to very small volumes of liquid. This increases sensitivity and throughput, reduces the experimental timescale, and allows almost total automation. For example, consider the widely used conventional separation method of 2DGE (Chapter 2). This requires relatively large sample volumes, separations can take more than 24 hours to complete, and the maximum sensitivity is somewhere in the low nanogram range. The gels are large and cumbersome, and they are difficult to integrate with downstream analysis procedures such as mass spectrometry. In contrast, on-chip separations can be carried out on tiny analyte volumes in less than 30 minutes, and 10 or more separations can be carried out simultaneously. Recent innovations allow on-chip protease digestion, peptide separation, and mass spectrometry using trypsin membranes and microscale capillary zone electrophoresis. Chips have been developed than can perform electrophoretic and/or chromatographic separations in two dimensions in a matter of minutes, and the elimination of hands-on sample processing means that there is no loss of analyte, providing picomole sensitivities.

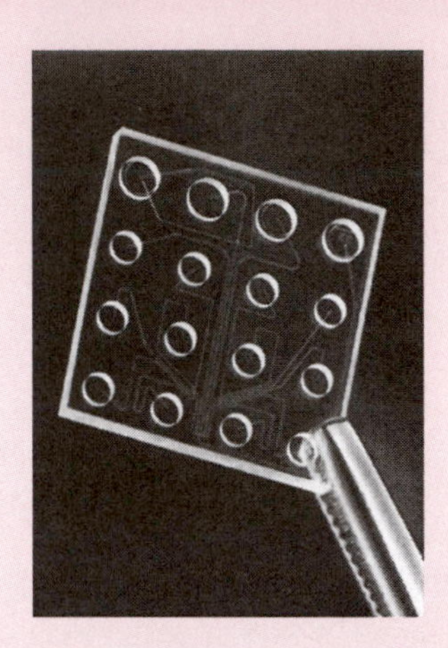

FIGURE 1 Prototype microfluidic channel system developed by Agilent Technologies, which allows 10 electrophoretic separations to be completed in less than an hour in combination with real-time viewing of separations. (Courtesy of Agilent Technologies, Inc., Santa Clara, CA.)

Figure 1 shows the prototype microfluidic channel system developed by Agilent Technologies, which allows 10 electrophoretic separations to be completed in less than an hour in combination with real-time viewing of separations.

Other protein chips are analogous to analytical microarrays in that their purpose is to capture proteins from an analyte. However, they do not contain arrays of specific capture agents, but instead possess various surface chemistries to capture broad classes of chemically similar proteins. For example, Ciphergen Biosystems Inc. produce a range of protein chips, marketed under the name ProteinChips, which retain different classes of proteins on a number of alternative chromatographic surfaces. Although relatively nonspecific compared with antibodies, complex mixtures of proteins can be simplified and then analyzed by mass spectrometry, as shown in Figure 2.

An advantageous feature of this system is the ease with which it is integrated with downstream MS analysis, since each chip doubles as a modified MALDI plate (p. 52). ProteinChips initially needed to be coated with matrix prior to MALDI analysis, but newer chips have the matrix compound incorporated into the surface chemistry. The chips can also be used to prepare conventional arrays (for example, antibody arrays), allowing direct MS analysis of captured proteins.

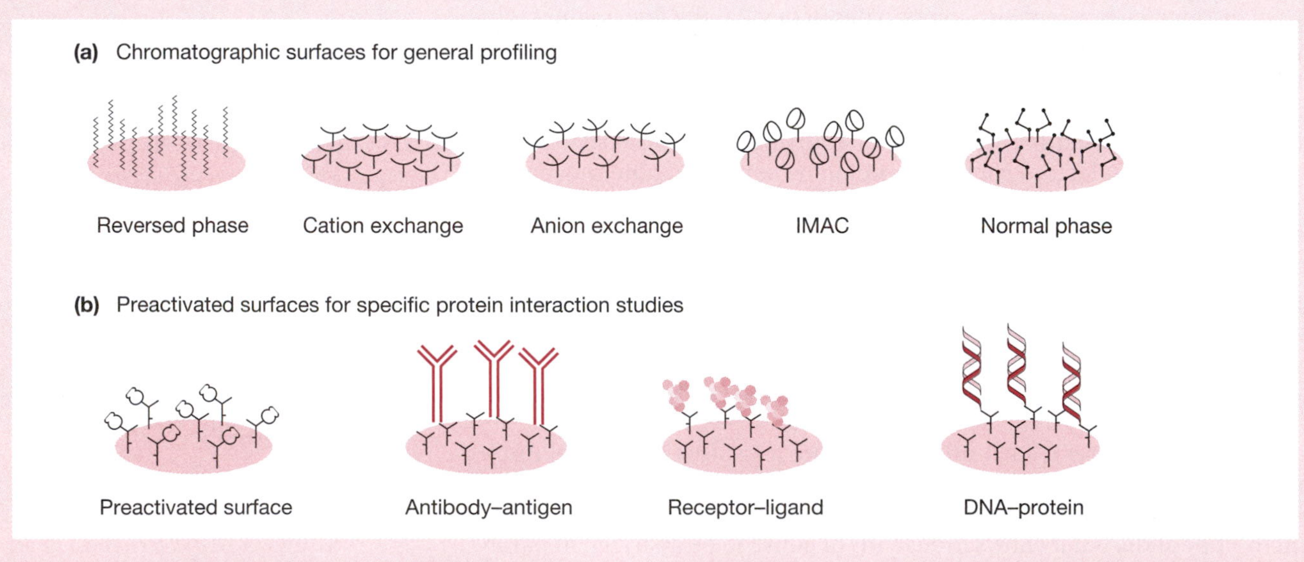

FIGURE 2 Various ProteinChip array surfaces. Both (a) chromatographic surfaces and (b) preactivated surfaces are illustrated. Chromatographic surfaces are composed of reversed-phase, ion exchange, immobilized metal affinity capture (IMAC), or normal-phase chemistries that function to extract proteins using quasi-specific affinity capture. Preactivated surfaces contain reactive chemical groups capable of forming covalent linkages with primary amines or alcohols. As such, they are used to immobilize specific capture molecules such as antibodies, receptors, or oligonucleotides often used for studying biomolecular interactions. (Courtesy of Scot Weinberger, Ciphergen Biosystems, Inc., Fremont, CA.)

microarray synthesis in many academic and industrial settings has made global expression profiling economical and practical, and the experiments yield immense amounts of useful data. Not surprisingly, there has been a strong drive toward the development of similar **protein microarrays** to reap the same benefits in the field of proteomics.

The concept of the protein array is not new. Miniaturized, multiplexed, solid-phase immunoassays were developed in the 1980s using protein microdots spotted manually onto nitrocellulose sheets and other solid supports. Such **multi-analyte immunoassay** formats were shown to be far more sensitive than standard immunoassays carried out in microtiter plates, because the sample volumes were much smaller. At about the same time, the first **gridded cDNA expression libraries** were developed, allowing the functions and binding capacities of large numbers of arrayed, immobilized proteins to be tested in parallel. The arrays were printed on sheets of nitrocellulose or nylon, which by current standards would be regarded as quite cumbersome. At the time, however, these were much more convenient to use than standard libraries taken from plate imprints (where the clones were distributed randomly) and also allowed data to be shared more efficiently between laboratories.

9.3 DIFFERENT TYPES OF PROTEIN MICROARRAYS

Analytical, functional, and reverse microarrays are distinguished by their purpose and the nature of the interacting components

Protein microarrays can be categorized according to their purpose, the nature of the arrayed targets, the nature of the analyte, the fabrication method, the detection method, and any number of other criteria, and this may appear confusing to those who are unfamiliar with the technology, particularly because different terms may be used to describe the same concept.

Generally, protein microarrays are divided into three major types according to their purpose:

1. **Analytical microarrays**. The defining feature of an analytical microarray is that the components arrayed on the surface are **affinity reagents** such as antibodies whose purpose is to capture proteins from the analyte, which is a complex mixture such as a cell lysate. The analytical array is used to detect/identify and quantify the components of the analyte. They are sometimes termed **capture microarrays**.

2. **Functional microarrays**. The defining feature of a functional microarray is that the components arrayed on the surface are diverse proteins (for example, representing an entire proteome) that can be used to assay a range of biochemical and molecular functions such as protein-protein interactions, protein interactions with nucleic acids or small-molecule ligands, enzyme activities, and interactions with antibodies. They are sometimes termed **target protein arrays**.

3. **Reversed-phase microarrays**. The defining feature of a reversed-phase microarray is that the components arrayed on the surface are complex mixtures such as cell lysates or serum samples, and these are probed with labeled affinity reagents to compare analytes from different sources —for example, to quantify the level of a particular protein in different samples.

Both analytical arrays and functional arrays are sometimes described as **forward arrays** because each address on the array contains a single component (affinity reagent or protein) and the analyte is complex—the converse of the reversed array, where the arrayed targets are complex and a simple probe such as an antibody is applied to the array to generate a signal.

Analytical microarrays contain antibodies or other capture reagents

Analytical microarrays typically contain high-density arrays of specific antibodies. The antibodies may be conventional full-size immunoglobulins produced in hybridoma cells, or recombinant derivatives such as single chain Fv (scFv) or Fab fragments expressed in bacteria. The chip is flooded with the analyte, allowing the antibodies to capture any antigens that are present, and then washed to remove unbound proteins. The proteins in the analyte may be labeled, allowing direct detection by fluorescence scanning or autoradiography, they may be detected using a label-free procedure such as surface plasmon resonance spectroscopy, or the bound, unlabeled antigen may be detected using a second, labeled antibody in a sandwich assay. The relative merits of these different signal detection methods are discussed in Section 9.6.

The two major technological problems limiting the development of antibody arrays are the difficulty in obtaining and expressing sufficient numbers of full-size antibodies for large-scale studies and the difficulty in obtaining antibodies of adequate specificity. An antibody microarray containing 810 cancer-specific antibodies was reported in 2010, but the most sophisticated commercial microarray contains 500 different antibodies and most contain approximately 100. Small numbers of antibodies are suitable for moderate-scale experiments, for example, monitoring the expression profiles of key proteins, but proteome-scale analysis would require at least a tenfold increase in complexity, and even this would only allow the analysis of relatively simple proteomes, such as those of bacteria and yeast.

Conventional hybridoma technology is too labor-intensive and time-consuming to achieve the scale of antibody production needed for the manufacture of whole-proteome analytical chips, so alternative methods are required. The best option at the current time is **phage antibody display** technology, which allows the production of complex libraries of single-chain antibodies (scFv) displayed on the surface of phage particles. This approach has yielded scFv arrays containing more than 18,000 probes.

Phage display not only allows the rapid generation of antibodies, but also facilitates the selection of high-affinity binders through several rounds of maturation and affinity panning. This may help to address the second limitation of current antibody microarrays, that of insufficient antibody specificity. The specificity problem reflects the fact that most antibody arrays are currently used to detect particular, restricted classes of proteins, often cytokines or other secreted factors that are released into the serum or culture medium. Such antibodies have been developed especially for serum profiling and in many cases have not been checked for broader cross-reactivity, for example, in cell lysates. In the few studies that have addressed this issue, the data suggest that up to 50% of antibodies cross-react with nontarget antigens (**Box 9.2**). The proteins in a typical analyte cover a broad dynamic range, so antibodies with high affinity for a scarce target antigen and low cross-affinity for an abundant non-target antigen might bind both antigens equally well. This would provide a completely false indication of the relative abundances of the two antigens in the analyte. It is therefore likely that many of the antibodies currently used for specific analytical applications will be unsuitable for proteome-wide applications. The problem of cross-reactivity is eliminated in the sandwich assay approach, because two noncompeting antibodies (that is, antibodies recognizing different epitopes of the antigen) would be required for each target protein. However, this in itself generates another volume problem because twice as many antibodies would be needed.

BOX 9.2 CASE STUDY.
Analytical protein chips with dual-color antibody arrays.

Dual labeling is widely used with DNA microarrays to compare gene expression levels across multiple samples. Typically, two samples are obtained, the mRNA or derived cDNA is universally labeled with one of two fluorescent molecules, and both samples are applied to the array. The array is scanned at the excitation wavelength of each fluorophore in turn, and the signals are read at the corresponding emission wavelengths. The images from each fluorophore are then rendered in false color and combined, providing a global snapshot of differential gene expression (Chapter 1).

Dual-labeling technology has been applied less frequently in proteomics because uniform labeling is more difficult to achieve (see Section 9.4). The technique of difference in-gel electrophoresis, which incorporates dual labeling, is discussed on page 75. There have also been several reports in which dual labeling has been used with protein microarrays. An early and extensive study into the feasibility of dual fluorescence microarray analysis was published by Brian Haab and colleagues in 2001 (see Further Reading). These investigators obtained 115 well-characterized antibody–antigen pairs from three commercial sources and transferred the antibodies onto glass slides that had been coated with poly-L-lysine. They then made various different preparations of the antigens, in which each protein varied in concentration from 1.6 μg/ml to 1.6 ng/ml. Six different preparations were made and each was labeled with the fluorescent molecule Cy5. The antigen preparations were then mixed with a reference preparation that had been labeled with Cy3, in which each antigen was present at a concentration of 1.7 μg/ml. The different combined samples were applied to the arrays and the levels of Cy3 and Cy5 fluorescence at each address were determined by laser scanning. The results showed that only about 18% of the antibodies were specific and accurate over the range of concentrations used in the experiment, but that these antibodies could detect target antigens at levels down to 100 pg/ml, which is suitable for clinical evaluation applications.

Sreekumar and colleagues (see Further Reading) applied the same dual labeling technology to protein profiling in colon cancer. They produced a microarray with 146 antibodies (1920 features in total) recognizing proteins involved in cell-cycle regulation, stress response, and apoptosis. VoLo carcinoma cells were irradiated with a cobalt-60 source and cultured for 4 hours before protein extracts were obtained and labeled with Cy3. Protein extracts from parallel cultures of nontreated cells were labeled with Cy5. The two samples were mixed and applied to the array. These experiments identified 11 proteins that were up-regulated in colon cancer, 6 of which were previously not known to be involved. Most of these proteins had roles in apoptosis, and increased apoptosis of the cells was observed after radiation treatment. Another protein, carcinoembryonic antigen, was shown to be down-regulated. The general approach for this type of experiment is shown in **Figure 1**.

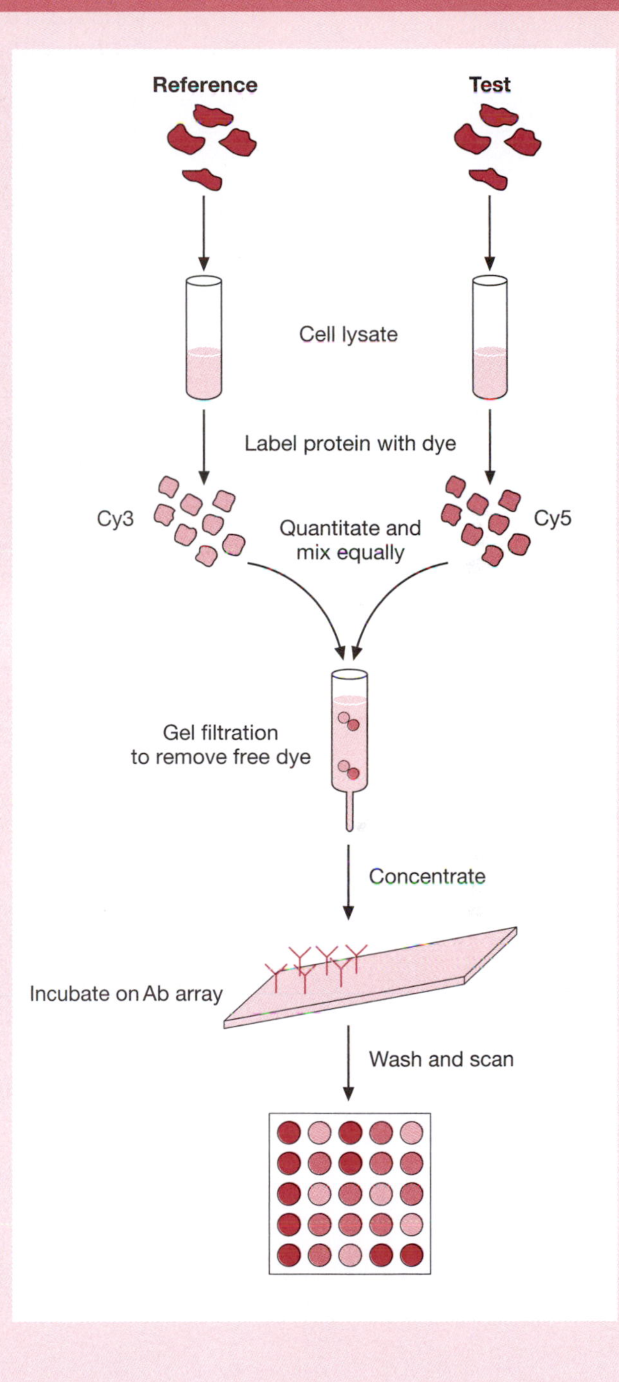

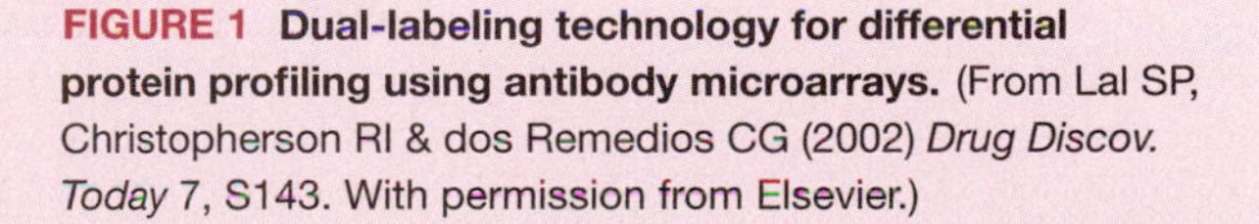
FIGURE 1 Dual-labeling technology for differential protein profiling using antibody microarrays. (From Lal SP, Christopherson RI & dos Remedios CG (2002) *Drug Discov. Today* 7, S143. With permission from Elsevier.)

Antibodies are not the only specific capture molecules that can be arrayed for the detection of proteins. Alternative reagents include **oligonucleotide** and **peptide aptamers**, which can be obtained from highly complex libraries and can be optimized for specific binding activity using *in vitro* evolution methods. There is great interest in peptide arrays not only because they are more stable than antibody arrays but also because they can be printed directly onto the chip surface using photolithographic techniques developed by Affymetrix Inc. for the manufacture of DNA microarrays (Chapter 1). An alternative strategy is to use completely synthetic **antibody mimics** or **protein scaffolds**, such as affibodies and Trinectin reagents, which are recombinant fibronectin structures. Other nonprotein molecules that can be used as affinity reagents include synthetic ligands identified by combinatorial chemistry, enzyme substrates, and ribozymes. A novel class of capture reagents known as **molecular imprinted polymers** (**MIPs**) could potentially be used to mimic any specific binding event. In this platform technology, a polymerization reaction is induced on the surface of a chip that has been flooded with a particular target molecule. The polymer forms around the target and becomes embossed with its shape. When the target molecule is removed, the imprint remains and can be used to capture similar molecules from solution. It is not known if these imprints will ever attain the sensitivity or specificity of antibodies, or if they can be developed on a scale suitable for proteomic applications. Other capture agents may be used to detect specific functional classes of proteins. Carbohydrate arrays, for example, can be used to capture lectins and other carbohydrate-binding proteins. In contrast, lectin chips can be used to capture glycoproteins and analyze their glycan structures (see Chapter 8).

Functional protein microarrays can be used to study a wide range of biochemical functions

Functional microarrays are arrayed with the proteins whose functions are under investigation. Unlike analytical microarrays, which are used for protein detection and expression profiling, functional microarrays can be used to investigate many different properties of proteins, including binding activity, the formation of complexes and biochemical functions (**Figure 9.1**).

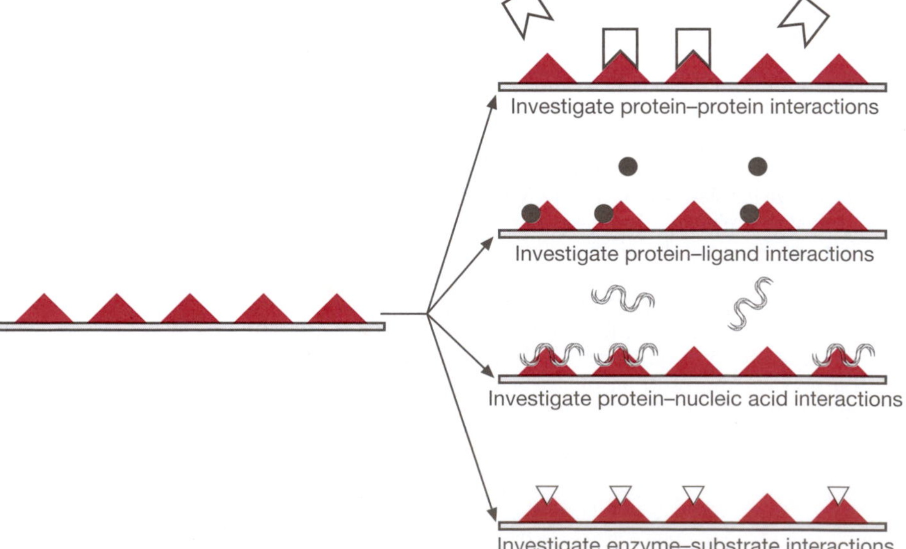

FIGURE 9.1 Some different uses for functional protein microarrays.

This is an evolution of microtiter plate assays for protein function (sometimes described as **functional proteomics** or **activity-based proteomics**) in which panels of proteins are systematically tested for biochemical activities. Such assays have also given rise to a more targeted approach known as **activity-based protein profiling** (**ABPP**) (**Box 9.3**).

As discussed above for antibody arrays, binding assays can involve the direct labeling of proteins in the analyte, sandwich assays, or label-free detection methods (Section 9.6). The two major problems limiting the development of functional microarrays are the difficulty in expressing large libraries of proteins and the difficulty in maintaining those proteins in an active state under a standard set of conditions. Functional microarrays containing up to 100 proteins have been available for several years and microarrays containing specific functional classes of proteins, for example, G-protein-coupled receptors, are used to screen for particular types of interaction. A **whole-proteome microarray** was developed for the yeast *Saccharomyces cerevisiae* in 2003 (the **Yeast ProtoArray**, containing about 5000 proteins) and a similar device for the bacterium *Escherichia coli* was reported in 2007 containing more than 4200 proteins. Some notable cases studies involving the use of functional microarrays are presented in **Box 9.4**.

9.4 THE MANUFACTURE OF FUNCTIONAL PROTEIN MICROARRAYS—PROTEIN SYNTHESIS

Proteins can be synthesized by the parallel construction of many expression vectors

The development of proteome-scale functional microarrays has demanded parallel advances in large-scale cloning, protein expression, and purification strategies. In the case of the yeast whole-proteome microarray discussed above, this was achieved by systematically cloning each gene in a vector that allowed the expression of a His_6-tagged glutathione-S-transferase (GST) fusion protein. About 95% of the yeast genome was successfully cloned in this manner and about 80% of the constructs were expressed as full-length, high-quality proteins. These proteins were isolated from cell lysates by GST pull-down (p. 140) and attached to nickel-nitrilotriacetate (Ni-NTA)-coated glass slides via the His_6 tags. Another popular approach has been to take advantage of the Gateway ligation-free cloning system in *E. coli* to generate many constructs in parallel. In addition to GST and His_6 tags, other researchers have used TAP cassettes (see Chapter 7) or photocleavable biotin tags to simplify the process of protein purification. The yeast homologous recombination and Gateway systems are compared in **Figure 9.2**.

Cell-free expression systems allow the direct synthesis of protein arrays *in situ*

The laborious process of preparing huge numbers of vectors and cell lines producing different recombinant proteins can be avoided by using **cell-free expression systems**, which allow the synthesis of proteins *in situ* from arrayed DNA constructs. Microarrays can also be produced with synthetic peptides, which can be spotted onto microscope slides using standard robotic techniques. There are also several methods for *in situ* peptide synthesis, including **photolithography**, which uses photolabile protection groups to selectively extend individual peptides, and the **SPOT method**, which involves the sequential addition of small volumes of activated amino acids.

For the construction of microarrays with full-length proteins, the original **protein *in situ* array** (**PISA**) or DiscernArray technology was a simple extension of the DNA microarray fabrication principle, in which DNA constructs produced by PCR are spotted on the activated surface of a glass slide

BOX 9.3 CASE STUDY.
Activity-based proteomics/protein profiling (ABPP).

An early example of activity-based proteomics was described using the term **biochemical genomics** by Martzen and colleagues in 1999. They produced 6144 yeast strains, in each of which a particular gene was expressed as a GST fusion, and grew the strains in defined pools in microtiter plates. The strains were assayed for a number of biochemical activities and then the pools were deconvoluted to identify strains expressing GST-tagged proteins with particular functions. This approach revealed the biochemical activities of three proteins whose functions were previously unknown.

The throughput and specificity of activity-based proteomics (**also termed activity-based protein profiling, ABPP**) can be increased by using specific functional probes and combining the technique with MS analysis. This can be achieved either in the liquid phase in approaches such as ABPP-MudPIT or using solid-phase assays such as a functional protein arrays.

The principle of **ABPP-MudPIT** is shown in **Figure 1**. The proteome is labeled with an ABPP probe that comprises a reactive group linked to a tag such as biotin (for capture) or rhodamine (for visualization). The reactive group is specific for a particular class of enzymes and becomes covalently linked to a nucleophilic residue in the active site. An example is fluorophosphonate (shown conjugated to the fluorescent label rhodamine in **Figure 2**), which is specific for the serine

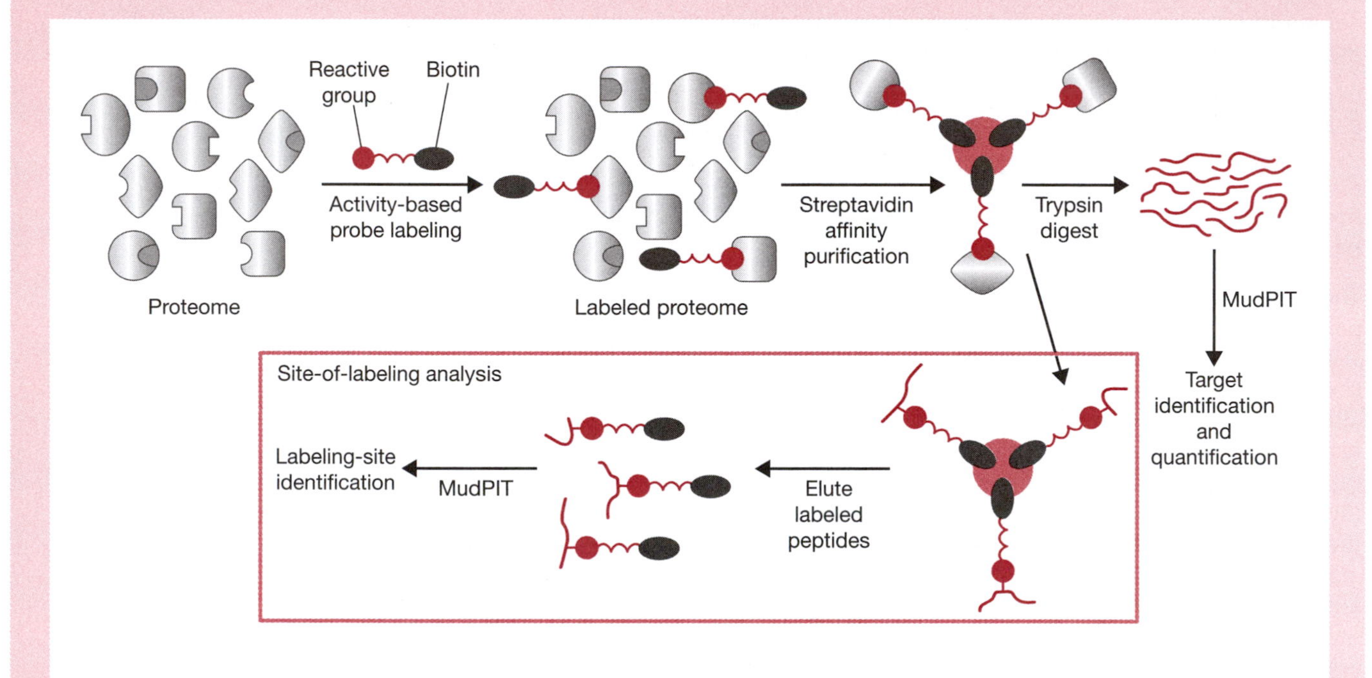

FIGURE 1 Activity-based protein profiling by MudPIT. Proteins identified by activity-related ligands are affinity purified and separated by multidimensional-ligand chromatography followed by MS/MS to identify the proteins. If trypsin digest is carved out before elution, the site of labeling can be determined. (From Speers AE & Cravatt BF (2009) *Curr. Protoc. Chem. Biol.* 1, 29. With permission from John Wiley & Sons, Inc.)

FIGURE 2 Fluorophosphonate (FP) conjugated to rhodamine.

hydrolase superfamily and binds irreversibly to the active-site serine nucleophile. However, ABPP probes have been developed for more than 20 enzyme classes. Fluorescence-labeled proteins can be separated by electrophoresis and identified individually by MALDI-TOF MS, whereas biotin-labeled proteins can be digested directly following affinity enrichment on streptavidin beads followed by MudPIT for identification and quantification (Chapter 3).

The use of bulky adducts is incompatible with *in vivo* labeling, but this can be achieved in a two-step process, first by interrogating enzyme activities with an ABPP probe comprising the reporter group linked to a small and latent chemical handle such as an azide group, which allows the probe to cross the cell membrane. After homogenization, a reporter tag can then by appended to the probe using Cu (I)-catalyzed azide–alkyne cycloaddition (click chemistry).

ABPP is similar in principle to chemical proteomics, as discussed in Chapter 10, but an important distinction is that chemical proteomics seeks small molecules that interact with proteins and could potentially be useful as drugs, whereas activity based proteomics seeks proteins that interact with reactive groups that select for active enzymes and do not interact with inactive mutants or even active enzymes that have been chemically inhibited.

to immobilize them. The innovation in the PISA method is that a second spotting operation is then used to overlay the DNA spots with the appropriate reagents for *in vitro* transcription and translation (for example, rabbit reticulocyte lysate) resulting in the synthesis of proteins *in situ*. The slides are also coated with a reagent that binds to an affinity tag on each protein (for example, glutathione for proteins tagged with GST or the HaloTag, a

BOX 9.4 CASE STUDY.
Functional protein chips.

Over the last decade, several studies have been published involving the use of small- to moderate- scale functional protein arrays. In 2000, one of the first reports described a universal protein array (UPA) comprising a 12 cm × 8 cm nitrocellulose sheet with 96 protein spots (Ge, see Further Reading). These represented 48 well-characterized human proteins (each spotted in duplicate to provide an internal control) including transcription factors, RNA processing proteins, and enzymes involved in DNA replication. The proteins were all expressed either in bacteria or using baculovirus vectors and were transferred to the sheet from a microtiter dish using a 96-well dot-blot arrayer. The array could be used to test protein interactions with proteins, nucleic acids, and small molecules. Among other observations, it was shown that the phosphorylated version of the transcriptional activator protein PC4 bound to DNA more strongly than the unmodified version.

Another notable study, published by MacBeath and Schreiber (see Further Reading), involved the printing of proteins onto derivatized glass slides with a high spatial density. The proteins were attached covalently to the slide surface but retained their biochemical activities and interaction potential. Several properties were tested, including protein–protein interactions (for example, IκBα with p50), protein–ligand interactions, and enzyme–substrate interactions. A rather different approach was used by Zhu et al. (2000) (see Further Reading) to study kinase activities in yeast. They produced a nanowell protein chip comprising 140 microwells in silicone elastomer sheets placed on top of microscope slides (Section 9.3). Sixteen chips were coated with different substrates and then each well was probed with a different protein kinase along with radiolabeled ATP (119 of the 122 known and predicted kinases were studied). The experimental format was thus distinct from the usual idea of arraying different targets on the chip, because, in this case, the chip was arrayed with identical targets and different substrates were applied at each address. These experiments confirmed many known kinase-substrate interactions and identified some new ones. The most surprising and provocative result, however, was that nearly a quarter of all the yeast kinases were capable of tyrosine phosphorylation even though they lacked the typical features of the tyrosine kinase family.

The most ambitious functional chip based analysis carried out to date involved a prototype Yeast ProtoArray (p. 197) that contained some 5800 separate protein features. As discussed in the main text, yeast genes were expressed as His_6-tagged fusion proteins and the proteins were immobilized on nickel-coated slides using the affinity tag (see Zhu et al., 2001; Further Reading). The chip was then probed with a selection of ligands, including calmodulin and a selection of phospholipids. This experiment revealed 6 of the 12 known calmodulin-binding proteins and 30 additional ones, and over 150 proteins that interacted with labeled liposomes containing specific phosphatidylinositides (PIs). Fifty of these proteins bound to specific classes of PIs and could be involved in PI-controlled signaling pathways.

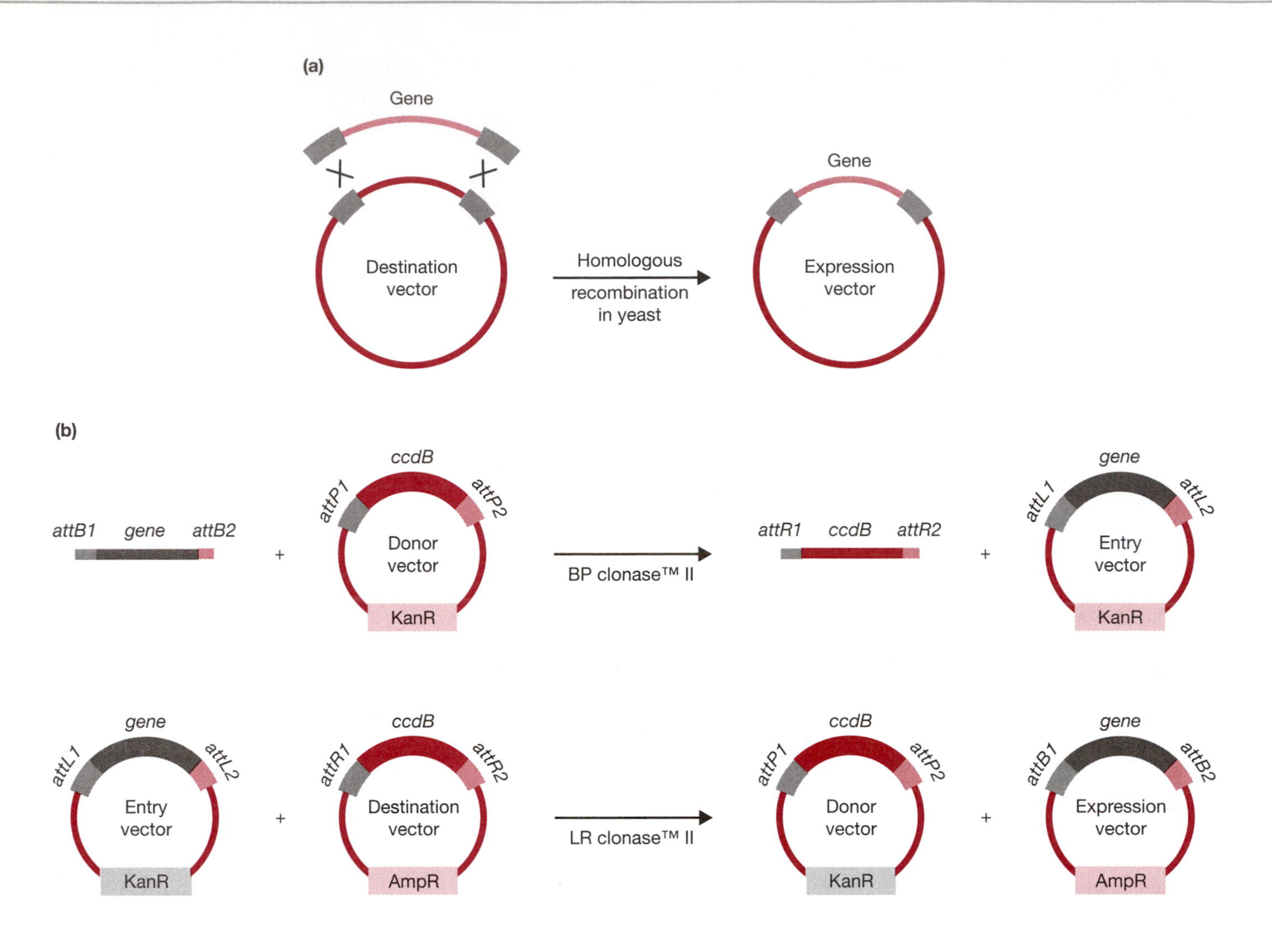

FIGURE 9.2 Comparison of the yeast homologous recombination and Gateway methods for the large-scale preparation of expression constructs, which are used to produce recombinant proteins for microarrays. (a) Yeast homologous recombination involves reciprocal exchange between the vector and an endogenous yeast DNA sequence to produce the expression vector. (b) In the Gateway Recombination System, the DNA fragment flanked with *att*B1 and *att*B2 is combined with a donor vector that contains *att*P1 and *att*P2 and the counterselectable marker, *ccdB*. The expression vector is produced by recombination between the entry vector and a specific destination vector that has the *att*R1 and *att*R2 sequences and the same counterselectable marker. (From Huang W, Wu L, Lu G & Lui S (2010) *Front. Biol.* 5, 331. With permission from Springer.)

33 kDa engineered derivative of bacterial hydrolase that binds tightly to surfaces activated with the HaloTag ligand). To streamline the process even further, the DNA and transcription/translation reagents can be spotted on the slide simultaneously.

An alternative method, first reported in 2004, is the **nucleic acid programmable protein array** (**NAPPA**) in which biotinylated cDNA plasmids encoding GST fusion proteins are printed onto an avidin-coated slide together with an anti-GST antibody. The addition of rabbit reticulocyte lysate results in protein expression, and the proteins are trapped *in situ* by the antibodies. This has been used to construct arrays containing more than 1000 different proteins.

The most recent development is the **DNA array to protein array** (**DAPA**) method, which is similar in principle to replica plating in that it can be used to create multiple copies of the same protein array without leaving a parallel DNA imprint. Cell-free protein synthesis is carried out on a semi-permeable membrane sandwiched between two microscope slides, one of which is arrayed with DNA constructs and the other of which is coated with a reagent to bind the recombinant proteins. The proteins diffuse across the membrane and become fixed to the coated slide, whereas the immobilized DNA is unable to cross the barrier. The proteins are immobilized in the same register as the DNA array, and the same DNA array can be used up to 20 times to generate replicas.

The PISA, NAPPA, and DAPA procedures are compared in **Figure 9.3**. There are also many emerging techniques for cell-free protein synthesis at various stages of development, including those based on inteins, ribosome display, and photocleavable tags.

9.5 THE MANUFACTURE OF FUNCTIONAL PROTEIN MICROARRAYS—PROTEIN IMMOBILIZATION

The performance of a protein microarray depends not only on the quality of the proteins attached to its surface, but also on the nature of the substrate, which determines how the proteins are attached and whether they maintain their structure and biochemical activity. The problem of maintaining proteins in an active state is difficult to address because of their diverse properties. DNA microarrays have been successful due to the similar properties of different DNA sequences, which makes them suitable for multiplex hybridization under a universal set of hybridization conditions. Nucleic acids can also be labeled uniformly, and can be tethered to solid substrates without interfering with their binding capabilities. In contrast, proteins are chemically and physically diverse, so it is difficult to envisage a universal set of conditions under which all proteins would adopt their native conformations, interact with their normal physiological ligands, and display their normal biological activities.

Most protein microarrays are based on coated glass slides, which are inexpensive and compatible with standard microarrayer contact printing apparatus. The surface of the slide is coated with a substance that either absorbs the arrayed proteins in a passive manner or allows proteins to be cross-linked to the reactive surface. Nitrocellulose, agarose, and poly-L-lysine fall into the former category, and nitrocellulose sheets can also be used as supports in their own right (**Table 9.1**). Cross-linking surfaces have been developed that provide aldehyde, amino, or epoxy groups that react with the primary

FIGURE 9.3 Comparison of the (a) PISA, (b) NAPPA, and (c) DAPA methods for *in situ* protein synthesis on microarrays. (From Huang W, Wu L, Lu G & Lui S (2010) *Front. Biol.* 5, 331. With permission from Springer.)

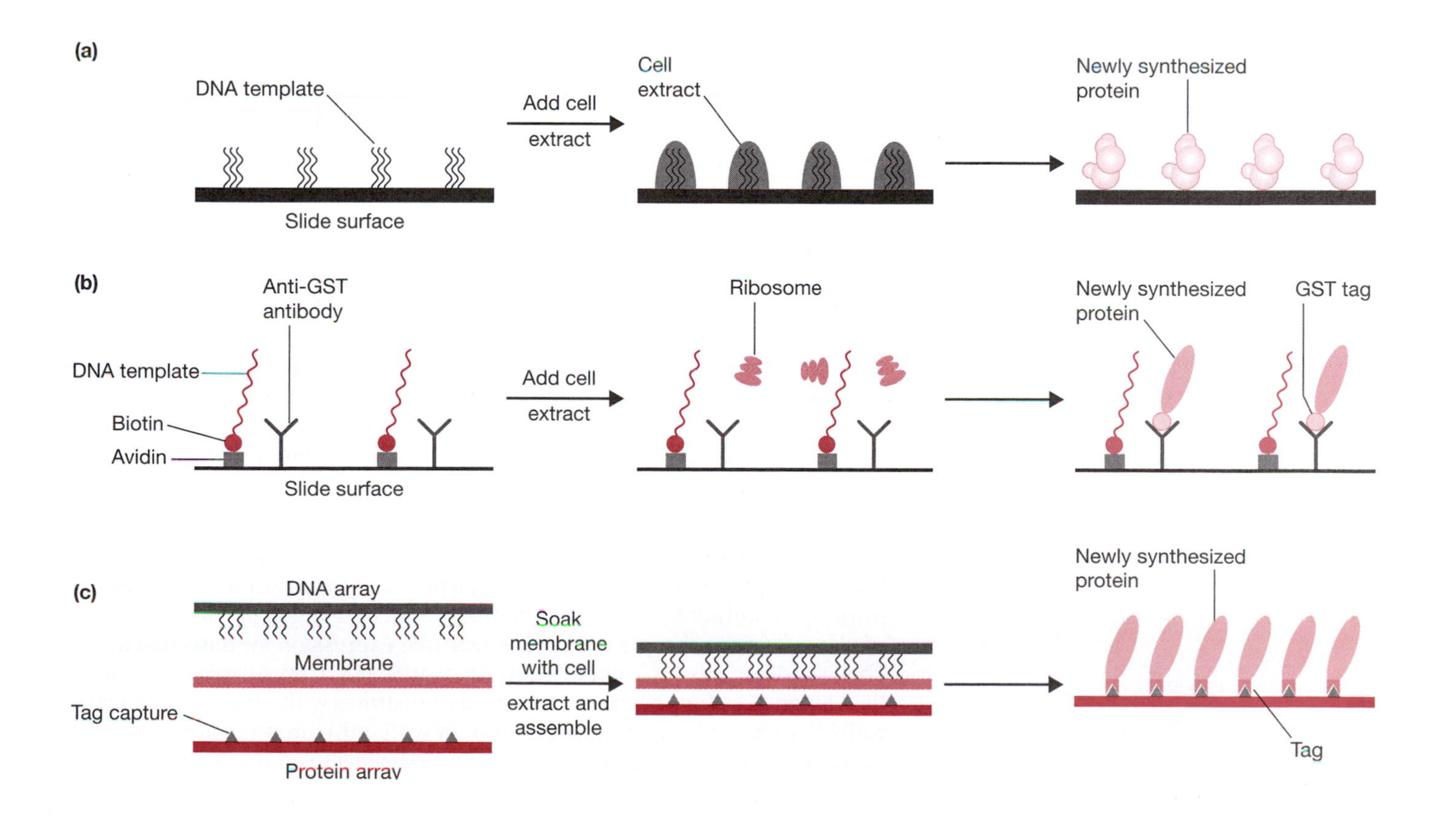

TABLE 9.1 PROPERTIES OF DIFFERENT SURFACE CHEMISTRIES USED FOR PROTEIN MICROARRAYS

Surface	Attachment	Advantage	Disadvantage
PVDF	Adsorption and absorption	No protein modification requirement, high protein-binding capacity	Nonspecific protein attachment in random orientation
Nitrocellulose	Adsorption and absorption	No protein modification requirement, high protein-binding capacity	Nonspecific binding, high background Low-density arrays
Poly-L-lysine coated	Adsorption	No protein modification requirement	Nonspecific adsorption
Aldehyde-activated	Covalent cross-linking	Strong and high-density protein attachment High-resolution detection methods available	Random orientation of surface-attached proteins
Epoxy-activated	Covalent cross-linking	Strong and high-density protein attachment High-resolution detection methods available	Random orientation of surface-attached proteins
Avidin-coated	Affinity binding	Strong, specific, and high-density protein attachment, low background	Proteins have to be biotinylated
Ni-NTA-coated	Affinity binding	Strong, specific, and high-density protein attachment, low background, uniform orientation of surface-attached proteins	Proteins have to be His_6-tagged
Gold-coated silicon	Covalent cross-linking	Strong and high-density protein attachment, low background. Can be easily coupled with SPR and mass spectrometry	Random orientation of surface-attached proteins, tough to fabricate, not commercially available
PDMS nanowell	Covalent cross-linking	Strong and high-density protein attachment, well suited for sophisticated biochemical analyses	Random orientation of surface-attached proteins
3D gel pad and agarose thin film	Diffusion	High protein-binding capacity, no protein modification requirement	Tough to fabricate, not commercially available
DNA/RNA-coated	Hybridization	Strong, specific, and high-density protein attachment, low background, uniform orientation of surface-attached proteins	Sophisticated *in vitro* production of labeled proteins

PVDF, polyvinylidene difluoride; NTA, nitrilotriacetate; PMDS, polydimethylsiloxane.
From Zhu H & Snyder M (2003) *Curr. Opin. Chem. Biol.* 7, 55. With permission from Elsevier.

amines of deposited proteins. Bifunctional thio-alkylene has been used to cross-link proteins to gold-coated slides, which are required for detection by surface plasmon resonance spectroscopy (see below). Proteins that adsorb passively or by cross-linking are arranged in a random manner and it is likely that a relatively large proportion of the immobilized proteins in both cases are either inaccessible or inactive due to linkages affecting important epitopes or functional residues. This problem can be avoided by using affinity tags to attach proteins to the surface. For example, in order to produce an antibody array with the antigen-binding domains of the antibodies exposed to the analyte, a recombinant staphylococcal protein A with five IgG-binding domains was covalently attached to the surface of a gold-coated slide. Antibodies bind to protein A via the Fc region; therefore, at least one antigen-binding domain should be accessible to the solvent. A similar principle applies to the use of His_6-tagged proteins, which can be immobilized in an oriented fashion to a chip surface coated with Ni-NTA resin (**Figure 9.4**).

One disadvantage of coated glass slides is the tendency for the protein solutions to evaporate during manufacture, a problem that must be addressed by printing in humidity-controlled chambers and/or using high concentrations of glycerol in the sample buffer. The cell-free expression systems discussed above address this problem to a certain degree, but array formats have been developed to reduce evaporation such as substrates with etched channels or wells (for example, the **nanowell chip**, in which the proteins are deposited into depressions in a polydimethylsiloxane surface fixed on the glass slide),

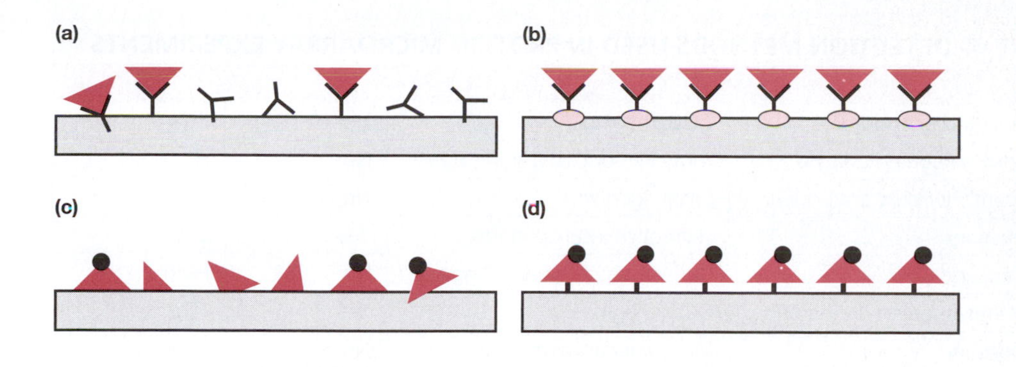

FIGURE 9.4 The importance of surface chemistry. (a) Antibodies will bind haphazardly to many surfaces via nonspecific bonds. Therefore, many of the antigen-binding domains are likely to be inactivated or obscured. (b) Coating the chip initially with staphylococcal protein A allows the antibodies to attach uniformly via the Fc regions, exposing the antigen-binding domain to the analyte. (c) Similarly, proteins on functional chips are often inactivated due to random binding. (d) The use of His_6-tagged proteins allows them to bind uniformly to a nickel-coated chip.

polyacrylamide **gel pads** printed on the surface of the slide (which provide a porous, three-dimensional matrix into which the protein can diffuse), and most recently bead- and particle-based arrays, which have the additional advantage that they can be sorted using microfluidics and analyzed by techniques such as fluorescence-activated cell sorting.

The techniques for applying proteins to microarray substrates have also advanced over the last 10 years, increasing the resolution of the devices and therefore allowing more proteins to be fitted onto smaller devices. As well as the typical robotics that create spots by **capillary transfer** (contact printing), more recent methods include **ink-jet printing** (a noncontact printing method in which small droplets of liquid are sprayed onto the substrate) and a piezoelectric method known as **tipnology**, which can accurately deliver as little as 333 picoliters of liquid. Proteins can also be deposited by **electrospray**, the same technique used to inject samples into ESI mass spectrometers. One of the most accurate methods is **dip-pen lithography**, in which an atomic force microscope probe is used to deliver protein solutions to nanometer-sized depressions generated by a focused particle beam.

9.6 THE DETECTION OF PROTEINS ON MICROARRAYS

Protein microarrays are used primarily to detect and characterize binding events, which may involve protein–protein, protein–nucleic acid, or protein–small-molecule interactions. In each case, the binding event generates a signal that must be detected and, ideally, quantified.

There are three broad classes of detection methods that can be used with protein microarrays (Table 9.2). In the first method, the analyte is labeled universally, either with a radioisotope or with an enzymatic or fluorescent conjugate, and the signal is detected directly from the bound analyte molecules on the array. In the second method, the analyte is not labeled, but a sandwich reaction is used to detect molecules bound to the array and the signal is produced by the labeled detection reagent. Finally, several label-free methods can be used to detect and/or identify proteins bound to capture reagents on the array surface. The relative merits of these different strategies depend on the sensitivity required and the effect of the labeling strategy on the interactions detected on the microarray.

Methods that require labels can involve either direct or indirect detection

Although radiolabels and colorimetric assays can be used with low-density protein microarrays, fluorescent labels are safer and more sensitive, and provide a greater spatial resolution, and they also provide the convenience of multiplex analysis using fluorescent labels with different spectral qualities. **Organic fluorophores** are widely used, but fluorescent inorganic **quantum**

TABLE 9.2 SUMMARY OF CURRENT DETECTION METHODS USED IN PROTEIN MICROARRAY EXPERIMENTS

Detection	Probe labeling	Data acquisition	Real time	Resolution
ELISA	Enzyme-linked antibodies	CCD imaging	No	Low
Isotopic labeling	Radioisotope-labeled analyte	X-ray film or phosphorimager	No	High
Sandwich immunoassay	Fluorescently labeled antibodies	Laser scanning	No	High
SPR	Not necessary	Refractive index change	Yes	Low
Noncontact AFM	Not necessary	Surface topological change	No	High
Planar waveguide	Fluorescently labeled antibodies	CCD imaging	Yes	High
SELDI	Not necessary	Mass spectrometry	No	Low
Electrochemical	Metal-coupled analyte	Conductivity measurement	Yes	Medium

ELISA, enzyme-linked immunosorbent assay; CCD, charge-coupled device; SPR, surface plasmon resonance; SELDI, surface-enhanced laser desorption/ionization.
From Zhu H & Snyder M (2003) *Curr. Opin. Chem. Biol.* 7, 55. With permission from Elsevier.

dots have also been conjugated to antibodies and are much brighter and more resistant to photobleaching. A fluorescent label can be incorporated directly into the analyte or into a secondary detection reagent that is applied once the microarray has been washed to remove unbound proteins. There are advantages and disadvantages to both methods.

The advantages of **direct labeling** are that protein detection and quantitation can be carried out in a single-step reaction and multiplex analysis is possible if required (Box 9.1). One disadvantage of direct labeling is that it lacks sensitivity compared with sandwich assays, although this can be addressed by introducing an intrinsic amplification step—for example, with avidin and biotin. A further disadvantage of direct labeling is that not all proteins are labeled with the same efficiency, and the label itself can alter the structure of some proteins and interfere with their binding capabilities. Direct labeling is therefore used most often with reversed-phase microarrays, where the complex substrate is immobilized and the labeled detection reagent is a single antibody whose behavior with and without the attached label can be verified. In forward assays, with individual proteins arrayed on the device and a complex analyte, the unpredictable consequences of labeling a large number of diverse proteins favors the indirect sandwich assay, particularly if quantitative data are required. However, the main disadvantage of sandwich assays is the requirement for two antibodies recognizing different epitopes for each antigen captured on the microarray, a problem that becomes more prevalent if there are hundreds or even thousands of proteins to detect.

A variation of the sandwich assay is the **immuno-RCA technique**, which involves a tertiary level of detection by **rolling circle amplification**. The principle is that a protein, captured by an immobilized antibody, is recognized by a second antibody in a sandwich assay as above, but the second antibody has an oligonucleotide covalently attached to it (**Figure 9.5** and color plates). In the presence of a circular DNA template, a strand-displacing DNA polymerase, and the four dNTPs, rolling circle amplification of the template occurs, resulting in a long concatemer comprising hundreds of copies of the circle, which can be detected using a fluorescent oligonucleotide probe.

Label-free methods do not affect the intrinsic properties of interacting proteins

Label-free methods use the intrinsic properties of proteins to report binding events on protein microarrays. Ciphergen's ProteinChips (Box 9.1) can be

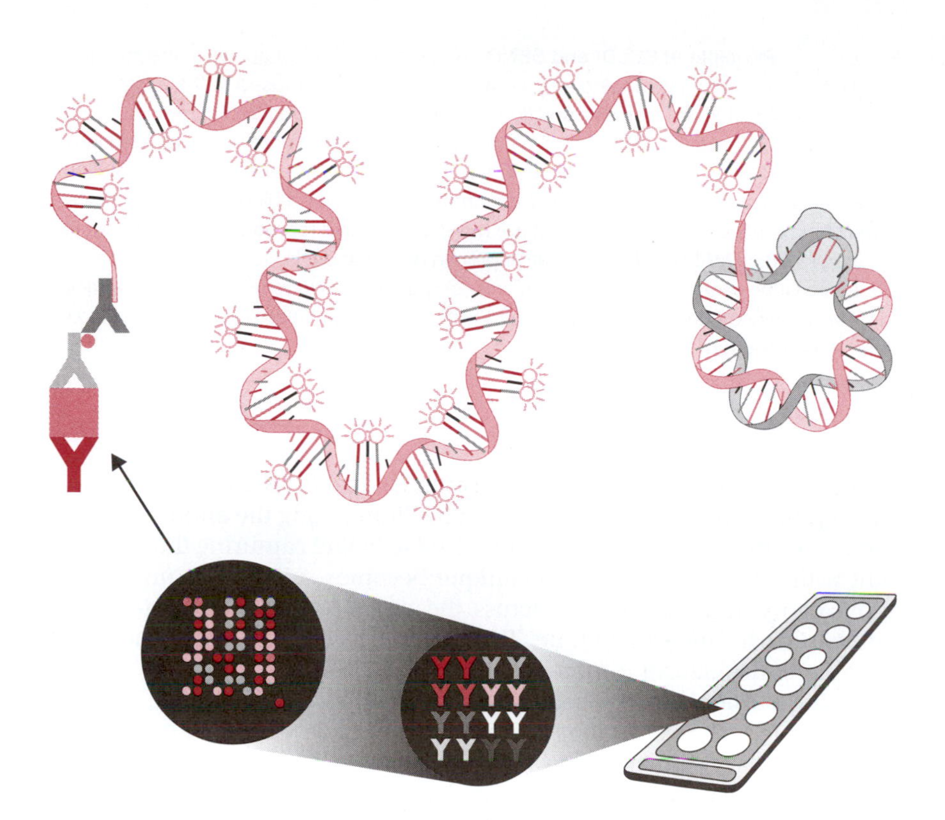

FIGURE 9.5 Sensitive protein detection using the RCA antibody chip. The chip is divided into 16 Teflon wells, each containing an array of 256 antibodies as probes. When a protein, represented by the *pink* square, is captured by one of the probes (*red*), it can be recognized using a second, biotinylated antibody (*light gray*), which is subsequently detected by a tertiary universal antibody (*dark gray*) connected to a circular oligonucleotide. A strand-displacing DNA polymerase can use this circular template, generating a long concatemer. See also color plates. (From Kingsmore SF & Patel DD (2003) *Curr. Opin. Biotechnol.* 14, 74. With permission from Elsevier.)

used as MALDI plates and scanned directly by MALDI-TOF mass spectrometry to reveal bound proteins and, if possible, identify them by peptide mass fingerprinting (p. 58). The ionization of proteins bound to ProteinChips is enhanced by the properties of the chip surface, leading to more uniform mass spectra than can be obtained with standard MALDI-MS analysis, a phenomenon described as **surface-enhanced laser desorption/ionization (SELDI)** (**Figure 9.6**). The quality of mass spectra is improved even further by incorporating the matrix compound into the chip surface, hence **surface-enhanced neat desorption (SEND)**.

Biacore Inc. produces a range of devices called Flexchips on which protein interactions can be detected by changes in **surface plasmon resonance (SPR)**. This is an optical effect that occurs when monochromatic polarized light is reflected from thin metal films. Some of the incident light energy interacts with the **plasmon** (the delocalized electrons in the metal), which results in a slight reduction in reflected light intensity. The angle of incidence at which this shadowing effect occurs is determined by the material adsorbed onto the metal film, which in this case would be one or more proteins (**Figure 9.7**). There is a direct relationship between the mass of the immobilized molecules and the change in resonance energy at the metal surface, which can be used to study interactions in real time. Put more simply, when light is shone on a gold-coated glass chip from underneath, the angle of incidence that induces SPR will change when molecules bind to the chip surface, and the change will reflect the size of the interacting molecule. A related technique is **ellipsometry**, in which the polarization state of the reflected light is altered due to changes in dielectric property or refractive index of the sample surface. This is often measured as the **oblique incidence reflectivity difference**.

The direct coupling of SPR spectroscopy and mass spectrometry allows the characterization of interaction kinetics and then the identification of

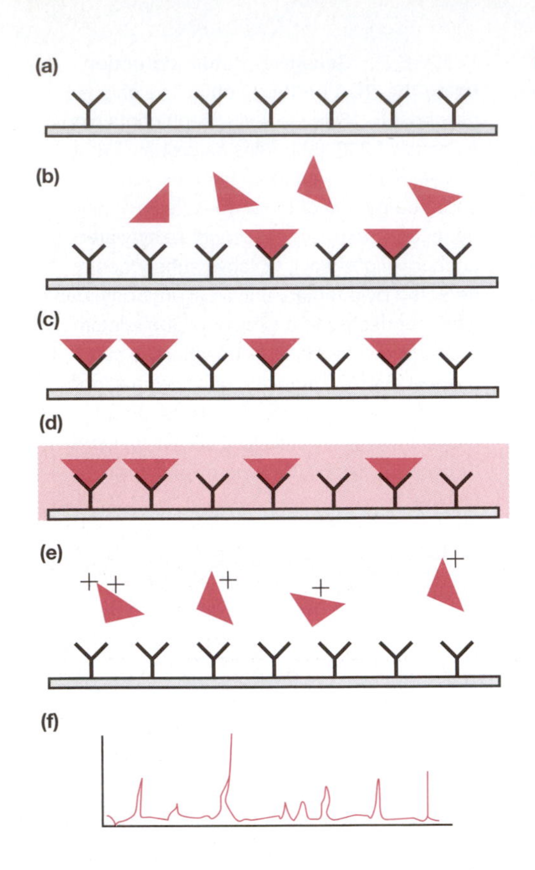

FIGURE 9.6 Principle of SELDI and SEND analysis. (a) In SELDI analysis, the protein chip (in this case preactivated and arrayed with antibodies) is exposed to the analyte (b) and captures antigens from the solution (c). The chip is then washed to remove unbound proteins and is coated with the matrix compound (d) before being inserted into the MALDI source of a mass spectrometer. A pulsed laser beam causes the captured proteins to ionize (e), producing a mass spectrum (f). The uniform binding of the proteins to the chip produces mass spectra that are more uniform and reproducible than possible with conventional MALDI, allowing relative protein quantitation. In SEND analysis, the procedure is the same except that the matrix compound is included in the chip's surface chemistry. Therefore there is no need to add further matrix and matrix ions do not appear in the mass spectrum.

interacting proteins. The typical SPR setup is ideal for the analysis of individual protein–protein interactions, but by illuminating the entire chip with a broad beam of monochromatic, polarized light and capturing the reflected light with a CCD camera, the technique becomes suitable for microarray applications because changes across the entire surface can be recorded simultaneously and in real time. This approach is sometimes termed **SPR imaging** to distinguish it from conventional SPR. The effect can be amplified using **nanohole arrays** because the periodic nanoholes couple incident photons into surface plasmons. A variant of this method uses the evanescent field of titanium pentoxide (Ti_3O_5) film to excite the fluorophores of labeled proteins bound to an array surface, which allows the detection of bound proteins without a washing step. Another emerging label-free detection method is **interferometry**, which measures the transformation of wave-front phase differences into observable intensity fluctuations known as interference fringes. Among many variants of this approach, **spectral reflectance imaging biosensing** (**SRIB**), **on-chip interferometric backscatter detection**, and **arrayed imaging reflectometry** are the most applicable to protein microarrays.

Finally, **atomic force microscopy** can detect protein interactions, albeit indirectly, if a binding event on the surface of a protein chip causes a change in surface topology. For example, this method has been used to detect the binding of an antibody immobilized on a thin gold film to a complementary

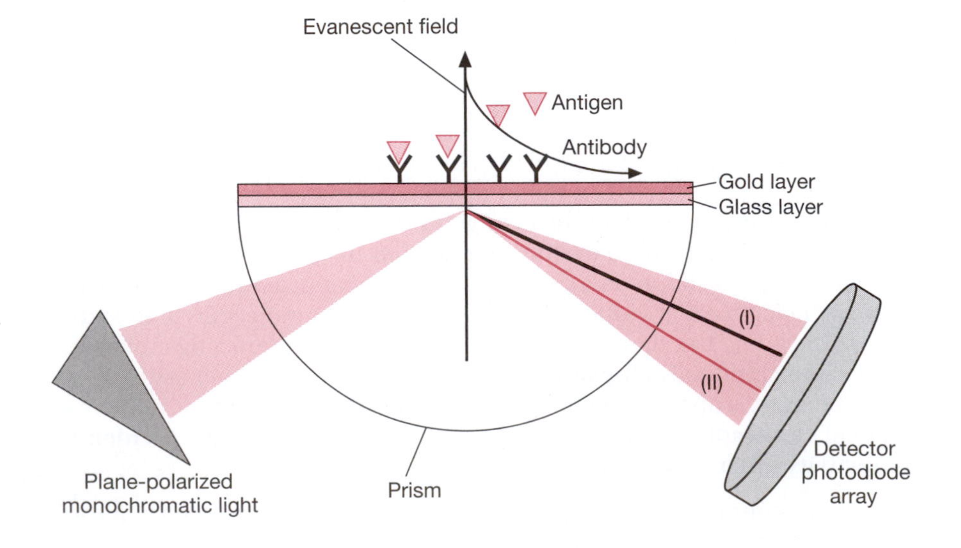

FIGURE 9.7 The principle of surface plasmon resonance spectroscopy. Plane-polarized light is incident on a gold-coated glass chip containing immobilized antibodies (or other capture agents). A change in mass at the surface, caused by antigen binding, causes a change in the refractive index and thus the resonance state. This is reported by a change in the angle of the reflected light (I to II), which can be detected using a photodiode array. (From Leonard P, Hearty S, Brennana J et al. (2003) *Enzyme Microb. Technol.* 32, 3. With permission from Elsevier.)

anti-IgG by virtue of the change in height. A scanning Kelvin probe force microscope has also been used to detect regional variations in the surface potential of a protein chip.

9.7 EMERGING PROTEIN CHIP TECHNOLOGIES

Bead and particle arrays in solution represent the next generation of protein microarrays

The protein arrays discussed thus far in the chapter are two-dimensional devices that are manufactured with a fixed number of probes in specific positions. Each individual probe is identified by its spatial address on the array. A disadvantage of this format is the lack of flexibility—the only way to incorporate new probes is to fabricate a new array. One way around this problem is to release the array from its two-dimensional format and instead use beads in solution as the probes. **Bead arrays** (or **solution arrays**) have all the advantages of solid-phase arrays in terms of throughput and sensitivity, but have improved solution kinetics and are much more flexible in terms of the number of probes that can be used. The question is how to identify specific probes when they are free in the solution. Several answers have been put forward, including the use of fluorescence-encoded beads and barcoded metal particles.

Fluorescence encoding makes use of multiple fluorescent dyes at different concentrations to provide a unique **spectral fingerprint** for each bead. For example, the use of 10 different fluorescent dyes would provide 10 unique labels, but 10 different dyes at 10 different concentrations would provide 100 unique labels. Over 100 different fluorescent dyes are currently available and, by using a range of different concentrations, the system becomes rapidly scalable. An example of this approach is the Qiagen/Luminex LiquiChip **protein suspension array**.

Barcoding is a similar strategy in which metal particles, functionalized to accept antibodies or other proteins, are produced with a unique set of gold, silver, and platinum stripes. The stripes are generated by carrying out electrochemical reductions with different metals and the stripe width can be varied by controlling the current. This system also has potentially unlimited scale, the decoding step can be automated, and the particles can be combined with downstream analysis by fluorescence imaging or mass spectrometry.

Cell and tissue arrays allow the direct analysis of proteins *in vivo*

A lot of useful functional information is lost when proteins are extracted into solution for further analysis, including their spatial distribution and subcellular location. As discussed in Chapter 7, large-scale studies of protein localization have been carried out using a procedure in which adherent mammalian cells are grown on microarrays of cDNA expression constructs that have been treated with a lipid transfection reagent to promote DNA uptake. On such **cell microarrays**, cells growing immediately above each DNA spot are able to express the cDNA and produce the protein, whose behavior in terms of subcellular distribution, oscillation in the context of the cell cycle, and response to external stimuli can be observed by immunocytochemical staining and fluorescence microscopy (Figure 9.8).

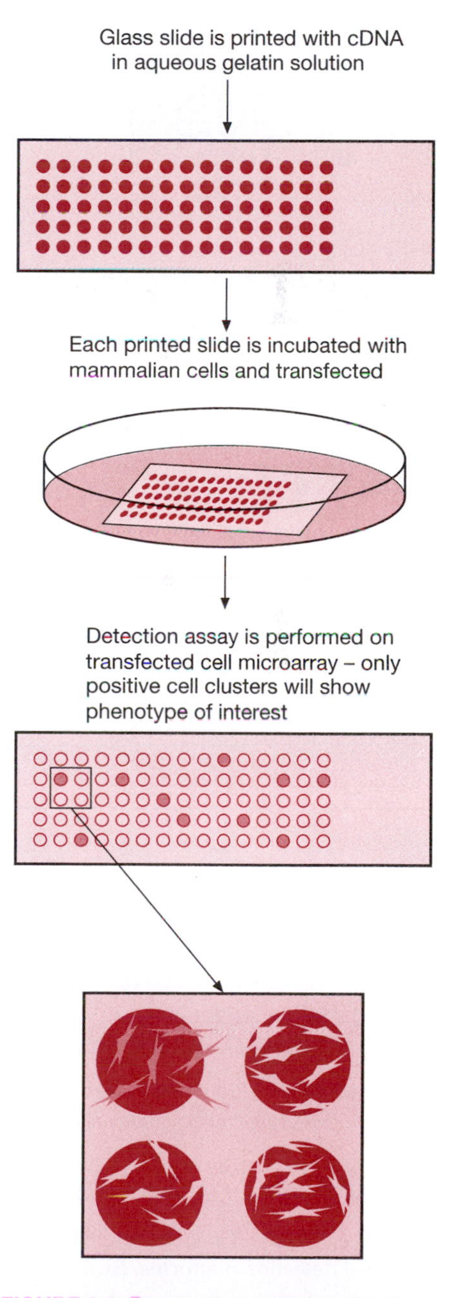

FIGURE 9.8 Preparation and use of cell arrays. (From Howbrook DN, van der Valk AM, O'Shaughnessy MC et al. (2003) *Drug Discov. Today* 8, 642. With permission from Elsevier.)

Information about a protein's spatial distribution can also be obtained by conventional histological examination, but this takes a long time and cannot be applied in a high-throughput manner. **Tissue microarrays** resolve this problem by placing large numbers of tissue samples on microscope slides and allowing highly parallel *in situ* detection methods to be applied under a constant set of conditions. Tissue microarrays are constructed by taking **needle biopsies** of different specimens and embedding these long, cylindrical tissue samples into blocks of paraffin in an array structure. The block of paraffin is then cut into sections (up to 300 per block), producing many copies of the same array (**Figure 9.9**). Advantages of this approach include the use of small analyte volumes and the fact that minimal damage is caused to the original specimen, which can also be subjected to conventional analysis if required.

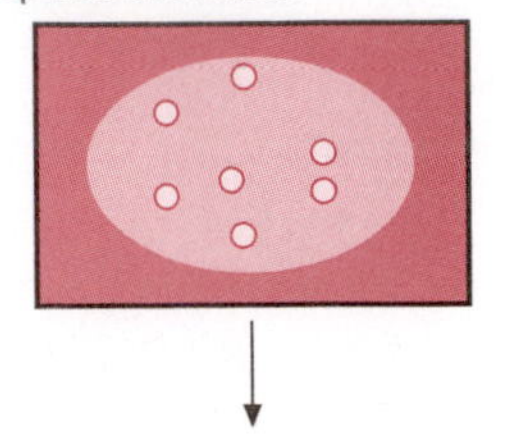

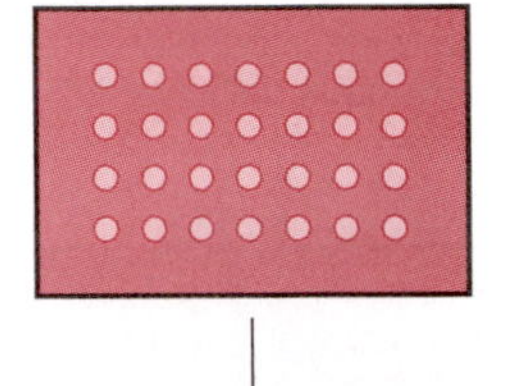

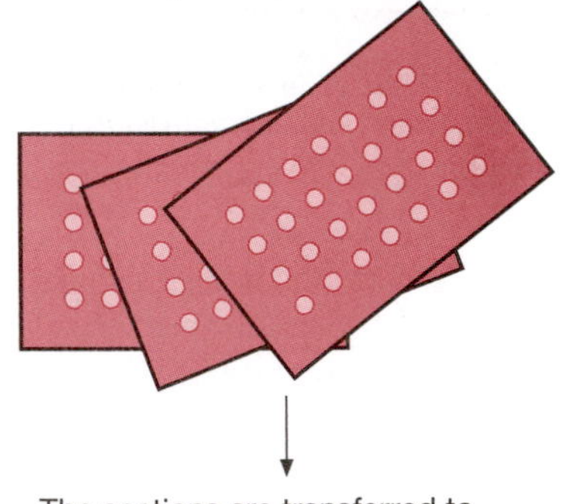

FIGURE 9.9 Preparation and use of tissue arrays. (From Howbrook DN, van der Valk AM, O'Shaughnessy MC et al. (2003) *Drug Discov. Today* 8, 642. With permission from Elsevier.)

FURTHER READING

Austin J & Holway AH (2011) Contact printing of protein microarrays. *Methods Mol. Biol.* 785, 379–394.

Braun P & LaBaer J (2003) High throughput protein production for functional proteomics. *Trends Biotechnol.* 21, 383–388.

Campbell DA & Szardenings AK (2003) Functional profiling with affinity labels. *Curr. Opin. Chem. Biol.* 7, 296–303.

Chandra H, Reddy PJ & Srivastava S (2011) Protein microarrays and novel detection platforms. *Expert Rev. Proteomics* 8, 61–79.

Chandra H & Srivastava S (2010) Cell-free synthesis-based protein microarrays and their applications. *Proteomics* 10, 717–730.

Chen R & Snyder M (2010) Yeast proteomics and protein microarrays. *J. Proteomics* 73, 2147–2157.

Cretich M, Monroe MR, Reddington A et al. (2012) Interferometric silicon biochips for label and label-free DNA and protein microarrays. *Proteomics* 12, 2963–2977.

Cutler P (2003) Protein arrays: the current state-of-the-art. *Proteomics* 3, 3–18.

de Wildt RMT (2000) Antibody arrays for high throughput screening of antibody antigen interactions. *Nat. Biotechnol.* 18, 989–994.

Fang Y, Lahiri J & Picard L (2003) G protein-coupled receptor microarrays for drug discovery. *Drug Discovery Today* 8, 755–761.

Ge H (2000) UPA, a universal protein array system for quantitative detection of protein-protein, protein-DNA, protein-RNA and protein-ligand interactions. *Nucleic Acids Res.* 28, e3i–e3vii.

Gershon D (2003) Proteomics technologies: probing the proteome. *Nature* 424, 581–587.

Gonzalez-Gonzalez M, Jara-Acevedo R, Matarraz S et al. (2012) Nanotechniques in proteomics: protein microarrays and novel detection platforms. *Eur. J. Pharm. Sci.* 45, 499–506.

Haab BB, Dunham MJ & Brown PO (2001) Protein microarrays for highly parallel detection and quantification of specific proteins and antibodies in complex solutions. *Genome Biol.* 2, 0004.1–0004.13.

Hahm JI (2011) Polymeric surface-mediated, high-density nano-assembly of functional protein arrays. *J. Biomed. Nanotechnol.* 7, 731–742.

Hall DA, Ptacek J & Snyder M (2007) Protein microarray technology. *Mech. Ageing Dev.* 128, 161–167.

He M, Stoevesandt O & Taussig MJ (2008) *In situ* synthesis of protein arrays. *Curr. Opin. Biotechnol.* 19, 4–9.

Hu S, Xie Z, Qian J et al. (2011) Functional protein microarray technology. *Wiley Interdiscip. Rev. Syst. Biol. Med.* 3, 255–268.

Kingsmore SF & Patel DD (2003) Multiplexed protein profiling on antibody-based microarrays by rolling circle amplification. *Curr. Opin. Biotechnol.* 14, 74–81.

Liu B, Huang L, Sihlbom C et al. (2002) Towards proteome-wide production of monclonal antibody by phage display. *J. Mol. Biol.* 315, 1063–1073.

Lopez MF & Pluskal MG (2003) Protein micro- and macroarrays: digitising the proteome. *J. Chromatogr. B Analyt. Technol. Biomed. Life Sci.* 787, 19–27.

MacBeath G (2002) Protein microarrays and proteomics. *Nat. Genet.* 32 (Suppl), 526–532.

MacBeath G & Schreiber SL (2000) Printing proteins as microarrays for high-throughput function determination. *Science* 289, 1760–1763.

Martzen MR, McCraith SM, Spinelli SL et al. (1999) A biochemical genomics approach for identifying genes by the activity of their products. *Science* 286, 1153–1155.

Mehan MR, Ostroff R, Wilcox SK et al. (2013) Highly multiplexed proteomic platform for biomarker discovery, diagnostics, and therapeutics. *Adv. Exp. Med. Biol.* 735, 283–300.

Niu Y & Jin G (2011) Protein microarray biosensors based on imaging ellipsometry techniques and their applications. *Protein Cell* 2, 445–455.

Phizicky E, Bastiaens PI, Zhu H et al. (2003) Protein analysis on a proteomic scale. *Nature* 422, 208–215.

Ray S, Mehta G & Srivastava S (2010) Label-free detection techniques for protein microarrays: prospects, merits and challenges. *Proteomics* 10, 731–748.

Sreekumar A, Nyati MK, Varambally S et al. (2001) Profiling of cancer cells using protein microarrays: discovery of novel, radiation-regulated proteins. *Cancer Res.* 61, 7585–7593.

Stoevesandt O, Taussig MJ & He M (2009) Protein microarrays: high-throughput tools for proteomics. *Expert Rev. Proteomics* 6, 145–157.

Voshol H, Ehrat M, Traenkle J et al. (2009) Antibody-based proteomics: analysis of signaling networks using reverse protein arrays. *FEBS J.* 276, 6871–6879.

Wiese R, Belosludtsev Y, Powdrill T et al. (2001) Simultaneous multianalyte ELISA performed on a microarray platform. *Clin. Chem.* 47, 1451–1457.

Wilson B, Liotta LA & Petricoin E 3rd (2010) Monitoring proteins and protein networks using reverse phase protein arrays. *Dis. Markers* 28, 225–232.

Zhu H, Bilgin M, Bangham R et al. (2001) Global analysis of protein activities using proteome chips. *Science* 293, 2101–2105.

Zhu H, Bilgin M & Snyder M (2003) Proteomics. *Annu. Rev. Biochem.* 72, 783–812.

Zhu H, Klemic JF, Chang S et al. (2000) Analysis of yeast protein kinases using protein chips. *Nat. Genet.* 26, 283–289.

CHAPTER 10

Applications of proteomics

10

10.1 INTRODUCTION

In the preceding chapters, we have discussed a range of different technologies that allow proteins to be studied on a large scale. These technologies can be applied in a number of different ways, some of which increase our basic scientific knowledge whereas others can be applied in fields such as medicine, agriculture, and industrial biotechnology.

The first type of application involves the acquisition of knowledge. This includes enumerating and identifying the proteins in cells and organisms as well as the determination of their sequences, structures, functions, interactions, modifications, expression profiles, subcellular localizations and modifications. It also includes linking the resulting proteomic datasets with equivalent large datasets delivered by the analysis of genomes (genomics), gene expression (**transcriptomics**), metabolite profiles (**metabolomics**), and the impact of perturbations caused by mutation, interference technologies, and environmental effects on phenotypes (**phenomics**). Together, these datasets constitute the empirical core of **systems biology**, in which computer modeling is used to propose specific testable hypotheses about complex biological systems (from simple genetic circuits through to entire cells, organisms, and even ecosystems) that can be experimentally validated and refined using the quantitative data from such large-scale biological datasets.

A second type of application involves the use of proteomics for detection and diagnosis. Here the aim is to find markers in individual samples or differences between related samples that provide indicators of a relevant process, for example, changes in protein abundance, localization, modification, or interactions that mirror the onset or progression of a disease. It also includes the detection of antigens and antibodies in blood samples, environment and food monitoring and the identification of microbes.

A third type of application involves the deployment of proteomics to facilitate discovery and invention. Here the aim is to develop a new product or process using evidence from proteomics to validate the approach. For example, proteomics is widely used in the development of new drugs because it can help to identify new drug targets and the most suitable pathogen proteins to target with vaccines. It can also help to develop new products for the protection of crops and food stocks, and new industrial processes, for example, the discovery of novel enzymes that improve the production of chemicals, fuels, and materials.

In this chapter, we briefly explore how proteomics can be applied to solve real problems, focusing on disease diagnosis and drug development because it is here that high-throughput studies of protein abundance and activity have made the largest impact. Indeed, most of the current investment in proteomics technology and research is funded by pharmaceutical companies, which see proteomics as a short cut to the development of novel drugs and diagnostics. Proteomics can play an important role throughout the pharmaceutical research and development value chain, and is particularly powerful when used in combination with other biological datasets (genome sequences, transcriptional profiles, single nucleotide polymorphism (SNP) catalogs, and mutant libraries). Proteomics can also be used to provide novel genetic markers for the rapid mapping of uncharacterized genomes. In this way, we come full circle from genomes to proteomes and back to genomes once again.

10.2 DIAGNOSTIC APPLICATIONS OF PROTEOMICS

Proteomics is used to identify biomarkers of disease states

A **biomarker** is a biological feature of a cell, tissue, or organism that corresponds to a particular physiological state. In a medical context, the most important biomarkers are those that appear or disappear specifically in the disease state (disease biomarkers) and those that appear or disappear in response to drugs (toxicity biomarkers, see later). There are many different types of disease biomarker, including the presence of particular pathogenic entities (**Box 10.1**), disease-specific cytological or histological characteristics, gene or chromosome mutations, the appearance of specific transcripts or proteins, new post-translational variants, or alterations in the level of mRNA or protein expression. Molecular biomarkers, such as mutations, transcripts, and proteins, are the most useful because they tend to appear well before the symptoms of the disease manifest, allowing early detection and prompt treatment. Furthermore, different biomarkers can sometimes be used to monitor the progress of a disease or its treatment.

As discussed in Chapter 1, proteins are advantageous biomarkers because the direct analysis of proteins can reveal characteristics, such as post-translational modifications, that cannot be identified by DNA sequencing or mRNA profiling. Perhaps more importantly, protein biomarkers can be assayed in body fluids, among which serum is the most valuable because it is in contact with all parts of the body and its composition is influenced by secretions or leakage from cells that are damaged by disease. Potential serum biomarkers for many types of disease have been discovered using different proteomics methods (**Table 10.1**). The ideal biomarker should highly specific for a certain disease condition, a feature that can only be established by extensive validation in a broad population. Unfortunately, such biomarkers are rare and most candidate biomarkers are found in many different types of disease, perhaps with different expression levels in each case. A combination of relatively nonspecific biomarkers can, in some cases, provide a more specific disease index, and proteomics is useful in this context since it allows the expression profiles of hundreds of proteins to be studied in parallel. Many licensed tests that use proteins for disease diagnosis are ELISA-based systems that exploit protein biomarkers found in easily accessible fluids so that the assay is noninvasive. In most cases, however, these assays have been developed after the fortuitous discovery of individual proteins that are overexpressed or ectopically expressed in the disease state. What proteomics can offer is the opportunity to compare the protein profiles of samples from healthy people and those with a given disease to identify protein biomarkers in a systematic fashion. Thus far, very few biomarkers discovered via proteomics have been introduced into the clinic, but this mainly reflects the long period of testing required for full validation.

BOX 10.1 CASE STUDY.
Identifying pathogens by mass spectrometry.

Typical proteomics experiments involve the careful preparation of samples and the analysis of proteins that have been isolated, solubilized, and digested *in vitro*. However, in parallel to the development of conventional proteomics, many of the techniques described in this book have also been used as diagnostic methods in their own right. For example, 2DGE and many of the associated bioinformatics methods were widely used to analyze clinical samples years before proteomics became a mainstream discipline, and similarly MALDI-TOF mass spectrometry has a long history of diagnostic use in microbiology in parallel to its development as a mainstream proteomics technology. This predominantly reflects its use in the development of **intact-cell mass spectrometry (ICMS)**, also known as **whole-cell mass spectrometry (WCMS)**.

The ICMS technique exploits the ability of MALDI-TOF MS to produce diagnostic fingerprints that are specific for different microbial species and strains without extensive preparation and without the need for protein purification and separation. Cells are cultured using conventional methods and then colonies are transferred to a MALDI plate and overlaid with the matrix compound for direct analysis.

Whereas conventional proteomics is often tailored to identify differences between similar samples (the dynamic proteome), ICMS instead focuses on the stable proteome, that is, the components that remain unchanged under different conditions. This is generally achieved by selecting an appropriate mass range for analysis (for example, 2–20 kDa selects for structural proteins, which tend to be stable, rather than enzymes and signaling proteins which, are more variable). Therefore, significant differences between samples represent static differences between structural components, which highlight differences between species and strains (**Figure 1**).

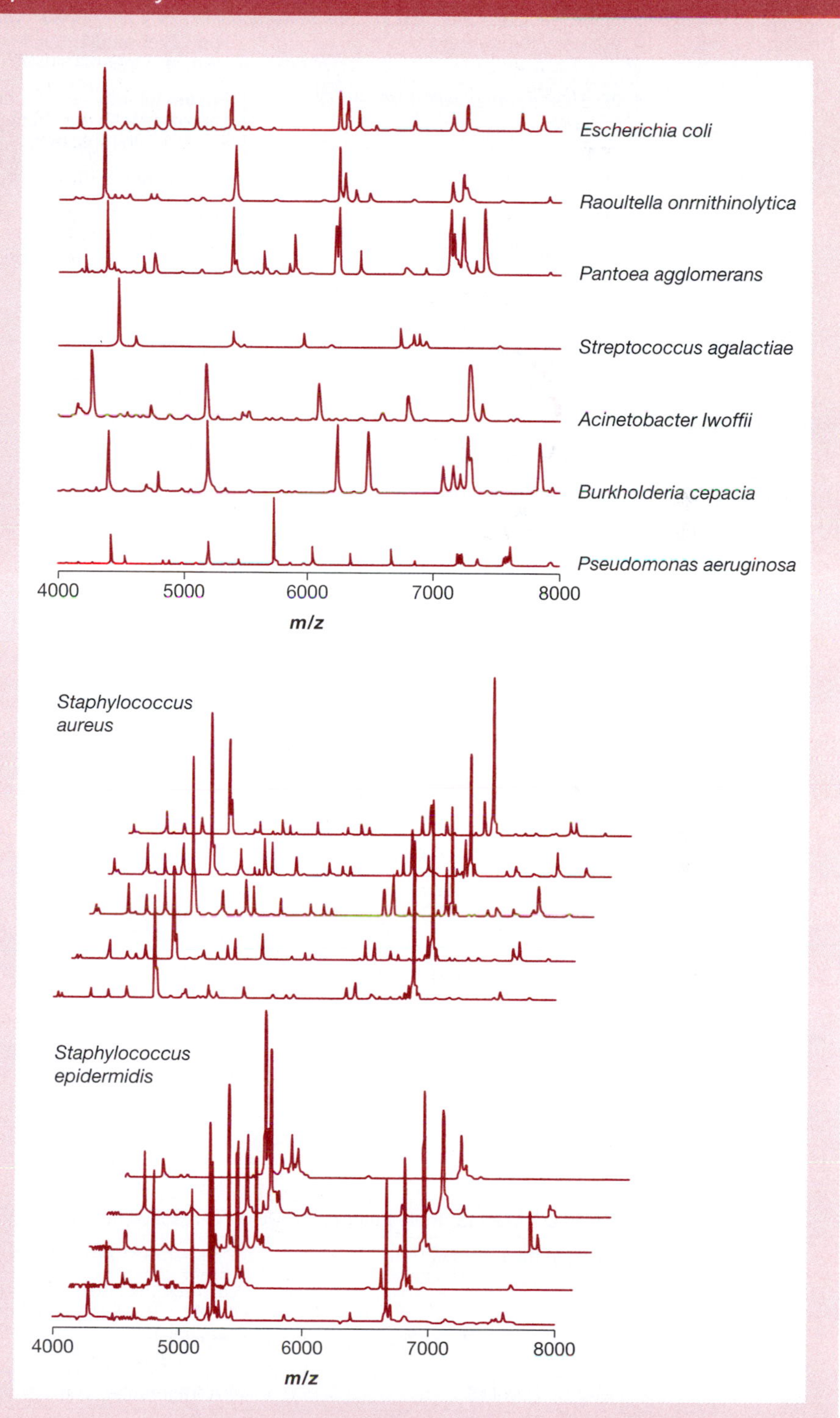

FIGURE 1 Distinguishing bacterial species and strains by MALDI-TOF mass spectrometry. The intact-cell spectra of different bacterial species are highly discriminatory, whereas those of different bacterial strains identify components that are common to the species and that differ according to the strain. (From Welker M (2011) *Proteomics* 11, 3143. With permission from John Wiley & Sons, Inc.)

TABLE 10.1 SERUM BIOMARKERS

Disease	Proteomic technique(s)	Serum biomarkers identified/detected
Cancers		
Prostate cancer	Antibody microarrays	Von Willebrand factor, immunoglobulin M (IgM), α_1-antichymotrypsin, villin, and IgG
	2D DIGE coupled with MALDI-TOF/TOF-MS	Epidermal fatty acid-binding protein 5, methylcrotonoyl-CoA carboxylase 2, palmitated protein A2, ezrin, stomatin-like protein 2, and smooth muscle 22
Colorectal cancer	2D DIGE	S100A8 and S100A9 (calgranulin A and B)
	2D DIGE	Transaldolase 1 and thyroid receptor interactor
	LC-MS/MS	Growth/differentiation factor 15 and trefoil factor 3
Liver cancer	Capillary-HPLC analysis, stable-isotope dilution-multiple reaction monitoring-MS	Clusterin and vitronectin
	2DGE coupled with MALDI-TOF/TOF	Vimentin
	2DGE coupled with MALDI-TOF/TOF	Heat shock protein 90
Pancreatic cancer	2DGE-MS	Cyclin I and Rab GDP dissociation inhibitor β
	Protein microarray	Phosphoglycerate kinase 1 and histone H4
Brain tumors	SELDI-TOF-MS, protein chips	Gliomas amplified sequence 64 and brain my035 protein
	SELDI-TOF-MS	α_2-Heremans–Schmid glycoprotein
Breast cancer	Nucleic acid programmable protein microarray (NAPPA)	TP53-specific antibodies
	MALDI-TOF-MS	CEA, CA15-3, cytokeratin fragment 21.1, leptin, and osteopontin
	Antibody microarray	Epidermal growth factor, soluble CD40 ligand and proapolipoprotein A1, kininogen, soluble vascular cell adhesion molecule 1, plasminogen activator inhibitor 1, vitamin D-binding protein, and vitronectin
Ovarian cancer	Protein arrays	Upstream stimulatory factor, cathepsin G, HLA-B-associated transcript 4, and zinc finger- and BTB domain-containing protein 22
	ESI MS/MS, reversed-phase protein microarray	S100A6 (calcyclin)
	2DGE, MS/MS, reversed-phase protein arrays	FK506-binding protein, Rho G-protein dissociation inhibitor, and glyoxalase I
	MALDI orthogonal TOF-MS	Complement C3 and inter-α (globulin) inhibitor H4, and single peptides from complement C4-A, transthyretin, and fibrinogen
Lung cancer	2DGE coupled with MALDI-TOF/TOF	Haptoglobin, transthyretin, and tumor necrosis factor superfamily member 8
Gastric cancer	SELDI-TOF-MS, antibody microarray	IPO-38 (H2B histone)
Autoimmune diseases		
Systemic lupus erythematosus	Protein microarrays	Cytokines, chemokines, growth factors, and soluble receptors
Rheumatoid arthritis	2D LC-MS/MS	C-reactive protein and six proteins from the S100 calcium-binding protein family
	Nano-LC-MS/MS, triple quadrupole MS	C-reactive protein
	Antigen microarrays	Tumor necrosis factor α, interleukin (IL)-1β, IL-6, IL-13, and IL-15
Multiple sclerosis	2DGE, MS	α_1-acid glycoprotein 1, α_1-B-glycoprotein, transthyretin, apolipoprotein C-III, serum amyloid P component, complement factor I, clusterin, gelsolin, hemopexin, kininogen-1, hCG1993037 isoform, and vitamin D-binding protein
	MALDI-TOF-MS	A fragment of complement C4 (1741 Da peptide)
Bowel diseases CD, ulcerative colitis	Nano-LC-MS	Complement C3 and C4A, fibrinogen α-chain, fibrinopeptide A, and apolipoprotein E
	SELDI-TOF-MS and antibody-based assay	Platelet aggregation factor 4, myeloid related protein 8, fibrinogen-binding peptide, and haptoglobin α_2
Ankylosing spondylitis	NAPPA	Autoantigens; glypican 3 and 4, connective tissue growth factor, osteonectin, melanocortin 4 receptor, chondromodulin1, matrix Gla protein, purinergic receptor, and extracellular matrix protein–SPARC-related modular calcium-binding protein 1

Disease	Proteomic technique(s)	Serum biomarkers identified/detected
Infectious diseases		
Tuberculosis	SELDI-TOF-MS	Amyloid A and transthyretin
Leprosy	2DGE, MALDI-TOF-MS/MS	Isoform of α_2-chain of haptoglobin
SARS	2DGE, MALDI-TOF-MS	Peroxiredoxin II
	2DGE, MALDI-TOF-MS	α_1-antitrypsin
Hepatitis	2DGE, LC/MS-MS	Haptoglobin β- and α_2-chains, apolipoprotein A-I and A-IV, α_1-antitrypsin, transthyretrin, and DNA topoisomerase IIβ
	2DGE, LC/MS-MS	α_2-macroglobulin, haptoglobin, albumin, complement C4, retinol-binding protein, and apolipoprotein A-I, and A-IV
Others		
Stroke	2DGE, MALDI-TOF-MS	Heart fatty acid-binding protein
	SELDI-TOF-MS, nano-LC/MS-MS	Apolipoprotein C-I and C-III, serum amyloid A, and antithrombin III fragment
Nonalcoholic fatty liver disease	LC/MS-MS	Fibrinogen B-chain, paraoxonase 1, prothrombin, complement C7, and serum amyloid P component
Diabetic nephropathy	2DGE, ESI-Q-TOF MS/MS	Extracellular glutathione peroxidase and apolipoprotein
Down syndrome	2D-DIGE, MALDI-TOF/MS, MudPIT-LC/LC-MS/MS	α_1-acid glycoprotein,α_2-antiplasmin precursor, antithrombin III, α_2-macroglobulin precursor, serum amyloid, inter-α-trypsin inhibitor heavy chain H4, afamin, and ceruloplasmin
	iTRAQ	Igλ-chain C region, serum amyloid P component, amyloid β A4, γ-actin, and titin
Sarcoidosis	SELDI-TOF-MS, 2DGE, MALDI-TOF	α-chain of haptoglobin
Schizophrenia	Label-free LC-MS/MS	Apolipoprotein A-I, and zinc finger- and BTB domain-containing protein 38

From Ray S, Reddy PJ, Jain R, et al. (2011) *Proteomics* 11, 2139. With permission from John Wiley & Sons, Inc.

Biomarkers can be discovered by finding plus/minus or quantitative differences between samples

Biomarker discovery was probably the first application envisaged for proteomics. As early as 1982, it was suggested that two-dimensional gels could be used to detect quantitative differences in protein profiles between healthy individuals and those suffering from particular diseases, although at the time there was no easy way to identify the differentially expressed proteins that were discovered. This all changed in the early 1990s with the advent of mass spectrometry techniques that allowed proteins to be identified by correlative database searching (Chapter 3). The combination of 2DGE and mass spectrometry soon became the standard way to find potential new protein biomarkers. An initial strategy was to compare silver-stained gels by eye or using visual analysis software. Spots that were present on one gel and absent on another, or spots that showed obvious quantitative differences between gels, were picked and analyzed by mass spectrometry. The proteins contained within the spots were thus identified, and their relative abundance in different samples was confirmed using other methods (**Figure 10.1**). This led to the discovery of numerous potential disease biomarkers, many of which offered the prospect of diagnosis for different forms of cancer, and also for cardiovascular disease, neurological disease, autoimmune and inflammatory diseases, and infectious diseases such as hepatitis.

Cancer has been the primary target for proteomic analysis because it is relatively easy to obtain matched samples of disease and healthy tissue from the same patient in sufficient amounts to carry out 2DGE (**Figure 10.2**). Good examples of this approach include the pioneering studies of Sam Hanash and colleagues that identified various biomarkers suitable for the diagnosis

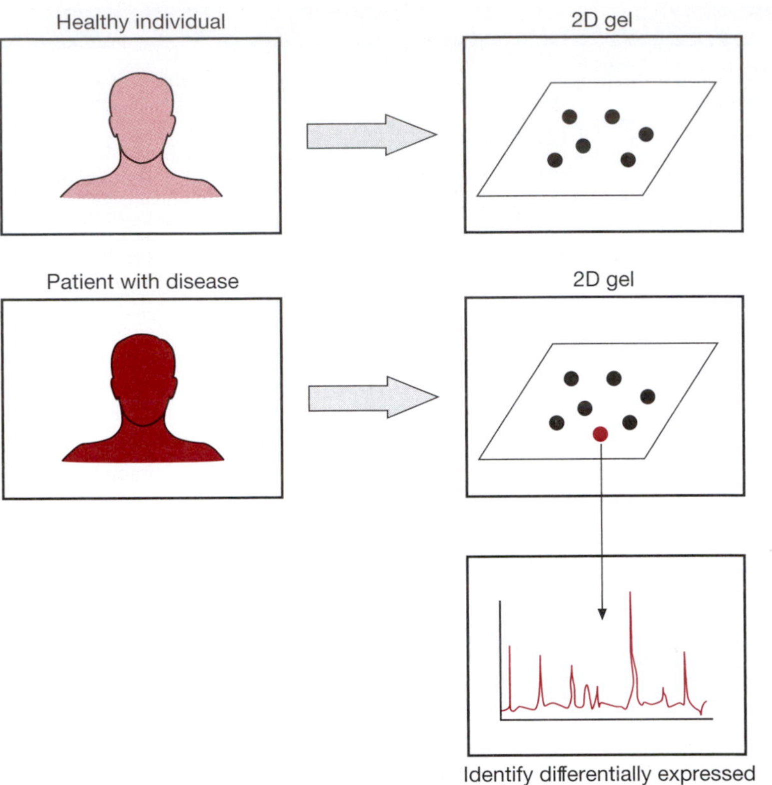

FIGURE 10.1 Identification of disease-specific biomarkers by 2DGE and mass spectrometry. After separation, proteins that are unique to the disease sample or significantly more abundant in the disease sample are selected for further characterization.

and classification of different forms of leukemia (see Further Reading). One such study identified the protein stathmin (otherwise known as oncoprotein 18), which functions as an intracellular signal relay in the transduction of growth factor signals, as a reliable biomarker for childhood leukemia. The interesting feature of this particular protein is that only the phosphorylated

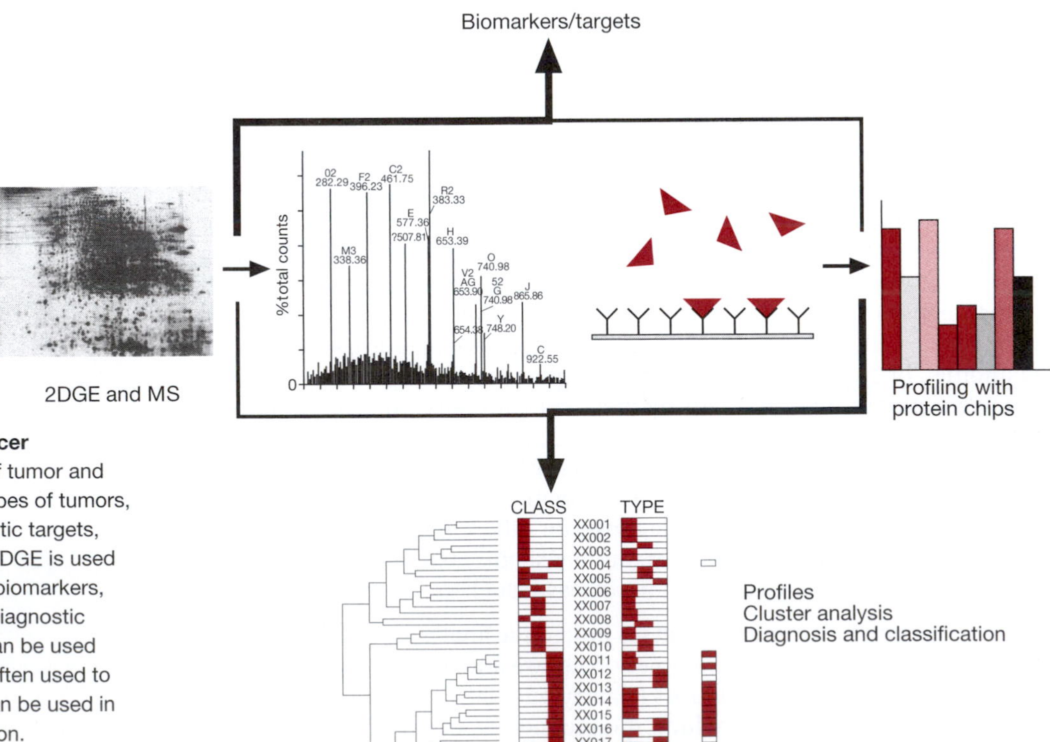

FIGURE 10.2 Overview of cancer proteomics. The comparison of tumor and non-tumor tissue, or different types of tumors, can reveal biomarkers, therapeutic targets, or diagnostic protein patterns. 2DGE is used predominantly to identify novel biomarkers, which are then developed into diagnostic immunoassays. Protein chips can be used to identify biomarkers, but are often used to generate pattern profiles that can be used in diagnosis and tumor classification.

form is implicated in the disease. Pioneering work was also carried out by Julio Celis and colleagues, who initially used 2DGE to study the changes in protein expression that occurred as cultured cells underwent growth transformation. The knowledge gained from this series of investigations was later applied to the analysis of bladder cancer, resulting in the discovery of several markers, including different forms of keratin, which can be used to follow the progression of the disease from normal epithelium through the early transitional epithelium stage to the late squamous cell carcinoma. Another protein, called psoriasin, is shed into the urine of squamous cell carcinoma patients and thus has the potential to be developed as a validated biomarker for disease diagnosis. Breast cancer has also received much attention, particularly since proteins can be isolated from nipple aspiration fluid, allowing noninvasive diagnosis. Several potential biomarkers have been identified through the comparative 2DGE analysis of bilateral matched samples of fluid taken from women with unilateral breast cancer (**Figure 10.3**).

Despite the many successes that have been reported, 2DGE has a number of disadvantages for biomarker discovery, including its low sensitivity and the requirement for relatively large samples. As discussed in Chapters 2 and 4, the information content of 2DGE can be improved through multiplex analysis (difference in-gel electrophoresis; p. 76) and the sensitivity can be increased through the use of novel protein stains or by pre-fractionation of the sample prior to separation. Various strategies for pre-fractionation have also been tested in biomarker discovery projects, including approaches that select a particular component of the proteome for analysis or eliminate a certain fraction of the proteome during analysis. The selection of cell surface proteins on cancer cells by labeling the extracellular portion of cell surface proteins on intact cells with a hydrophilic biotin reagent is an example of the first approach (**Figure 10.4**). An example of the second approach is the use of narrow pH range gels or simple chromatographic procedures that select proteins with particular physicochemical properties. In these cases, however, it is beneficial to use even larger amounts of the starting material to provide enough of the protein sample to facilitate the identification

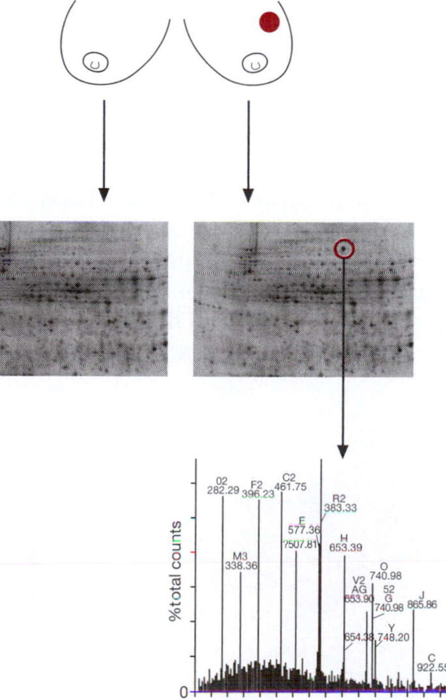

FIGURE 10.3 Novel biomarkers for breast cancer can be identified by the comparative analysis of nipple aspiration fluid from affected and unaffected breasts in women with unilateral tumors.

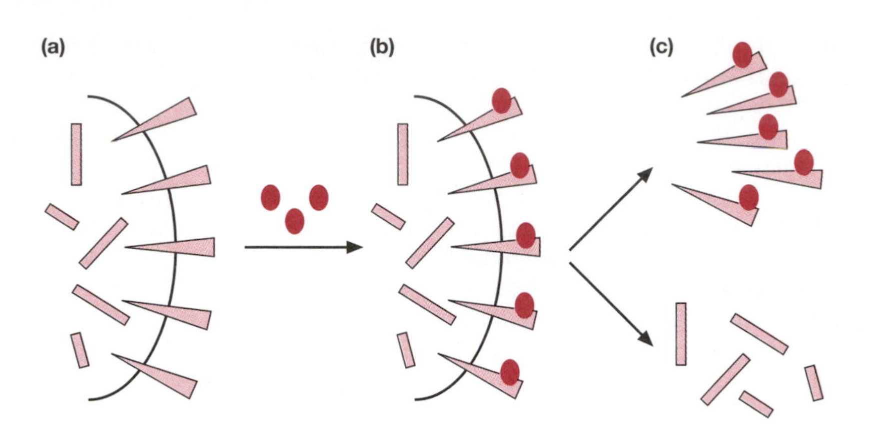

FIGURE 10.4 The cell surface subproteome can be isolated using affinity reagents that cannot penetrate the plasma membrane. (a) The cellular proteome comprises both cell surface and internal proteins. (b) A hydrophilic affinity reagent derived from biotin is added to intact cells, resulting in universal labeling of cell surface proteins. (c) After disruption of the cells, the surface proteins can be captured using affinity chromatography with streptavidin beads, while the internal proteins are recovered in the eluate.

of low-abundance proteins. Unfortunately, most clinical samples are small and heterogeneous, and are surrounded by contaminating normal tissue, which makes the detection of useful biomarkers much more difficult. One way to address the problem of contamination is to use **laser capture microdissection** (**LCM**), a technique in which particular cell populations can be isolated under direct microscopic visualization (**Figure 10.5**). Although this combination of methods has led to the discovery of several novel cancer biomarkers (for example, glyoxalase-1 and FK506BP, which are more abundant in invasive ovarian cancer cells), it remains difficult to produce sufficient amounts of starting material.

More sensitive techniques can be used to identify biomarker profiles

As discussed in Chapter 2, the sensitivity of 2DGE methods is limited by the ability to detect differences between patterns on gels. Alternative methods such as LC-MS and the use of protein microarrays are less biased and can process samples at a higher throughput, thereby increasing the potential for biomarker discovery. Such techniques help to identify not only individual biomarkers (see **Box 10.2,** which discusses the detection of biomarkers for rheumatoid arthritis using MS and MS/MS methods) but also diagnostic fingerprints or profiles comprising different proteins that together provide a reliable indicator of disease. For example, Bateman and colleagues (Further Reading) analyzed cancer epithelial cells microdissected from 25 breast cancer patients using LC-MS and found 113 proteins that differed in abundance between stages 0 and III, including known biomarker candidates and several novel proteins that were indicative of the disease stage. Protein microarrays have been used for the identification of individual biomarkers and profiles, for example, the dual fluorescence labeling of antibody arrays to profile colorectal cancer (p. 195) and the identification of diagnostic protein profiles by SELDI mass spectroscopy (p. 205), which has helped to identify several defensins that are induced by viruses in CD8 cells (**Figure 10.6**). Disease diagnosis is achieved by looking at the SELDI spectra

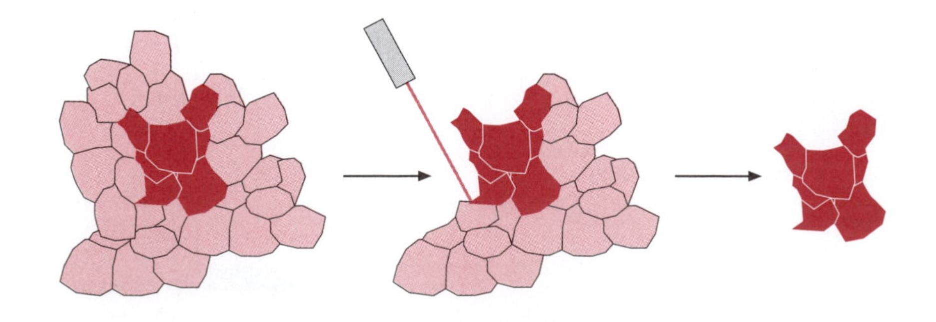

FIGURE 10.5 Laser capture microdissection uses a finely focused laser beam to select individual cells in a heterogeneous tissue sample. In this example, tumor cells (*dark*) are being captured from a heterogeneous biopsy, which also includes normal cells.

BOX 10.2 CASE STUDY.
Biomarkers for rheumatoid arthritis.

There are no reliable early markers for rheumatoid arthritis, and although the presence of the disease correlates well with changes in erythrocyte sedimentation rate and levels of C-reactive protein (CRP), these do not give any prognosis, that is, prediction of severity. However, the analysis of serum and synovial fluid using proteomics methods has revealed that a number of potential biomarkers corresponding to inflammatory activity in the joints (for example, stromelysin-1), cartilage integrity (for example, CS846, a marker of aggrecan turnover in cartilage), and the breakdown of bone and connective tissue (various proteases and neoepitopes of collagen) are elevated in chronic rheumatoid arthritis. Several cytokines have also been proposed as biomarkers, for example, interleukin-15 (IL-15) for juvenile rheumatoid arthritis and tumor necrosis factor α (TNF-α), IL-12, IL-15, and IL-18 for general rheumatoid arthritis. These can be tested in either serum or synovial fluid, the latter appearing more reliable. The application of MS/MS to serum and synovial fluid has also revealed the potential of matrix metalloproteinase 13, myeloid-related protein 8, and calgranulins A-C, the latter appearing to correlate well with severity. Importantly, the sensitive detection of specific biomarkers for rheumatoid arthritis in serum and synovial fluid helps to differentiate between different forms of the disease and also distinguish it from other joint diseases such as osteoarthritis.

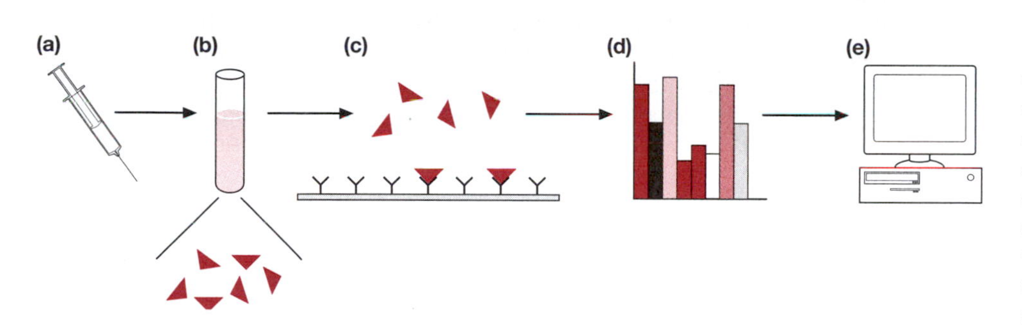

FIGURE 10.6 Protein patterns in disease diagnosis. A blood sample (a) contains many proteins (b), which can be captured and quantified on an analytical protein chip (c). The relative abundances of the proteins provide a unique signature, or fingerprint (d), which can be detected by pattern-matching algorithms (e) and used to diagnose disease and classify different forms of tumor.

produced by different samples, and the peaks on the spectra are diagnostic in themselves without necessarily identifying the proteins to which they correspond. In theory, SELDI spectra could provide higher sensitivity than the analysis of single biomarkers, because the latter are often expressed in multiple diseases, making a precise diagnosis more difficult.

A useful example of SELDI pattern profiling is the early diagnosis of ovarian cancer, a disease that is usually detected at the late stage when cancer cells have already spread and the prognosis is poor. In the original study, mass spectra derived from the serum samples of women with ovarian cancer and from unaffected controls were used as a training set for a pattern-matching algorithm. A discriminatory pattern was identified, which was applied to another set of samples. This resulted in the correct diagnosis of all ovarian cancers (including 18 stage I cancers, where the prognosis is good because the neoplastic cells are still contained within the ovary) and a false-positive rate of only 5%. Similar algorithms have been used to diagnose breast and prostate cancers. In each case, the sensitivity of the pattern profiling method appears to be significantly higher than that of single biomarkers and can achieve correct diagnosis in up to 100% of cases in some diseases.

10.3 APPLICATIONS OF PROTEOMICS IN DRUG DEVELOPMENT

Proteomics can help to select drug targets and develop lead compounds

The marketing of a new drug marks the end of a complex, lengthy, and expensive process (**Figure 10.7**). It has been estimated that every drug candidate entering the clinic costs $100 million to develop and requires 250 employee years during the development process. Since many drugs fail at late clinical stages, the real cost of bringing a new drug to market is probably in the region

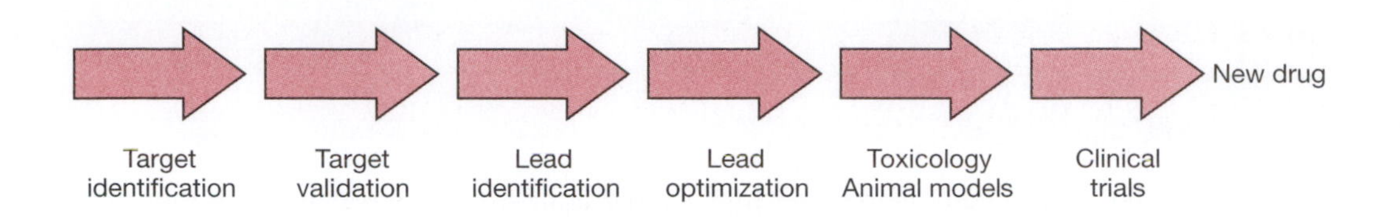

FIGURE 10.7 The major stages in drug development.

of $1 billion. Very few drugs recapture in sales the amount of money spent on their development. The pharmaceutical industry has therefore been quick to embrace any new technology that increases the pace of drug discovery and reduces the **attrition rate** (the number of products abandoned during development), since attrition contributes most of the development costs. Such technologies include combinatorial chemistry, high-throughput screening, *in silico* screening, genomics, and now proteomics.

Disease biomarkers represent proteins that are expressed specifically or preferentially in either the disease state or the healthy state. In many cases, such proteins appear or disappear merely as a consequence of the disease, but it is also possible that their presence or absence could contribute to the disease symptoms. Where comparative proteomic analysis identifies proteins that are expressed only in the disease state, such proteins might represent not only useful biomarkers but also likely therapeutic targets. Biomarker discovery therefore provides fuel for the target identification stage of drug development (**Figure 10.8a**). Furthermore, where specific proteins are depleted in the disease state, these proteins could themselves be developed as potential drugs. Comparative proteomic analysis can therefore provide leads toward novel therapeutic proteins (Figure 10.8b).

As well as the direct comparison of healthy and disease tissues, another useful strategy in target identification is the analysis of protein profiles when particular cells, tissues, or perhaps animal models have been treated with a regulatory molecule, such as a growth factor or cytokine. Since these molecules stimulate different forms of cell behavior, including proliferation, differentiation, and adhesion, they can be useful in modeling the outcome of diseases. Examples of the above include the discovery of cytokine-regulated proteins in intestinal epithelial cells, the discovery of targets regulated by phorbol esters in erythroleukemia cells, and the identification of phosphoproteins induced by epidermal growth factor treatment in HeLa cells. The last example is interesting because antibodies were used to

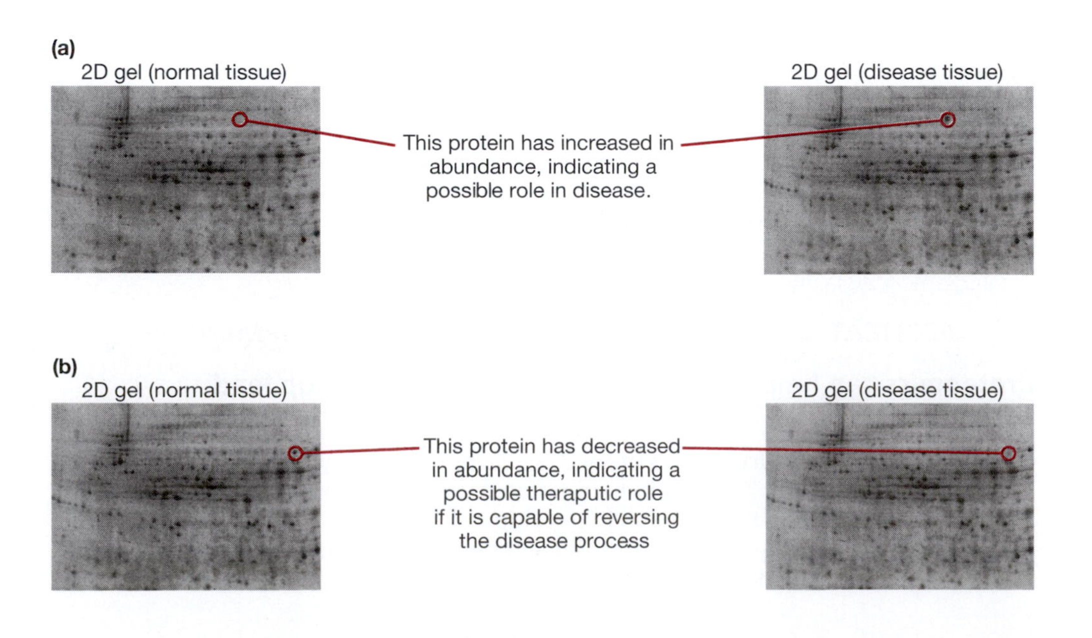

FIGURE 10.8 The value of proteomics in (a) target identification and (b) the identification of potential protein therapeutics.

isolate phosphotyrosine-containing proteins, and comparative analysis was therefore facilitated by the selective purification of the phosphoproteome (Chapter 8). Potential drug targets in lymphocytes treated with interferon-α and interleukin-2 have also been identified in this manner.

Proteomics not only provides methods for the identification of new therapeutic targets based on human disease-specific proteins, but also offers many strategies for the identification of targets in the pathogen proteome, thereby providing leads in drug and vaccine development for infectious diseases (**Table 10.2**). The characterization of cell surface proteins or secreted proteins that are involved in infection or pathogenesis provides a rapid route for the identification of novel drug or vaccine targets. Several bioinformatics methods can be used to analyze pathogen protein sequences and thus identify proteins that are likely to be secreted or integrated into the plasma membrane and exposed to the environment, including those detecting secretion signals and/or hydrophobic transmembrane regions. Such methods can be tailored to different species or taxonomic groups (for example, Gram-positive or Gram-negative bacteria). Other techniques focus on enriching the proteome for surface proteins. One approach is to separate the pathogen proteome by 2DGE and use hyperimmune sera from different patients to identify the immunodominant proteins. This subset of the proteome, the "immunome," is likely to be a good source of drug targets as well as vaccine targets (since sites on these proteins are proven to be exposed to the host immune system). This strategy can also be used to identify proteins that are expressed during particular stages of the pathogen life cycle, as has been demonstrated in the case of the malaria parasite *Plasmodium falciparum*. Various methods have been developed to isolate the cell surface protein fraction from microbial cultures, including **cell surface labeling** with dyes, ^{18}O, or affinity tags (see Figure 10.4 for an analogous approach in human cells), cell wall digestion to release surface proteins into the medium, and **cell surface shaving**, which involves the digestion of intact cells with proteases. Trypsin is advantageous because the released peptides are directly compatible with MS, but trypsin may not cleave shorter peptide sequences projecting from the membrane. Therefore, mild digestion with less specific proteases (for example, proteinase K) can be useful to identify these peptides.

Alternative strategies to find microbial drug targets include the comparative proteomic analysis of pathogenic and nonpathogenic isolates of the same organism to identify pathogenesis-related proteins, or the comparative analysis of host and/or pathogen tissues before and after infection, to

TABLE 10.2 A SELECTION OF 2DGE DATABASES FOR MICROBES AND PLANTS

Database	Contents	URL
Department of Medical Microbiology, University of Aberdeen	*Haemophilus influenzae*, *Streptococcus pneumoniae*, and *Neisseria meningitidis*	http://www.abdn.ac.uk/~mmb023/2dhome.htm
European Bacteria Proteome Project (EBP)	*Chlamydia pneumoniae*, *Helicobacter pylori*, and *Mycobacterium tuberculosis*	http://www.mpiib-berlin.mpg.de/2D-PAGE/
SWICZ: Swiss–Czech Proteomics Server	*Caulobacter crescentus* and *Streptomyces coelicolor*	http://proteom.biomed.cas.cz/
Various	*Borrelia garinii*, *Francisella tularensis* LVS, and *Mycoplasma pneumoniae* M129	http://www.mpiib-berlin.mpg.de/2D-PAGE/microorganisms/index.html
COMPLUYEAST-2DPAGE	*Candida albicans* and *Saccharomyces cerevisiae*	http://compluyeast2dpage.dacya.ucm.es/
Yeast Protein Map	*Saccharomyces cerevisiae*	http://www.ibgc.u-bordeaux2.fr/YPM/
CNRS/INRA/Bayer CropScience	*Arabidopsis* seed proteome	http://www.seed-proteome.com/
The *Arabidopsis* Mitochondrial Proteome Project	*Arabidopsis* mitochondrion	http://www.gartenbau.uni-hannover.de/genetik/AMPP

identify proteins specifically involved in host–pathogen interactions. The agent responsible for tuberculosis, *Mycobacterium tuberculosis*, has been compared with its nonpathogenic relative *M. bovis* BCG to identify proteins that are specific to the virulent strain. Comparative 2DGE analysis revealed 56 proteins specific to *M. tuberculosis* out of 96 spot differences, and 32 of these proteins were identified by mass spectrometry and are currently being investigated as novel vaccine targets. *In vitro* models of infection can be useful as long as they are physiologically relevant. For example, several models of tuberculosis infection have been established, including the Wayne model (where the bacterium persists in a nonreplicative state under conditions of reduced oxygen) and a disease model involving *in vitro* macrophage infection. In the latter case, 16 proteins were shown to be induced by infection and 28 were shown to be repressed.

Proteomics is also useful for target validation

Proteomics and other large-scale technologies have generated a boom in **target discovery**, but this has resulted in a bottleneck at the **target validation** stage. This is where supporting evidence is generated to show that interfering with the activity of target protein will alter the course of the disease in a beneficial way. Compared with target identification, target validation is a low-throughput enterprise, since it can take 1–2 years to validate each novel target protein.

A wide variety of genomic and proteomic methods can be employed at the validation stage, including structural analysis, the investigation of protein interactions, expression studies to see if the protein generates an informative phenotype when overexpressed or suppressed, and the analysis of genetic variation in the target population. Targets with high levels of polymorphism within the population are usually unsuitable because the different variants are likely to show different responses to candidate drugs. In such cases, genetic and biochemical interaction studies may reveal potential interacting proteins, involved in the same metabolic or signaling pathway, which show less polymorphism and would represent more suitable targets. One way that proteomics can be applied in target validation is the use of affinity-based probes to select proteins on the basis of their ability to bind particular small molecules (**Table 10.3**). Target proteins identified in this type of **chemical proteomics** screen are, in effect, pre-selected as susceptible to inhibition by small molecules, and are therefore more likely to respond favorably to candidate drugs. Affinity-based probes can be incorporated into cell-based and animal testing programs and, as discussed below, can also be used as a starting point in the development of novel chemical entities.

Chemical proteomics can be used to select and develop lead compounds

During the chemistry stages of drug development, certain proteomic technologies can provide a high-throughput approach for the identification and optimization of suitable **lead compounds**. For example, methods that identify protein–protein interactions can be used to screen lead compounds for interfering activity that would suggest useful physiological effects in the body. Functional protein chips can be used to assay protein–protein interactions *in vitro* in the presence and absence of potential lead compounds, allowing the rapid identification of molecules that prevent normal binding events. These molecules would have a significantly higher chance of interfering with protein interactions in living cells. When suitable leads have been identified, the same strategy can be used for lead optimization. In this case, protein interactions could be tested in the presence of selected chemical derivatives of the lead compound, to identify those with the most potent effects. Similar experiments could be carried out *in vivo*, using the yeast reverse two-hybrid system and its derivatives (p. 152) to test for the disruption of protein interactions.

TABLE 10.3 A SELECTION OF FUNCTION-SPECIFIC AFFINITY REAGENTS THAT CAN BE USED TO IDENTIFY FUNCTIONAL CLASSES OF PROTEINS IN GELS, ON MEMBRANES, AND ON MICROARRAYS

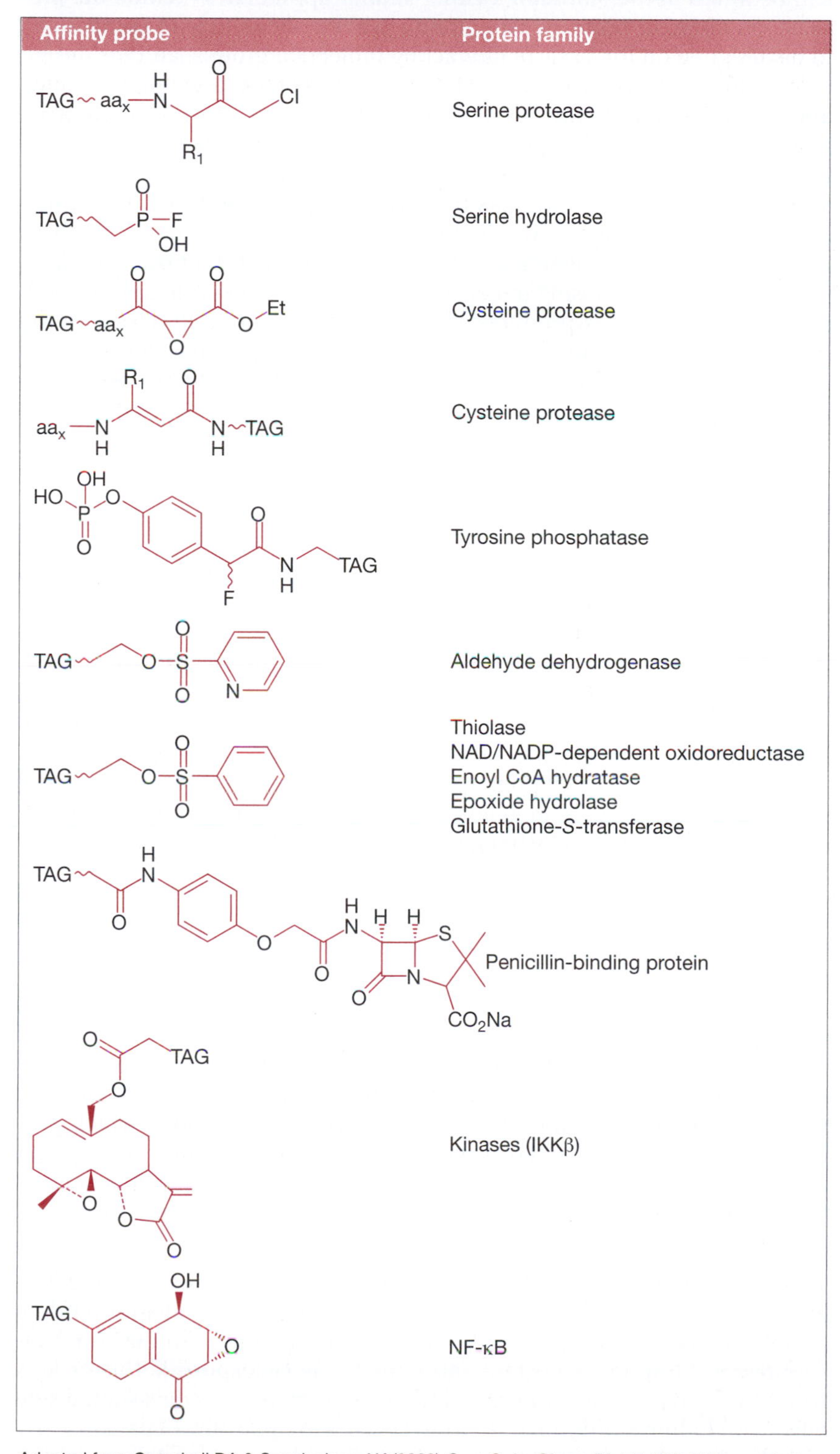

Adapted from Campbell DA & Szardenings AK (2003) *Curr. Opin. Chem. Biol.* 7, 296. With permission from Elsevier.

Structural proteomics also plays an important role in lead optimization, because the structural analysis of a potential drug target allows putative protein–small-molecule interactions to be modeled *in silico* prior to experimental screening. Armed with structural information, researchers can use powerful computer programs to search through databases containing the structures of many different chemical compounds. The computer can select those compounds that are most likely to interact with the target, and these can be tested in the laboratory. Other similar approaches include the preselection of target proteins with affinity probes, allowing lead compounds to be designed on the basis of interacting molecular groups, and the design of lead compounds on the basis of known ligands of the target protein. Such approaches eliminate the need for high-throughput screens of complex chemical libraries, and in principle allow lead compounds to be perfected using computers before any chemical screening is necessary. This is known as rational drug design, and has been successful in the development of a number of current drugs. The first drug produced by rational design was zanamivir (trade name Relenza), which is used to treat influenza. Zanamivir was developed by choosing molecules that were most likely to interact with neuraminidase, a virus-produced enzyme that is required to release newly formed viruses from infected cells. Many of the recent drugs developed to treat HIV infections (for example, ritonavir and indinavir) were designed to interact with the viral protease, the enzyme that splits up the viral proteins and allows them to assemble properly.

Proteomics can be used to assess drug toxicity during clinical development

Proteomics can be applied in studies that investigate mechanisms of drug activity and toxicity, both of which provide valuable data during the clinical stages of drug development. Adverse drug responses are the largest source of litigation in the United States, and cost the pharmaceutical industry millions of dollars every year. Such effects often occur because the drug accumulates to toxic levels or is broken down into a toxic derivative, and in each case the clinical symptoms reflect unanticipated interactions between the drug or its metabolic byproducts and nontarget proteins in the cell. The result of drug toxicity is often a change in gene expression or protein abundance, which can be detected as a toxicity biomarker.

The study of protein profiles altered in response to drugs, using either traditional 2DGE or protein chips, has provided a great deal of data about the biochemical basis of drug activity and the pathways and networks upon which drugs act. Many different systems have been studied, including acute promyelocytic leukemia cells before and after treatment with retinoic acid and Burkitt lymphoma cells before and after treatment with 5′-azacytidine. This approach can be used not only to study human proteins or those in animal models of disease, but also the proteomes of pathogens to see how they respond to antibiotics. Although expression proteomics shows which proteins are induced or repressed by particular antibiotics, interaction maps are useful in pinpointing hubs and redundant pathways. This can help to predict which combinations of drugs, acting on separate targets, will interfere most destructively with the pathogen life cycle.

Toxicity biomarkers have been discovered predominantly using 2DGE and mass spectrometry. A good example is calbindin, a calcium-binding protein that is depleted in the kidneys of patients treated with the immunosuppressant drug cyclosporin A (also known as ciclosporin). This drug is widely used to prevent organ rejection in children, but has devastating side effects, including kidney toxicity, which occurs in up to 40% of patients. The toxicity is associated with the loss of calcium in the urine and the resulting calcification of the kidney tubules. A comparison of two-dimensional gels from treated/untreated samples of rat and human kidneys showed a striking difference in the abundance of calbindin, and the loss of this protein following drug treatment provides a mechanistic explanation for the toxicity effect. Unlike humans, monkeys do not suffer cyclosporin A toxicity effects and proteomic analysis of monkey kidneys shows that there is no calbindin depletion following drug treatment. Studying the way in which monkeys metabolize the drug may therefore provide insight into novel ways to prevent side effects in humans.

10.4 PROTEOMICS IN AGRICULTURE

Proteomics provides novel markers in plant breeding and genetics

Although biomedical applications dominate proteomics and are likely to do so for the foreseeable future, proteomics also has a long history in agriculture (see **Box 10.3**). Since the early 1980s, two-dimensional gels have been used to study the extent of genetic variability in natural plant populations, and more recently the same techniques have been applied in the analysis of genetically modified (GM) crops. Proteomics has also been used to study plant development, physiology, and interactions with other organisms, helping to identify proteins involved in defense and stress responses and those expressed in improved agricultural varieties.

Among the earliest proteomic studies in plants were those performed by de Vienne and colleagues to distinguish between different wheat varieties (see Further Reading). Such experiments have been carried out in several important crops in addition to wheat, including rice, barley, sugarcane, and

BOX 10.3 CASE STUDY. Biomarkers for stress tolerance in plants.

One of the greatest constraints on agricultural productivity is the loss of yields due to biotic stress factors (weeds, pests, diseases) and abiotic stress factors (climate, temperature, poor soil quality, salinity). For example, the two major cereal crops that feed the human population (rice and maize) are strongly affected by high temperatures (which cause sterility) and by salinity (which reduces yields and inhibits growth and development). Plants have mechanisms that allow them to respond to stress and ultimately tolerate it, and those with the best stress tolerance are good targets for breeding. Proteomics can be used to identify proteins that are upregulated or downregulated under stress; for example, heat stress in wheat leads to the induction of low-molecular-weight heat shock proteins and transcription factors responsible for activating the corresponding tolerance genes. Such proteins have been shown to confer heat stress tolerance on plants when introduced by genetic engineering. Protein–protein interaction studies have shown that certain heat shock proteins interact with calmodulin-binding protein, providing insight into the signaling pathways that induce stress tolerance and therefore offering an opportunity to induce heat tolerance by applying chemicals externally.

Salinity stress is a major target for proteomics because large amounts of potential agricultural land are affected by dehydration and/or salinity, and the development of salt-tolerant crops would expand the amount of land available for agriculture. Plants respond to drought by closing their stomata and reducing water loss, which is controlled by the hormone abscisic acid (ABA). Several proteomic studies have shown that plants under drought or salinity stress induce overlapping sets of proteins, many of which also respond to ABA signaling, again providing insight into the underlying mechanisms and potential strategies to induce drought tolerance. Many forms of stress also result in the production of reactive oxygen species, so enzymes such as superoxide dismutase (which remove free radicals) are often induced by stress, as are enzymes that synthesize antioxidant metabolites such as ascorbic acid.

Proteomics is also useful to address biotic stress because it can be used to identify proteins induced in the plant by contact with pests and pathogens, and proteins induced in the pests/pathogens that enable attack. Both can then be used as targets to improve resistance (such as designing the plants to express antibodies or dsRNA targeting proteins that are essential for attack. For example, 2DGE has been used to identify 41 wheat proteins induced by the fungus *Fusarium graminearum*, many of which are involved in jasmonic acid signaling, pathogen defense, and nitrogen/amino acid metabolism pathways, whereas photosynthesis-related proteins have been shown to be repressed. Up-regulated fungal proteins included those responsible for the production of antioxidants (presumably to counter the "oxidative burst" response) and the acquisition of carbon.

pepper, as well as a number of tree species. Proteomics has also been used to study interspecific variety among cultivated wheats and other cereals. In the initial studies, several wheat species were analyzed by 2DGE, and the similarities and differences in the distribution and abundance of protein spots was used to calculate similarity indices. This allowed a phylogenetic tree to be constructed, which were found to be in excellent agreement with trees based on DNA sequences and classical taxonomy. As is the case for medical proteomics, a number of databases have been set up to catalog the proteomes of various plant species, including many of agricultural importance (Table 10.2).

A large number of proteomic studies have been carried out to investigate physiological processes in plants, often with the aim of identifying proteins corresponding to useful agronomic traits. In some cases, such studies have been used to compare mutant and wild-type varieties to characterize downstream effects of the mutation at the whole-proteome level. For example, 2DGE has been used to study global differences between near-isogenic pea lines differing only at the classical Mendelian locus R, which determines seed morphology. A comparison of RR seeds (round) and rr seeds (wrinkled) revealed extensive differences, affecting the abundance of about 10% of the proteome, agreeing well with the numerous biochemical and physiological differences that have been observed between these seeds in previous studies.

Proteomics has also been exploited to identify polymorphisms that can be used as genetic markers. Polymorphism occurs at three levels, described as position shifts (PS), presence/absence polymorphisms (P/A), and quantitative polymorphisms (sometimes called protein quantity loci, or PQLs). The first two types of polymorphism are often useful genetic markers if they represent variant forms of the same polypeptide chain differing in mass or charge, but they may also represent different post-translational variants, which are less suitable because they do not represent an underlying genetic variation at the same locus. Such markers have been used in wheat, maize, and pine, for example, to generate comprehensive genetic maps also containing DNA markers. Quantitative polymorphisms are useful for identifying quantitative trait loci (QTLs) that cannot be pinned down by traditional map-based cloning or candidate gene approaches. The basis of the method is that protein quantitative polymorphisms that colocalize with QTLs can be used to validate candidate genes. An example is provided by de Vienne et al. (see Further Reading). These investigators identified a candidate gene on chromosome 10 of maize for a QTL affecting drought response. The candidate gene was *ASR1*, known to be induced by water stress and ripening. Verification of the association was possible because a protein quantitative polymorphism found during the comparison of two-dimensional gels from control and drought-stressed maize plants mapped to the same region; the quantitative polymorphism reflected different levels of the ASR1 protein under different drought stress conditions.

Most proteomic studies in plant biology have involved the comparison of healthy plants with those infected with pathogens, or stable plants with those exposed to particular forms of stress. Rice (*Oryza sativa*) and the model plant *Arabidopsis thaliana* have received most of the attention because the genomes of both species have been sequenced and extensive EST resources are available, facilitating the identification of proteins by peptide mass fingerprinting and fragment ion analysis (Chapter 3). Since rice is the most important food crop in the world, representing the staple diet for over 50% of the population, proteomics has been used to identify proteins that might affect traits of agronomic value in this species. For example, rice leaves treated with jasmonic acid, a known fungal elicitor, were compared with untreated leaves and 12 jasmonate-induced proteins were detected. Nine of these were similar to known pathogenesis-related proteins. The abundance of 21 proteins was found to change when rice plants were infected with the bacterial blight pathogen (*Xanthomonas oryzae* pv. *oryzae*), whereas 17 proteins were induced by infection with the blast fungus *Magnaporthe grisea*. Comparison of wild-type rice seeds and the seeds of a semi-dwarf variety revealed the specific expression of different forms of the storage protein glutelin in each type of seed. All these studies revealed proteins that might have specific roles in defense or agricultural improvement.

Protein interaction analysis has also been useful to predict functions for plant proteins. For example, flowering in *A. thaliana* is controlled by a closely related group of transcription factors whose DNA-binding domain is known as a MADS box. Similar transcription factors have been identified in other plants, but it is often difficult to determine their precise roles without painstaking analysis of mutant phenotypes. One shortcut is to assign functions based on conserved protein interactions. For example, the DAL13 protein of the Norway pine is a MADS box protein that could conceivably be the functional equivalent of any of the *Arabidopsis* proteins. In interaction screens, however, DAL13 interacted specifically with the *Arabidopsis* protein PISTILLATA, suggesting that DAL13 and PISTILLATA are functionally homologous (orthologs). A similar interaction screen, shown in **Figure 10.9**, has identified *Arabidopsis* SEPELLATA3 (SEP3) as the functional ortholog of petunia FBP2 and FBP5, since a conserved set of interactions is observed.

Proteomics can be used for the analysis of genetically modified plants

One of the major applications of plant biotechnology is the development of GM crops. In some cases, the crops are used as bioreactors for the production of valuable proteins or metabolites, but other crops are modified to improve their agronomic traits and are intended to be used as food or feed. Concerns about the safety of GM foods revolve around the concept of substantial equivalence, that is, whether the process of genetic modification has introduced any unanticipated (and undesirable) changes in the host plant. Such changes, resulting from the random integration of DNA into the plant genome and the influence this may have on endogenous genes, could include metabolic disruption and therefore changes in the concentrations of nutrients and potentially toxic compounds. Strategies for the comparative analysis of GM and non-GM crops often focus on specific compounds,

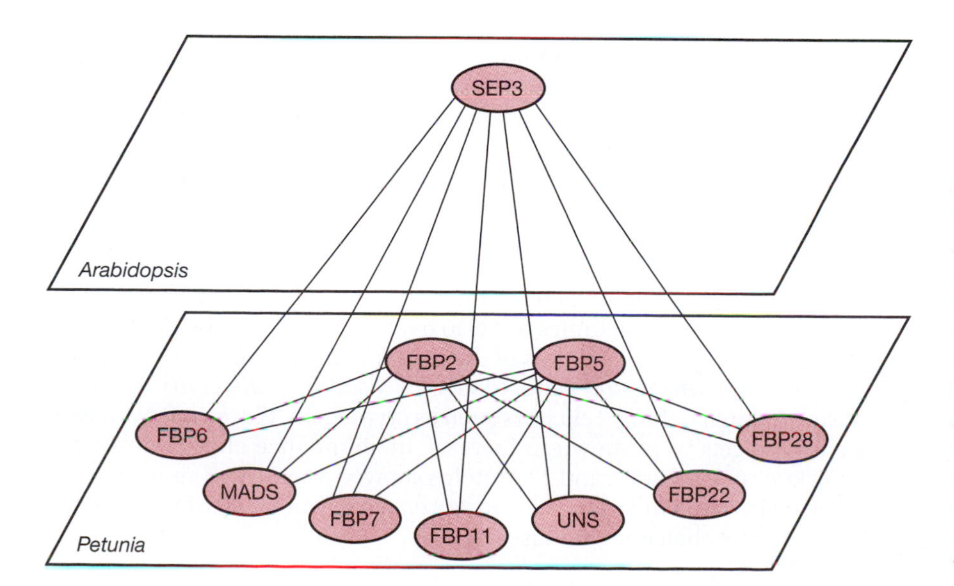

FIGURE 10.9 Heterologous interactions between SEPALLATA3 (SEP3) from *Arabidopsis* and petunia MADS-box proteins identified by the yeast two-hybrid system. The interaction partners of SEP3 are identical to those of the FBP2 and FBP5 proteins from petunia, indicating that SEP3 and FBP2 and FBP5 are orthologs. Abbreviations: FBP, FLORAL BINDING PROTEIN; pMADS, petunia MADS-box protein; SEP, SEPALLATA; UNS, UNSHAVEN. (From Immink RG & Angenent GC (2002) *Trends Plant Sci.* 7, 531. With permission from Elsevier.)

such as particular nutrients, metabolites, or toxins. However, in order to fully characterize any changes that may have occurred, it is necessary to use global and nontargeted approaches, including transcriptional profiling, proteomics, and metabolomics (see Chapter 1).

Proteomics potentially has an important role to play in food safety assessment because it allows changes in protein abundance and post-translational modifications to be identified, which cannot be detected using other global analysis methods. For example, the UK Food Standards Agency has funded projects in which GM and non-GM crops are compared using multidimensional liquid chromatography and ICAT-based quantitative mass spectrometry. An important component of these studies is the evaluation of natural proteomic variation within the model crops that make GM-specific differences harder to identify.

10.5 PROTEOMICS IN INDUSTRY—IMPROVING THE YIELD OF SECONDARY METABOLISM

Plants produce a wide variety of chemical substances with complex structures. These substances are known as secondary metabolites because they are not part of core metabolism, but they perform many useful functions, including defense against pathogens and attraction of pollinators. Secondary metabolites have been of interest to humans for centuries because they can be used as drugs, dyes, fragrances, nutritional supplements, and flavors. Unfortunately, many of the most beneficial secondary metabolites are produced in such low quantities that their commercial extraction is unfeasible. Therefore, researchers have studied secondary metabolic pathways in an attempt to produce more of these desired compounds. A further hurdle is that secondary metabolic pathways are extremely complex, involving many enzymatic steps, extensive branching and cross-talk, and the compartmentalization of different steps in different cell types or organelles. Traditional methods to study secondary metabolism involve the step-by-step characterization of individual reactions using feeding experiments and labeled intermediates. Proteomics can accelerate discovery in plant secondary metabolism because it can identify not only enzymes, but also regulatory proteins and proteins involved in the shuttling of intermediates between compartments.

One of the best model systems for the production of useful secondary metabolites is the Madagascar periwinkle (*Catharanthus roseus*), which produces hundreds of alkaloids, including the potent anti-neoplastic drugs vinblastine and vincristine (**Figure 10.10**). Proteomic analysis of *C. roseus* cell cultures under culture conditions known to affect alkaloid production has revealed five proteins whose abundance mirrors the level of alkaloid accumulation, suggesting either a catalytic or regulatory role. The common regulation of many of the genes involved in terpenoid indole alkaloid biosynthesis in *C. roseus* suggests that a useful strategy for increasing the levels of alkaloids in this system would be to identify and manipulate transcription factors that control the expression of these genes. One such transcription factor, known as ORCA2, has been identified using the yeast one-hybrid system (p. 141), and another has been found in a screen involving insertional mutagenesis (p. 10). Proteomics has also been used to study the phytoalexin synthesis pathway in cell cultures of the common bean (*Phaseolus vulgaris*), chickpea (*Cicer arietinum*), and tobacco (*Nicotiana tabacum*) following exposure to fungi or fungal elicitors such as cryptogein. Eleven spots on two-dimensional gels were shown to increase in abundance after treatment of the chickpea cell cultures, and 23 proteins showed a change of abundance in tobacco cells. These included the two key enzymes phenylalanine ammonia lyase (PAL) and chalcone synthase (CHS).

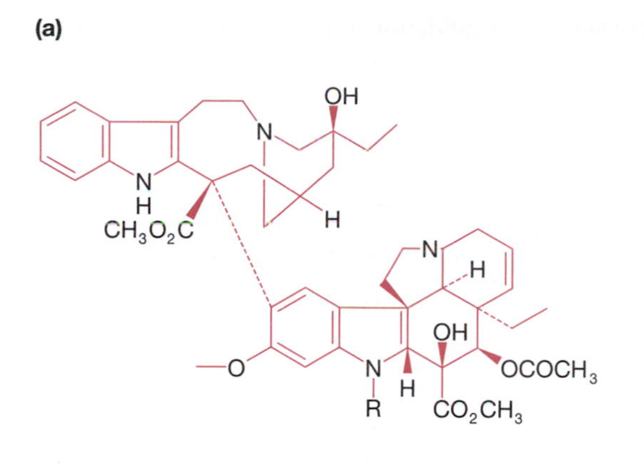

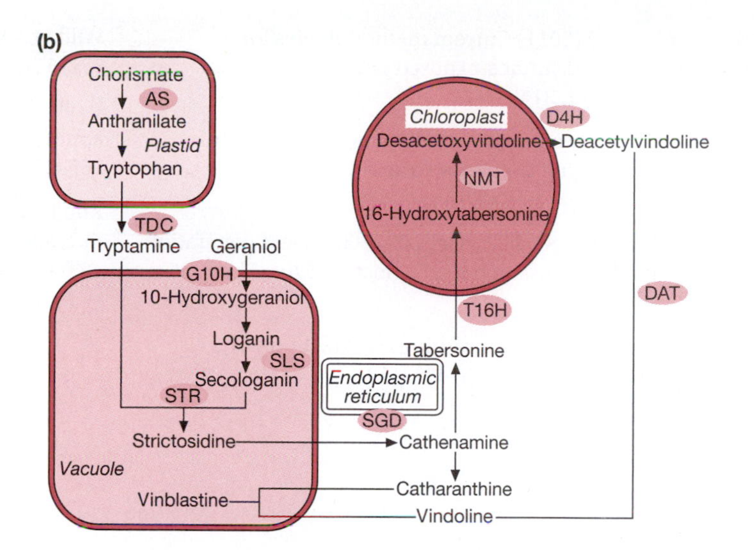

FIGURE 10.10 Overview of the TIA pathway in *C. roseus*. (a) The generic chemical structure of the alkaloid antineoplastic drugs vinblastine and vincristine (R = $-CH_3$ in vinblastine and –CHO in vincristine). (b) An abbreviated representation of the very complex and highly compartmentalized biosynthetic pathway for these compounds.

FURTHER READING

Alaiya AA, Roblick UJ, Franzen B et al. (2003) Protein expression profiling in human lung, breast, bladder, renal, colorectal and ovarian cancers. *J. Chromatogr. B* 787, 207–222.

Bateman NW, Sun M, Bhargava R et al. (2011) Differential proteomic analysis of late-stage and recurrent breast cancer from formalin-fixed paraffin-embedded tissues. *J. Proteome Res.* 10, 1323–1332.

Bumpus SB, Evans BS, Thomas PM et al. (2009) A proteomics approach to discovering natural products and their biosynthetic pathways. *Nat. Biotechnol.* 27, 951–956.

Celis JE, Kruhoffer M, Gromova I et al. (2000) Gene expression profiling: monitoring transcription and translation products using DNA microarrays and proteomics. *FEBS Lett.* 480, 2–16.

de Vienne D, Leonardi A, Damerval C & Zivy M (1999) Genetics of proteome variation for QTL characterization: application to drought-stress responses in maize. *J. Exp. Bot.* 50, 303–309.

Eldakak M, Milad SI, Nawar AI & Rohila JS (2013) Proteomics: a biotechnology tool for crop improvement. *Front. Plant Sci.* 4, 35.

Hanash S (2003) Disease proteomics. *Nature* 422, 226–232.

Hanash SM, Madoz-Gurpide J & Misek DE (2002) Identification of novel targets for cancer therapy using expression proteomics. *Leukemia* 16, 478–485.

He QY & Chiu JF (2003) Proteomics in biomarker discovery and drug development. *J. Cell Biochem.* 89, 868–886.

Hewitt SM (2009) Tissue microarrays as a tool in the discovery and validation of tumor markers. *Methods Mol. Biol.* 520, 151–161.

Lee JM, Han JJ, Altwerger G & Kohn EC (2011) Proteomics and biomarkers in clinical trials for drug development. *J. Proteomics* 74, 2632–2641.

Mattoon D, Michaud G, Merkel J & Schweitzer B (2005) Biomarker discovery using protein microarray technology platforms: antibody-antigen complex profiling. *Expert Rev. Proteomics* 2, 879–889.

Moseley FL, Bicknell KA, Marber MS & Brooks G (2007) The use of proteomics to identify novel therapeutic targets for the treatment of disease. *J. Pharm. Pharmacol.* 59, 609–628.

Petricoin EF, Ardekani AM, Hitt BA et al. (2002) Use of proteomic patterns in serum to identify ovarian cancer. *Lancet* 359, 572–577.

Petricoin EF & Liotta LA (2004) SELDI-TOF-based serum proteomic pattern diagnostics for early detection of cancer. *Curr. Opin. Biotechnol.* 15, 24–30.

Petricoin EF, Zoon KC, Kohn EC et al. (2002) Clinical proteomics: translating benchside promise into bedside reality. *Nat. Rev. Drug Discov.* 1, 683–695.

Ray S, Reddy PJ, Jain R et al. (2011) Proteomic technologies for the identification of disease biomarkers in serum: advances and challenges ahead. *Proteomics* 11, 2139–2161.

Sahab ZJ, Semaan SM & Sang QX (2007) Methodology and applications of disease biomarker identification in human serum. *Biomark. Insights* 2, 21–43.

Shin BK, Wang H, Yim AM et al. (2002) Global profiling of the cell surface proteome of cancer cells uncovers an abundance of proteins with chaperone function. *J. Biol. Chem.* 278, 7607–7616.

Solis N & Cordwell SJ (2011) Current methodologies for proteomics of bacterial surface-exposed and cell envelope proteins. *Proteomics* 11, 3169–3189.

Van Eyk JE (2002) Proteomics: unraveling the complexity of heart disease and striving to change cardiology. *Curr. Opin. Mol. Therapeut.* 3, 546–553.

Voelkerding KV, Dames SA & Durtschi JD (2009) Next-generation sequencing: from basic research to diagnostics. *Clin. Chem.* 55, 641–658.

Welker M (2011) Proteomics for routine identification of microorganisms. *Proteomics* 11, 3143–3153.

Williams M (2003) Target validation. *Curr. Opin. Pharmacol.* 3, 571–577.

Wulfkuhle JD, Liotta LA & Petricoin EF (2003) Proteomic applications for the early detection of cancer. *Nat. Rev. Cancer* 3, 267–275.

Zhu H & Qian J (2012) Applications of functional protein microarrays in basic and clinical research. *Adv. Genet.* 79, 123–155.

Glossary

%C
In polyacrylamide gels, the proportion of the monomer mass represented by the linking agent bis-acrylamide.

%T
In polyacrylamide gels, the total concentration of monomer as a percentage of the gel volume before it is cast.

ABPP-MudPIT
Activity-based protein profiling (the use of chemical probes that interact with mechanistically related enzymes) combined with multidimensional chromatography and mass spectrometry for protein identification.

absorption peak
A peak on a chromatogram representing the maximum absorption (for example, of UV light) by a component of the sample.

activation trap
An integrating DNA construct that contains a strong, outward-facing promoter to activate nearby genes.

activity-based protein profiling (ABPP)
The functional analysis of proteins using probes that react with mechanistically related enzymes.

activity-based proteomics
See **activity-based protein profiling**.

affine gap penalty
In sequence comparisons, a gap penalty comprising an initial penalty for opening a gap and a smaller penalty for extending it.

affinity capture
The capture of particular molecules from solution using a specific binding ligand.

affinity chromatography
A chromatography format based on affinity capture.

affinity depletion
The use of affinity chromatography to remove superabundant proteins from a mixture, allowing the analysis of the less abundant proteins.

affinity pull-down
The isolation of specific proteins in solution by binding to ligand-coated beads.

affinity purification
The use of affinity chromatography to purify a specific target protein.

affinity reagents
Ligands that are used to capture specific proteins.

alignment score
A numerical value that shows the quality of a sequence alignment.

amino acid
An organic molecule that includes amine and carboxylic acid functional groups along with a specific side chain.

amino acid composition
The proportion of each different amino acid in a protein.

analogous
Describing sequences that are structurally and/or functionally related but not derived from a common ancestor, namely, a product of convergent evolution.

analytical microarray
A protein microarray comprising immobilized capture agents such as antibodies, which is used to investigate the abundance of proteins in solution.

angle restraints
In NMR spectroscopy, restraints based on the torsion angles of chemical bonds.

anion exchange (AEX) chromatography
Chromatography mode based on the exchange of negatively charged ions on cationic resins, for example, quaternary amine groups.

anomalous scattering
A change in the phase of X-ray diffraction caused by the presence of atoms (usually heavy metals) that cause strong X-ray absorbance.

antibody mimics
Proteinaceous molecules that mimic the structural properties of antibodies.

anti-phosphotyrosine antibodies
Antibodies that specifically recognize phosphotyrosine residues on proteins.

array
Analytical targets arranged in a grid, usually provided in a miniature format (*see* **microarray**).

arrayed imaging reflectometry
Method for the detection of proteins binding to a non-reflective surface, because binding causes the surface to reflect light in a manner that is proportional to the amount of bound protein.

atomic force microscopy
A high-resolution method for investigating the properties of surfaces using a small tip attached to a cantilever.

attrition rate
The proportion of drugs that fail during development.

autoactivation
In the yeast two-hybrid system and derivatives, the ability of a protein to activate an expression cassette without an interaction partner, often because it is a transcription factor.

automated crystallization workstations
Highly parallel automated devices for testing many different crystallization strategies simultaneously.

bait
In the yeast two-hybrid system and derivatives, the protein that is used to identify interacting partners, usually expressed as a fusion with the DNA-binding domain of transcription factor.

barcoding
(1) The use of a specific oligonucleotide to identify a construct in high-throughput, assays for example, allowing detection by hybridization or high-throughput sequencing. (2) The use of unique sequences of stripes to identify the individual beads in a bead array.

Basic Local Alignment Search Tool
A rapid sequence alignment algorithm that is heuristic and not based on dynamic programming.

bicinchoninic acid (BCA) assay
An assay that measures the total protein concentration in a solution.

bead arrays
Microarrays that are not constrained to a grid format but are instead presented as freely diffusing beads.

bimolecular fluorescence complementation (BiFC)
Assays used to confirm protein interactions by assembling two nonfunctional components of a fluorescent reporter protein into a functional complex.

binary interactions
Interactions between two components.

biochemical genomics
The large-scale analysis of gene function by studying the biochemical characteristics of the corresponding proteins.

bioinformatics
The field of computational biology involving the storage, retrieval, organization, and analysis of large biological datasets.

bioluminescence resonance energy transfer (BRET)
Energy transfer between the fluorophores of two bioluminescent proteins, which confirms their close proximity and suggests they interact.

biomarker
Any indicator of biological state, such as a disease or drug response.

biphasic column
A chromatography column containing two distinct stationary phases with different chemical properties allowing orthogonal separation in one column.

BLAST
See **Basic Local Alignment Search Tool.**

BLOCKS
A database of short, highly-conserved functional sequences in proteins.

BLOSUM (blocks substitution matrix)
A substitution matrix based on amino acid substitutions observed in blocks of aligned sequences with a certain level of identity.

bottom-up proteomics
Any proteomics strategy involving the identification of proteins based on the behavior of peptides.

Bradford assay
An assay that measures the total protein concentration in a solution.

***b*-series and *y*-series ions**
Peptide ions generated by fragmenting the peptide bond, with the charge remaining on the N-terminal fragment (*b*-series) or C-terminal fragment (*y*-series).

capillary electrophoresis
Electrophoresis carried out in thin glass capillaries.

CATH (class, architecture, topology, homologous superfamily)
Strategy for the structural classification of proteins, and the associated database.

cathodic drift
The tendency of analytes to move toward the cation during electrophoresis.

cation exchange (CAX) chromatography
Chromatography mode based on the exchange of positively charged ions on anionic resins, for example, sulfate and carboxylate groups.

cell microarrays
Microarrays containing discrete groups of cells rather than immobilized nucleic acids or proteins.

cell surface labeling
A strategy to label extracellular proteins by using dyes that cannot cross the cell membrane.

cell surface shaving
A strategy to isolate extracellular proteins by cleaving them from the cell surface using proteases without disturbing the contents of the cell.

cell-free expression systems
Expression systems that utilize purified transcriptional and translational components *in vitro* rather than recombinant cells.

census sequencing
Any method for the analysis of gene expression that involves counting the occurrence of different DNA sequences representing messenger RNAs.

charge density
The concentration of charges on the surface of a protein.

charge-coupled device (CCD)
A device that is capable of moving electrical charges to specific sites and converting them into digital values, for example, for imaging.

chemical derivatization
The conversion of a specific chemical group into a derivative, for example, to improve stability or to achieve labeling.

chemical proteomics
The high-throughput analysis of protein functions by testing proteins against libraries of small chemical compounds.

chemical shift
In NMR spectroscopy, a shift in the resonant frequency of a nucleus compared with a standard, revealing the presence of adjacent chemical groups.

chip
A small device, usually the size and shape of a microscope slide, used for microanalytics.

Chou–Fasman method
An empirical method for the prediction of secondary structures in proteins based on the propensity of different amino acids to occur in particular secondary structures.

chromatofocusing (CF)
A protein separation method using ion exchange resins and pH elution gradients, which allows proteins to be separated according to their isoelectric point.

chromatogram
The visual output of chromatography comprising a series of absorption peaks representing different components.

chromatography
Any separative process in which molecules are partitioned between a solid stationary phase and a mobile liquid phase.

circular dichroism (CD)
The differential absorption of left and right circularly polarized light, which helps to identify protein secondary structures.

co-immunoprecipitation
The immunoprecipitation of a soluble protein by virtue of its ability to bind to specific antibody, which is captured by affinity to protein A/G or by cross-linking.

collision cell
The part of a mass spectrometer that allows the gas-phase ions to collide with a neutral gas, thus generating fragments (product ions).

collision induced dissociation (CID)
The fragmentation of intact peptide ions by collision with a neutral gas stream in the collision cell of a mass spectrometer.

collision induced dissociation (CID) spectrum
Mass spectrum produced by collision-induced dissociation.

combined fractional diagonal chromatography (COFRADIC)
A quantitative proteomics method that distinguishes between natural N-termini and those generated by proteases.

comparative genomics
The study of relationships between the genomes of different species.

comparative modeling
See **homology modeling**.

complete interpretation
The derivation of a complete amino acid sequence from a mass spectrum.

complete orthogonality
A hypothetical separation in which properties affecting the separation of proteins in one dimension do not overlap with those affecting separation of proteins in another dimension, resulting in maximum peak capacity in both separations. In practice, there is always some overlap between properties and complete orthogonality is never achieved.

complex
An interacting group of proteins forming a single multimeric structure.

complex interactions
Interactions between proteins within the complex, not necessarily involving only two partners.

complexome
All the protein complexes in a cell or organism.

comprehensive mutant library
A library of cells or organisms containing mutants representing every gene in the genome.

consensus sequences
A sequence representing the most common amino acids at each position in a collection of aligned sequences.

conservative substitutions
Amino acid substitution in which one amino acid is replaced with another that has similar chemical properties.

context-based searching
A pattern recognition strategy that improves the sensitivity of sequence alignment by calculating the alignment score based on context, not just matches at the single amino acid level.

context-specific BLAST (CS-BLAST)
A derivative of BLAST that includes 12 amino acids of surrounding sequence context to identify more distant relationships.

context-specific iterated BLAST (CSI-BLAST)
A derivative of PSI-BLAST that includes 12 amino acids of surrounding sequence context to identify more distant relationships.

core glycan
The glycan structure formed on glycoproteins when they enter the endoplasmic reticulum, which is decorated and modified in other parts of the secretory pathway.

correlative database searching
Database searching based on correlations between empirical data and theoretical data derived from the database for example, peptide masses derived by mass spectrometry and theoretical peptide masses based on database sequences.

COSY (correlation spectroscopy)
A standard form of two-dimensional NMR that identifies spins which are coupled to each other (namely, chemical bonds).

cross-correlation
See **correlative database searching**.

***c*-series and *z*-series ions**
Peptide ions generated by fragmenting the N–C bond, with the charge remaining on the N-terminal fragment (*c*-series) or C-terminal fragment (*z*-series).

database pollution
The propagation of incorrect data in a database due to the automatic annotation of new proteins using existing incorrect annotations.

daughter ion scanning
The use of two mass analyzers in series to separate intact ions and then their fragments (also known as product ion scanning).

Dayhoff matrices
See **PAM**.

denatured conformations
Any protein conformation that is not a typical native state, usually caused by incorrect folding or unfolding due to the presence of heat or chemicals.

densitometer
A device that measures optical density.

densitometric resolution
The resolution to which a densitometer can separate degrees of optical density.

deuterated
Molecules in which hydrogen atoms are replaced with the heavy hydrogen isotope deuterium.

Dicer
The enzyme responsible for cleaving messenger RNAs during RNA interference.

difference in-gel electrophoresis (DIGE)
A method for the simultaneous separation of different protein samples by two-dimensional electrophoresis, based on the labeling of each sample with different fluorescent dyes.

differential display PCR
Method in which pools of cDNA from different samples are amplified using the same primers and displayed side-by-side in adjacent gel lanes to identify differences in the appearance or abundance of particular bands.

diffraction patterns
The patterns on a reflector generated by the scattering of X-rays by a protein crystal.

dimethylation-TAILS
A quantitative proteomics method that distinguishes between natural N-termini and those generated by proteases.

direct analysis of large protein complexes (DALPC)
A method in which isolated protein complexes are digested into peptides and analyzed using the MudPIT biphasic chromatography–MS approach.

direct detection
Detection of specific proteins using labeled probes, in contrast to sandwich assays in which the initial probe is not labeled and a secondary detection agent is required.

direct labeling
The detection of proteins by labeling those proteins rather than detection with a separate labeled probe.

discontinuous system
Any orthogonal separation system in which an operator collects eluted samples and feeds them into the next separation stage manually. For example, samples may be collected as they emerge from one chromatography column and fed into another.

discovery driven
Research driven by the desire to discover the components of the system rather than to test a particular hypothesis.

discovery proteomics
A proteomics approach that is not based on any specific hypothesis but instead on the desire to enumerate the components of the system.

distance restraints
In NOESY NMR spectroscopy, restraints based on the proximity between two nuclei.

distributed computing
A computing task that is solved by a network of computers coordinating their actions rather than by a single central facility.

disulfide bridges
Intermolecular or intramolecular bonds formed between cysteine residues.

DNA array to protein array (DAPA)
A method for the manufacture of protein microarrays that involves the use of DNA arrays as templates for the cell-free expression of proteins in the same pattern.

DNA microarray
A miniature device containing nucleic acid probes that are used to detect nucleic acids in solution by hybridization.

domain
A component of a protein that is structurally, functionally, and evolutionarily independent.

domain enhanced lookup time accelerated BLAST (DELTA-BLAST)
A BLAST search that also incorporates information from a database of conserved protein domains.

domain fusion method
A functional annotation method that compares multi-domain proteins in one species with single-domain proteins in other species to infer functional relationships.

dual baits
Any system for testing differential protein interactions between a single prey protein and two different baits.

dynamic light scattering (photon correlation spectroscopy)
A label-free method for measuring protein interactions by reporting changes in the hydrodynamic radius of molecules in solution.

dynamic programming
A programming method for solving complex problems by breaking them down into simpler problems.

edge detection filters
Image scanning software that identifies regions with a sharp change in pixel intensity signifying the edge of a spot.

Edman degradation
A method for protein sequencing based on the progressive removal and identification of amino acids from the N- terminus.

electron capture dissociation (ECD)
A method for the fragmentation of peptides based on the capture of a thermal electron by a multiprotonated peptide or protein cation.

electron density map
A three-dimensional image of the electron clouds within a molecule produced by X-ray diffraction.

electron diffraction
A method for structural determination similar to X-ray diffraction, which is particularly useful for two-dimensional arrays such as flat protein sheets.

electron tomography
An extension of transmission electron microscopy that allows the detailed three-dimensional structure of macromolecular objects to be determined.

electron transfer dissociation (ETD)
A method for the fragmentation of peptides based on the transfer of an electron from a radical anion to the multiprotonated peptide or protein.

electro-osmotic flow
The bulk of movement of solvent toward an electrode during electrophoresis.

electrophoresis
The movement of a charged molecule in solution when an electric field is applied.

electrospray ionization
A soft ionization method in which the liquid containing the analyte is dispersed into an aerosol through a fine needle.

ellipsometry
A technique for detecting protein interactions on chips by measuring changes in the polarization state of reflected light due to the changing refractive index of the sample surface (the oblique incidence reflectivity difference).

enhancer mutations
Second mutations that increase the severity of the phenotype caused by a primary mutation.

enhancer trap
An insertion construct that comprises a reporter gene downstream of a minimal promoter, which is therefore activated when it integrates near to an endogenous enhancer.

enzyme
A protein that catalyzes a biochemical reaction.

enzyme-linked immunosorbent assay (ELISA)
A solid-phase assay for the detection and quantitation of proteins that uses antibodies carrying an enzymatic or fluorescent label as detection reagents.

epistasis
A genetic interaction in which a mutation in one gene masks the phenotype of a mutation in a second gene.

epistatic miniarray profiles (E-MAPs)
A high-throughput method involving the systematic mating of haploid yeast cells to screen for epistatic interactions.

ESI
See **electrospray ionization**.

exoglycosylase
An enzyme that removes sugars from the ends of oligosaccharides.

expressed sequence tags (ESTs)
Partial cDNAs generated by single-pass sequencing.

expression proteomics
The branch of proteomics that deals with the expression levels and abundance of proteins.

extracted ion chromatogram (XIC) method
The measurement of precursor ion signal intensity by isolating m/z values representing one or more individual analytes on a chromatogram.

false negative
An experimental result that falsely indicates the absence or inactivity of a particular component that is actually present or active, reflecting the failure or limitations of the experimental setup or method.

false positive
An experimental result that falsely indicates the presence or activity of a particular component that is actually absent or inactive, reflecting the failure or limitations of the experimental setup or method.

family
A group of evolutionarily related genes or proteins. There is no precise quantitative definition, and the description of families, subfamilies, and superfamilies is context dependent.

far-UV circular dichroism spectrophotometry (CDS)
The measurement of absorption spectra from proteins in solution exposed to far-UV circularly polarized light.

FASTA
A heuristic algorithm used to find sequence alignments.

fingerprints
(a) *See* **peptide mass fingerprinting**; (b) groups of short, ungapped and highly-conserved sequences (the latter described as motifs or blocks, depending on the database).

fluorescence resonance energy transfer (FRET)
The transfer of energy between two fluorophores that occurs when the emission wavelength of one overlaps with the excitation wavelength of the other.

fold
The three-dimensional structure of a protein.

fold library
A database of three-dimensional protein structures.

fold recognition
See **threading**.

folding@home
A distributed program that uses spare computer resources to model protein folding.

forward arrays
A protein microarray comprising single components (proteins or affinity reagents) arranged in a grid, which are exposed to a complex analyte.

forward genetics
A strategy in genetics in which a mutation or protein is used as the starting point to isolate the corresponding gene.

FSSP
Families of structurally similar proteins: one of three major databases of protein structures. *See* **CATH**, **SCOP**

FT-ICR
Fourier transform ion cyclotron resonance—a sophisticated form of mass spectrometry in which ions orbiting between two charged plates create an image current that is converted by Fourier transformation into the component frequencies and amplitudes of individual ions, corresponding to specific m/z values.

functional genomics
Any large-scale genomics-based strategy for the determination of gene functions.

functional microarray
A protein microarray containing proteins that are functionally tested, for example, for biochemical activity or interactions.

functional proteomics
The branch of proteomics dealing with the large-scale analysis of protein functions.

gap
A missing segment in a sequence alignment.

gap penalties
A reduction in an alignment score awarded for the introduction of a gap into sequence alignment.

Gaussian fitting
Any method that attempts to fit values to a normal distribution.

gel filtration chromatography
The separation of molecules in an analyte by molecular mass, based on the size-dependent exclusion of molecules from pores in the resin.

gel matching
The alignment of two-dimensional gels so that corresponding spots show perfect correlation.

gel pads
Small three-dimensional gel structures that are used to construct microarrays less likely to suffer from evaporation artifacts.

gene knockout
The creation of a null (complete loss of function) mutant by using homologous recombination to interrupt an endogenous gene or replace it with a nonfunctional sequence.

Gene Ontology
A bioinformatics initiative to unify the naming and annotation of genes and proteins across all species.

gene silencing
Any strategy to abolish the expression of a preselected gene.

gene trap
An insertion construct that contains a reporter gene downstream of a splice acceptor site so that it is expressed as a fusion protein when it integrates into an endogenous gene.

genome
The complete complement of DNA in a cell or organism.

genomics
The analysis of genomes, specifically through the application of genetics and molecular biology methods at the scale of the entire genome.

global distance test (GDT)
An accurate way to measure the similarity between protein structures.

global similarity
Similarity between sequences throughout their entire length.

glycan
An oligosaccharide or polysaccharide, often used to describe the carbohydrate portion of a glycoprotein.

glycoforms
Structural variants of a protein based on the presence of different glycans.

glycoproteins
Proteins with attached glycans.

glycoproteome
The sum of all glycoproteins in a cell or organism.

glycosidase
An enzyme that digests glycosidic linkages to release sugars.

glycosylation
The addition of glycans to proteins, a form of post-translational modification.

glycosylphosphatidylinositol (GPI) anchor
A glycolipid attached to the C-terminus of a protein allowing it to be inserted into a membrane.

GOR (Garnier–Osguthorpe–Robson) method
An empirical method for the prediction of secondary structures in proteins based on comparisons with known structures and statistical relationships between neighboring residues.

gradient elution
The elution of bound molecules from a chromatography column by gradually changing the properties of the elution buffer.

green fluorescent protein
A bioluminescent protein widely used as a reporter molecule in molecular biology.

gridded cDNA expression libraries
A clone library in which the clones are arranged in a gridlike pattern on a membrane or filter, which can be thought of as a forerunner of the protein microarray.

GST pull-down
A method for analyzing protein interactions in which the bait protein is expressed as a fusion with glutathione *S*-transferase and then captured (along with any bound prey) on glutathione-conjugated beads.

guide tree
The order of similarity among multiple sequences in a progressive alignment.

guilt by association
The tentative annotation of a protein by virtue of its association or interaction with another protein whose function is already known.

Hart scores
A method for socioaffinity scoring based on the combined observations of different interaction screens.

heated capillary dissociation (HCD)
A method for the fragmentation of peptides based on the induction of thermal dissociation using a heated capillary.

helical propensity
The likelihood of a given amino acid residue appearing in a helix.

helical wheel
A diagram in which the positions of amino acids are plotted on a circle corresponding to the pitch of an ideal helix.

heteronuclear
NMR spectroscopy methods in which a protein is labeled with multiple isotopes.

hidden Markov model
A stochastic model of autonomous and partially observable systems that is widely used for pattern recognition.

High-energy C-trap dissociation
A method for the fragmentation of peptides that is used with ion trap mass spectrometers.

high mannose
One of three major glycan structures, featuring a branched terminus containing many mannose residues.

high performance liquid chromatography (HPLC)
Chromatography carried out at high operational pressures (50–350 bar) compared with normal chromatography, where the movement of the mobile phase through the stationary phase is determined by gravity.

higher-energy collision dissociation (HCD)
See **high-energy C-trap dissociation**.

high-scoring segment pairs
In BLAST, an extension of the perfect but short matching segment that initiates the alignment.

homologous
Related by common ancestry.

homologous recombination
Recombination between two closely related DNA sequences.

homology modeling
The modeling of protein structures using a closely related protein as a template.

homology transfer
The assumption that interactions between proteins in one organism also occur among the orthologs of those proteins in other species.

homonuclear
NMR spectroscopy methods in which proteins are labeled with one isotope.

HUPO Proteomics Standards Initiative
A Human Proteome Organization (HUPO) working group developing data standards for proteome research.

hydrazine-based solid-phase extraction
A widely used chemical method for the enrichment of glycoproteins.

hydrogen–deuterium exchange mass spectrometry (DXMS)
A mass spectrometry method that can be used to determine regions of flexibility or disorder within a protein.

hydrophilic interaction chromatography
A form of liquid chromatography with a polar stationary phase, which is used to separate molecules on the basis of their hydrophilic interactions.

hydrophobic collapse
The folding of a protein beginning with the initial compact arrangement of hydrophobic residues in the core.

hydrophobic interaction chromatography
A form of liquid chromatography with a hydrophobic stationary phase, which is used to separate molecules on the basis of their hydrophobic interactions.

hypothesis-driven
Research whose aim is to test a specific hypothesis.

hypothetical protein
A protein that is predicted to exist on the basis of a gene sequence or phenotype, in the absence of direct evidence at the protein level.

IDBOS scores (interaction detection based on shuffling)
A method for socioaffinity scoring that specifically looks for direct physical interactions and assumes that purified complexes can be randomly permutated.

identity
In sequence comparisons, the percentage of identical nucleotides or amino acids in an alignment.

identity matrix
A substitution score matrix where identical amino acids score 1 and non-identical amino acids score 0.

image acquisition
The process of digitizing an image for further analysis.

imino acid
A molecule containing an imino group and a carboxyl group, for example, proline.

Immobilines
Reagents that are used to generate a pH gradient during gel electrophoresis, and that remain immobile because they are conjugated to the gel matrix.

immobilized metal-affinity chromatography (IMAC)
A form of liquid chromatography in which metal ions are immobilized on the resin to capture negatively charged proteins and peptides.

immobilized pH gradient (IPG)
A pH gradient established in a gel by the arrangement of Immobiline reagents.

immunoaffinity enrichment
The enrichment of particular classes of proteins or peptides (such as phosphoproteins) using a specific antibody.

immunoassay
Any assay involving the use of an antibody for detection.

immunoblot
An immunoassay in which proteins are first transferred to a solid-phase support and immobilized.

immuno-RCA technique
A sandwich assay involving a tertiary level of detection by rolling circle amplification.

improved socioaffinity scores (ISAs)
A method for socioaffinity scoring that makes full use of repetitive purifications.

in-cell NMR
A noninvasive form of NMR spectroscopy that can be used to analyze protein structures in living cells.

indel
In sequence alignments, an indel represents an insertion in one sequence and a deletion in the other when it is unclear which sequence arose first.

indirect detection
A protein detection method in which the detection reagent is unlabeled and is in turn detected by a second, labeled reagent.

infrared multiphoton dissociation (IRMPD)
A method for the fragmentation of peptides based on induced collision with photons.

INSDC nucleotide sequence database
Any of the three major sequence databases operated by the International Nucleotide Sequence Database Collaboration.

insertional mutagenesis
The creation of a mutant allele by inserting a DNA sequence that inactivates the gene.

in-source CID
High-energy ionization, for example, using the MALDI method to induce multiple collisions between the peptide and matrix compound to fragment the peptide backbone.

integrated spot intensity
A value assigned to protein spots on a two-dimensional gel image to represent the overall intensity, accounting for shape and density variations, etc.

interaction hotspot
An area of protein surface that is critical for interactions with other proteins, often a valuable drug target.

interaction map
A topographical map showing links between proteins and protein complexes in the interactome.

interaction proteomics, interactomics
The branch of proteomics that deals with protein interactions.

interface
The localized area on the surface of a protein that interacts with other molecules.

interferometry
Any label-free detection method involving the measurement of phase differences between waves that cause intensity fluctuations known as interference fringes.

intermolecular comparison
A method for the comparison of protein structures in which two proteins are superimposed and the algorithm attempts to minimize the distance between superimposed atoms.

International Nucleotide Sequence Database Collaboration (INSDC)
An international consortium that provides access to the three major collaborative sequence databases.

InterPro
A database of protein family domains and functional sequences.

intramolecular comparison
A method for the comparison of protein structures involving side-by-side comparison. The algorithm measures the internal distances between equivalent atoms in each structure.

ion exchange (IEX) chromatography
A form of liquid chromatography with a polar stationary phase that is used to separate molecules on the basis of their charge density. *See* **anion exchange chromatography**, **cation exchange chromatography.**

ion packet
A group of ions with the same m/z value and cyclotron frequency in a uniform magnetic field.

ion trap
A chamber surrounded by a ring electrode that allows ions of a certain m/z value to cycle (and thus remain trapped) within the field.

isobaric tagging
A quantitative mass spectrometry technique in which proteins are labeled with chemical groups that nominally have the same mass but that yield reporter ions with different masses when fragmented.

isoelectric focusing (IEF)
A form of electrophoresis in a pH gradient that allows proteins to be separated according to their isoelectric point.

isoelectric point
The position in a pH gradient where a protein bears no intrinsic charge.

isomorphous crystals
Protein crystals with the same structure but incorporating different atoms to produce alternative diffraction patterns.

isothermal titration calorimetry
A method for detecting protein interactions based on changes in temperature that occur when proteins associate and disassociate in solution.

isotope-coded affinity tags (ICATs)
A protein tag widely used for quantitative proteomics, comprising a reactive group that labels cysteine residues, an isotopically coded linker, and an affinity reagent such as biotin for affinity isolation.

isotope-coded protein labeling (ICPL)
A technique similar to ICAT in which the reactive group labels free amino groups rather than cysteine residues.

iTRAQ (isobaric tags for relative and absolute quantification)
A quantitative mass spectrometry technique in which proteins are labeled with chemical groups that nominally have the same mass but yield reporter ions with different masses when fragmented.

knock-in
A homologous recombination technique that is used to replace one gene sequence with another.

***k*-tuples**
An ordered list of elements of length *k* that is used as the basis of heuristic sequence alignment methods.

lab-on-a-chip
Any small device that can carry out normal laboratory procedures such as electrophoresis and chromatography on a minute scale.

landmarks
Protein spots present on every gel during comparative 2DGE, and which can therefore be used as references for gel alignment.

Laplacian of Gaussian (LOG)
A mathematical method for the detection of spots on a digital image.

large-scale DNA sequencing
Automatic DNA sequencing carried out using highly parallel techniques.

laser capture microdissection (LCM)
A technique using laser beams to dissect small tissue samples and recover specific cells, for example, cancer cells from a larger biopsy.

lead compounds
Any chemical compound with pharmacological activity that is used as a starting point for chemical modifications to improve drug-like properties such as potency, selectivity, and efficacy.

lectin-affinity chromatography
A form of chromatography used to capture glycoproteins by exploiting their affinity for lectins.

lectin
A protein that interacts with carbohydrates, often with great specificity.

Levinthal paradox
The astronomically large number of potential conformations that theoretically can be adopted by a protein, indicating that protein folding does not occur by randomly sampling all different possible confirmations but must follow an ordered path.

line analysis
A method for spot detection on two-dimensional gels that involves the sequential analysis of vertical scan lines to identify density peaks.

liquid chromatography (LC)
Any form of chromatography that involves a solid stationary phase and a liquid mobile phase.

local similarity
Short regions of similarity within a longer sequence.

loss-of-function
A mutation that abolishes the function of a gene.

low-complexity sequences
Commonly occurring sequences that are not necessarily evolutionarily related, such as transmembrane helices.

Lowry assay
An assay that measures the total protein concentration in a solution.

LTQ (linear trap quadrupole)
A sophisticated mass spectrometer featuring a linear-field quadrupole and ion-trap mass analyzer that traps ions in a two-dimensional quadrupole field.

luciferase
A class of oxidative enzymes responsible for bioluminescence, widely used as reporters in molecular biology.

LUMIER (luminescence-based mammalian interactome mapping)
A high-throughput mammalian two-hybrid platform in which the bait and prey hybrids are fused to luciferase to facilitate detection, and to an epitope to facilitate the affinity-based capture of interacting proteins.

macroheterogeneity
Different protein glycoforms characterized by alternative glycan site occupancy.

MAD (multiple-wavelength anomalous dispersion)
The use of synchrotron radiation of different monochromatic wavelengths to generate alternative diffraction patterns from protein crystals containing heavy metal atoms by exploiting the phenomenon of anomalous scattering.

magnetic moment
The magnetic properties of a nucleus with non-integer spin.

MALDI
See **matrix assisted laser desorption/ionization**.

MALDI-TOF
See **matrix assisted laser desorption/ionization**, **time of flight**.

manual alignment
Sequence alignment carried out without computer assistance, or to improve quality following computer alignment.

MAPPIT (mammalian protein–protein interaction trap)
A high throughput mammalian two-hybrid platform that works by restoring the activity of the JAK–STAT signaling pathway.

marker ion
See **reporter ion**.

mark-up language
Any strategy to annotate a document in a way that is distinguishable from the document contents.

Mascot
A commonly used program for peptide mass fingerprinting and correlative database searching.

mass deviations
Differences in mass between experimental and theoretical peptides caused by the presence of isotopes, mass tags, or particular post-translational modifications.

mass filter
The component of a mass spectrometer that allows the selection of ions on the basis of their m/z ratio.

mass instability mode
The use of an ion-trap mass spectrometer to eject ions with an m/z ratio above a certain threshold.

mass label
Any adduct conjugated to a peptide or protein, or a isotopic substitution, which creates a defined mass difference that can be detected by mass spectrometry. (*Also see* **mass tag**).

mass spectrometer
A device that can produce, separate, and detect gas-phase ions to generate a mass spectrum.

mass spectrometry
An analytical technique that produces spectra representing the masses of molecules in a sample.

mass spectrum
The output of a mass spectrometer, showing peaks representing ions with different m/z ratios.

mass tag
A mass label used in quantitative mass spectrometry generated by chemical derivatization rather than the incorporation of stable isotopes.

mass/charge (m/z) ratio
The mass-to-charge ratio of an ion, which is the selective criterion for separation by mass spectrometry.

mass-coded abundance tag (MCAT)
See **mass tag**.

massively parallel signature sequencing (MPSS)
A large-scale sequencing method involving the parallel sequencing of many DNA molecules immobilized on beads.

matrix assisted laser desorption/ionization
A soft ionization technique in which the analyte is mixed with an aromatic matrix that absorbs laser energy and allows the desorption as gas-phase ions.

matrix interactions
Direct and indirect interactions between proteins in a complex.

matrix screening method
A systematic version of the yeast two-hybrid system in which panels of defined bait and prey are mated in an array format.

membrane proteomics
The branch of proteomics that deals with membrane-localized proteins.

membrane-based yeast two-hybrid
See **split ubiquitin system**.

metabolic labeling
The incorporation of an isotopic label into proteins while the sample is still metabolically active.

metabolic tagging
The chemical derivatization of proteins in living cells.

metabolome
The sum of all metabolites (small molecules) in a cell or organism.

metabolomics
The global analysis of all the small metabolic compounds in a cell or organism.

microarray
A miniature device in which analytical targets are arranged in a grid. *See* **DNA microarray, protein microarray**

microheterogeneity
Different protein glycoforms characterized by alternative glycan structures at the same site.

middle-down proteomics
Protein identification by mass spectrometry starting with large peptides.

Minimal Information About a Molecular Interaction Experiment (MIMIx)
A minimal set of standards for the presentation of experimental data relating to protein interactions.

Minimal Information About a Proteomics Experiment (MIAPE)
A minimal set of standards for the presentation of experimental proteomic data.

Minimum Information about a Biomedical or Biological Investigation (MIBBI)
An overarching standards initiative for all biomedical experiments.

Minimum Information About a Microarray Experiment (MIAME)
The prototype minimal set of standards initiative seeking a standardized approach to the presentation of DNA microarray data.

mobile phase
The non-fixed phase in a chromatography column, which may be gas or liquid, but is usually liquid in proteomics.

modification-specific antibodies
Antibodies that recognize particular types of post-translational modification.

molecular chaperones
Proteins whose function is to facilitate the folding or re-folding of other proteins.

molecular exclusion
The separative principle used in gel filtration chromatography, which is based on the inability of molecules above a certain mass to enter pores in the resin.

molecular imprinted polymers (MIPs)
A polymer that forms around another molecule and is left with the imprint of its shape.

molecular replacement
The creation of an electron density map from X-ray diffraction data using phases from related protein structures.

molten globule
A model of protein folding involving an intermediate folding state.

monosaccharide composition
A quantitative and qualitative description of the sugar components of a glycan.

morpholino antisense oligonucleotide
A type of stable nucleic acid analog that is used for gene silencing.

motifs
(a) Functionally related, short and conserved protein sequences that have been defined experimentally; (b) three-dimensional motifs in proteins comprising multiple secondary structures (for example, helix-turn-helix motif) or specific configurations of functional amino acids (for example, catalytic motif); (c) the name for a short ungapped highly conserved protein sequence in the PRINTS database; (d) natively unstructured regulatory modules in proteins, more fully known as short linear motifs (SLiMs) or MiniMotifs.

MOWSE score
A probability score showing the likelihood that an experimental peptide mass matches the theoretical peptide mass derived from a database sequence.

MS/MS
Tandem mass spectrometry—a form of mass spectrometry in which two mass analyzers operate in series, separated by a collision cell, allowing the analysis of fragment ions..

MS-BLAST
A variation of the BLAST algorithm that is used to search sequence databases with small protein tags derived by *de novo* mass spectrometry-based sequencing.

MS^n
Higher order tandem mass spectrometry, where n refers to the number of rounds of fragmentation and analysis.

MS-Tag
An algorithm used to search sequence databases with short peptide tags derived by *de novo* mass spectrometry-based sequencing.

multi-analyte immunoassay
Any immunoassay in which protein microdots are spotted onto a solid support. These can be considered forerunners of the protein microarray.

multidimensional liquid chromatography (MDLC)
Any form of multistage liquid chromatography that involves the sequential use of different separative principles.

multidimensional protein identification technology (Mud-PIT)
A key proteomics platform in which proteins or peptides are separated by multidimensional liquid chromatography using a biphasic column and then automatically transferred to a mass spectrometer for protein identification.

multiple isomorphous replacement (MIR)
A comparison of reflections generated by the X-ray diffraction of isomorphous crystals containing different heavy metal atoms.

multiple reaction monitoring (MRM)
A quantitative method in mass spectrometry involving the analysis of multiple transition pairs of precursor and product ions.

multiple sequence alignment
The alignment of more than two sequences.

multiplex hybridization
Hybridization experiments in which there is more than one target, often arranged in the form of a grid as in a DNA microarray.

mutability scores
The weighting of different amino acid changes in a substitution matrix based on the likelihood of such exchanges occurring in nature.

mutation data matrix
A substitution matrix generated by studying mutation rates in nature.

MYTH
See **split ubiquitin system**.

nanohole array
A microfluidic device integrated with surface plasmon resonance spectroscopy, which incorporates nanoscale holes in a metal film to increase sensitivity.

nanowell chip
Any chip-like device or microarray featuring nanoscale indentations to improve sensitivity and reduce surface evaporation.

native conformation
The conformation of the protein in its normal functional state.

needle biopsies
Solid biological samples taken with a thin-bore needle.

Needleman–Wunsch algorithm
A dynamic programming algorithm that aligns sequences by searching for global similarity.

neutral loss scan mode
A mass spectrometry mode in which the first analyzer scans the full mass range, the next acts as a collision cell, and the third scans the full mass range at different register to detect neutral losses from the product ions.

neutron diffraction
A method for determining protein structures by scattering neutrons from the nuclei, allowing the detection of hydrogen atoms (which cannot be detected by X-rays).

next-generation sequencing
Any of various ultra-high-throughput DNA sequencing methods that generate millions of short sequences in parallel.

***N*-linked glycosylation**
The addition of sugar chains to the amide nitrogen on the side chain of asparagine residues.

NMR spectroscopy
See **nuclear magnetic resonance (NMR) spectroscopy**.

NOESY (NOE spectroscopy)
A form of NMR spectroscopy that takes advantage of the nuclear Overhauser affect and allows the detection of nuclei that are close together in space but not connected.

non-equilibrium pH gradient electrophoresis (NEpHGE)
A form of gel electrophoresis in which proteins are separated while the pH gradient is forming.

nonlinear pH gradient
A pH gradient that is flattened to improve the separation of proteins with typical pI values, created by increasing the spacing between Immobiline reagents over a particular pH range.

N-terminome
The complete collection of N-terminal peptides representing all the proteins in the proteome.

nuclear magnetic resonance (NMR) spectroscopy
The measurement of radio waves emitted from nuclei exposed to resonant radio waves in an applied magnetic field, which yields data about the relative positions of the nuclei and hence the molecular structure.

nuclear Overhauser effect (NOE)
The transfer of nuclear spin polarization from one nuclear spin population to another via cross-relaxation, allowing the detection of interactions between nuclei that are close together in space but not directly connected.

nucleic acid programmable protein array (NAPPA)
A procedure for manufacturing protein microarrays in which biotinylated cDNAs encoding GST fusion proteins are printed onto avidin-coated slides together with an anti-GST antibody, followed by *in situ* protein synthesis such that the resulting proteins are captured by the antibodies.

null mutations
A mutation that causes a complete loss of gene function.

oblique incidence reflectivity difference
A property of the polarization state of the light that is measured by ellipsometry.

oligonucleotide and peptide aptamers
Short nucleic acid or protein sequences with selective affinity for particular target molecules that can therefore be used as capture agents on protein microarrays.

oligonucleotide chip
A device for analyzing gene expression that is constructed by synthesizing oligonucleotides *in situ* on the chip surface.

oligosaccharide
A short chain of sugar residues, usually fewer than ten.

***O*-linked glycosylation**
The addition of sugar chains to the hydroxyl oxygen on the side chain of hydroxylysine, hydroxyproline, serine, or threonine.

'omics'
Large-scale systematic analysis of a biological system defined by the suffix -ome (for example, proteomics is the analysis of the proteome).

on-chip interferometric backscatter detection
A form of interferometry that can be used to characterize protein interactions on microarrays.

one-third rule
A limitation of ion-trap mass spectrometers, based on the principle that the ratio between the precursor-ion m/z and the lowest m/z of the trapped fragment ion never increases above 0.3.

organellar proteomics
The branch of proteomics that deals with the protein content of specific organelles.

organic fluorophores
An organic molecule that can emit light at one wavelength following excitation by incident light at another wavelength.

orientation restraints
Restraints in NMR spectroscopy resulting from residual dipolar coupling.

orphan gene
A confirmed gene sequence whose function is unknown because it is unrelated to any functionally characterized gene sequence in the databases.

orthogonality
The degree to which properties affecting the separation of proteins in one dimension overlap with those affecting separation of proteins in another dimension.

orthologs
Genes (and the corresponding proteins) that have diverged by speciation and fulfill equivalent functions in different species.

PAM
Substitution score matrix based on the percentage of accepted mutations using an explicit evolutionary model.

PAM units
Evolutionary time required for one amino acid change in a sequence of 100 amino acids.

paralogs
Genes (and the corresponding proteins) that have diverged from a common ancestor within a species and are not necessarily functionally equivalent.

peak capacity
The number of chromatography peaks that can be separated from one another up to a predefined capacity ratio, used as a measure of resolving power in chromatography.

peptide mass fingerprinting (PMF)
A strategy for protein identification based on correlating experimentally determined peptide masses and theoretical masses derived from sequence databases.

peptide tag
A short *de novo* peptide sequence derived by mass spectrometry that can be used for database searching.

peptide-*N*-glycosidase F (P*N*Gase F)
An enzyme that cleaves a N_4-(acetyl-β-D-glucosaminyl) asparagine residue and thus removes *N*-linked glycans from proteins and peptides.

periodic acid/Schiff (PAS)
A gel staining method that works well with glycoproteins.

phage antibody display
Variant of the phage display system that is used to present antibodies and therefore select those with the most suitable properties.

phage interaction display
A variant of the phage display system that is used to isolate prey proteins which interact with baits expressed on the surface of phage particles.

phase problem
The inability to determine the phase angles of the reflections in a diffraction experiment.

phenocopy
A cell or organism with the appearance of a mutant phenotype that is caused by gene silencing rather than by mutation.

phenomics
The large-scale analysis of phenotypes.

phenyl isothiocyanate
A reagent used to label the N-terminal amino acid of a protein to facilitate sequencing by Edman degradation.

PHI-BLAST (pattern-hit initiated BLAST)
An algorithm similar to PSI-BLAST that allows the explicit inclusion of a sequence signature by the user.

phosphatase
An enzyme that adds phosphate groups to proteins.

phosphoforms
Different forms of a protein varying in the number and or position of phosphate groups.

phosphoproteins
Proteins that carry phosphate groups as post-translational modifications.

phosphoproteome
The sum of all phosphoproteins in a cell or organism.

phosphotyrosine-specific immonium ion scanning, PSI scanning
A mass spectrometry technique for the analysis of phosphoproteins, which detects the immonium ion by breaking the polypeptide backbone on either side of a phosphotyrosine residue.

photolithography
A technique for the production of microarrays involving *in situ* synthesis of nucleic acids or peptides.

phylogenetic profiling
A method for predicting protein interactions and functions based on their presence in phylogenetically related organisms.

plasmon
The collective oscillations of electrons. Plasmons on the surface of metal films can be used to study the kinetics of molecular interactions. *See* **surface plasmon resonance spectroscopy.**

polyacrylamide gel electrophoresis (PAGE)
The separation of charged molecules by electrophoresis using a polyacrylamide gel as a support matrix and to improve sieving by size.

pooled matrix screening method
A systematic yeast two-hybrid screening method in which arrays of bait proteins are tested against pooled prey proteins and then deconvoluted when interactions are detected.

Poppe plot
A kinetic plot that indicates the performance limits of separation systems by comparing peak capacity and speed of separation.

position-specific score matrix (PSSM)
A conserved sequence that includes the likelihood of different amino acids appearing at different positions.

post-source decay (PSD)
High-energy ionization using the MALDI method to induce multiple collisions between the peptide and matrix compound, resulting in the delayed fragmentation of the peptide backbone.

post-translational modification (PTM)
Any chemical modification of a protein after synthesis.

practical peak capacity
The peak capacity of multidimensional chromatography, taking incomplete orthogonality into account.

precursor ion scanning
A mass spectrometry method in which the first analyzer scans the full *m*/*z* range of the precursor ion and the second is set to detect specific reporter ions.

pre-focusing
The act of initiating electrophoresis before adding the sample to the gel in order to achieve a pH gradient.

prey
Proteins that are tested for interactions with specific bait.

primary sequence databases
Databases that store raw sequence data.

probability-based matching
A form of uninterpreted spectral analysis in mass spectrometry, in which virtual spectra are derived from relevant peptides and compared with observed fragmentation data.

product scan mode
A mass spectrometry method in which the first analyzer gates specific precursor ions and after collision the second analyzer scans the full *m*/*z* range of the product ions.

progressive alignment
A multiple alignment method beginning with the two most similar sequences and progressively aligning the remainder in order of decreasing similarity.

propagation
A method for matching spots on two-dimensional gels by starting at a known landmark and working outwards.

protein
A biopolymer comprising one or more chains of amino acids.

protein chip
See **protein microarray**.

protein complementation assay
Any assay for detecting protein interactions that works by reassembling a functional protein from two components which are fused to potential interactors.

protein complex
A multimeric structure comprising multiple proteins.

Protein Data Bank
The principal database of protein structures.

protein domain
A functionally and structurally independent region of a protein, typically able to function in isolation.

protein expression profiling
The analysis of protein expression and abundance in different cells or tissues.

protein fragment complementation
The principle underlying the yeast two-hybrid system and other protein complementation assays, which involves the assembly of a functional protein from two nonfunctional fragments.

protein localization traps
An insertion construct that identifies particular types of protein based on their localization in the cell.

protein microarray
A small device containing many proteins immobilized in the form of a grid, which (depending on the design) can be used to measure protein abundance, interactions, and activity.

protein scaffolds
Recombinant proteins that have been developed as specific affinity reagents.

protein sequence
The primary amino acid sequence of a protein or peptide.

protein signature databases
Databases containing highly-conserved short protein sequences.

protein suspension array
A protein microarray based on a bead or particle format suspended in a liquid medium.

protein–protein interactions
Interactions between two or more proteins to form a dimer or complex.

proteolysis
The digestion of a protein into amino acids.

proteome
The complete complement of proteins present in a cell, organ, tissue, organism, or alternative sample such as serum or whole blood.

proteome linkage maps
A topographical map showing the interactions between proteins in the proteome.

proteomics
The large-scale or global analysis of proteins.

Proteomics Standards Initiative (PSI)
A Human Proteome Organization (HUPO) working group developing data standards for proteome research.

PSI-BLAST (position-specific iterated BLAST)
A variant of BLAST that performs iterative searches and includes new hits in the query sequence after each round.

pulsed q dissociation (PQD)
A method for the fragmentation of low-mass fragment ions by activating the precursor ion for a short time with high energy.

purification enrichment score
A method for socioaffinity scoring that scores individual interactions separately.

QqLIT
A quadrupole linear ion trap mass spectrometer.

QqQ
A triple quadrupole mass spectrometer in tandem MS mode, with the second quadrupole acting as a collision cell.

QqTOF
A hybrid quadrupole time-of-flight mass spectrometer in tandem MS mode, with the second quadrupole acting as a collision cell.

quadrupole
A mass analyzer comprising four parallel metal rods, with a voltage applied across them such that they can filter ions on the basis of their m/z values.

quadrupole ion trap (QIT)
A quadrupole ion trap mass spectrometer, usually comprising an ion trap with a typical three-dimensional quadrupole field.

quantitative mass spectrometry
Any mass spectrometry approach that can yield the absolute or relative quantities of specific ions.

quantum dots
Miniature semiconductors with the ability to emit light due to quantum confinement, often used for imaging applications.

quenching
Interference between two fluorophores in close proximity which prevents the emission of light.

questionable
A putative gene whose existence is suggested by the sequence but whose expression and function has not been verified.

random library screening
In the yeast two-hybrid system, the use of libraries of random DNA fragments as baits and/or prey, instead of defined open reading frames.

random mutagenesis
Any nonselective mutagenesis strategy, for example, irradiation or chemical mutagenesis.

RAS recruitment system (RRS)
A method for detecting interactions among membrane proteins by making cell survival dependent on interactions with the critical membrane-bound regulator RAS.

reductionist approach
A scientific approach for the analysis of biological systems in which each system is broken down into its components, which are investigated individually.

reflections
The spots on a detector generated by the diffraction of X-rays through a protein crystal.

reflectron
A type of time-of-flight mass analyzer in which ion mirrors are used to reverse the direction of the ions and reduce the spread of flight times for ions with the same m/z ratio.

reporter ion
An ion that reveals the presence of a particular target analyte or class of target analyte.

resonance excitation method
A method used to eject ions from an ion trap mass spectrometer by oscillating the voltage of the end cap electrodes while varying the trapping voltage amplitude.

reverse genetics
A strategy in genetics in which a gene is used as the starting point to determine the function of a protein, for example, by mutation or overexpression.

reverse two-hybrid system
A method used to screen for mutations or chemicals that disrupt particular protein–protein interactions. It is based on the same principle as the yeast two-hybrid system, but uses a lethal reporter gene so that cells only survive if interactions are prevented. The system is often used for testing drugs that disrupt protein interactions.

reversed-phase microarray
Any microarray in which the arrayed features are complex mixtures, generally probed with specific affinity reagents.

reversed-phase (RP) chromatography
Any form of liquid chromatography using strongly hydrophobic (reversed-phase) resins.

RF-only mode
An operational mode for quadrupole mass spectrometers in which only a radio-frequency (RF) voltage is applied across the quadrupoles, thus allowing ions of any m/z ratio to pass through.

RNA-induced silencing complex (RISC)
A protein complex containing short interfering RNA that mediates the cleavage of target messenger RNA molecules during gene silencing by RNA interference.

RNA interference (RNAi)
A gene silencing method in which double-stranded RNA acts as a trigger to silence any gene with a homologous sequence.

RNA-Seq
Next-generation sequencing of cDNA.

rolling circle amplification
The generation of copies of a DNA sequence by the perpetual replication of a circular template.

root mean square deviation (RMSD)
A calculation used to measure the differences between observed values and the values predicted by a model.

Russian doll effect
The consequences of continuous structural variation within a protein family, which means that some members at the extremes of the continuum may appear structurally unrelated.

SAD (single-wavelength anomalous dispersion)
An X-ray crystallography method used to solve the phase problem, in which monochromatic X-rays are used to induce anomalous scattering from heavy metal atoms in a protein crystal.

salvage pathway
In structural genomics, any single technique or combination of techniques used to increase the likelihood of producing useful protein crystals.

sandwich assay
An immunological assay in which the target is captured by one antibody and detected by a second antibody recognizing a different epitope.

Sanger chain-termination method
A DNA sequencing method that dominated molecular biology from 1977 until 2005, involving the separation of nested DNA molecules with a common starting point but different endpoints to determine the sequence.

scanning mode
An operational mode for quadrupole mass spectrometers in which a variable voltage is applied across the quadrupoles thus sequentially selecting ions over a range of m/z values.

SCOP (structural classification of proteins)
One of three major databases for the structural classification of proteins. *See also* **CATH, FSSP.**

screening for interactions between extracellular proteins (SCINEX-P)
A method for the detection of protein interactions in the secretory pathway by exploiting the yeast unfolded protein response.

SDS-PAGE (sodium dodecylsulfate polyacrylamide gel electrophoresis)
A form of polyacrylamide gel electrophoresis in which the highly anionic detergent sodium dodecylsulfate is added to the gel to ensure that each protein carries a uniform charge enabling separation by size.

secondary antibody
An antibody used to detect a primary antibody that in turn is bound to the target.

selected ion monitoring (SIM)
A sensitive mass spectrometry mode in which only one m/z value is monitored.

selected reaction monitoring (SRM)
A highly selective MS/MS mode in which the parent and product ions are analyzed at specific m/z values to detect particular fragmentation reactions.

selenomethionine
A naturally occurring amino acid containing selenium, which is useful for X-ray diffraction experiments because the selenium atom causes anomalous scattering.

separative transport
The separation of molecules based on different rates of migration.

sequence identity
See **identity**.

sequence patterns
Representation of conserved sequences showing alternative amino acids at each position but not their weighting.

sequence profiles (gapped weight matrices)
Position-specific score matrices for longer sequences (*see* **position-specific score matrix**).

Sequence Read Archive (SRA)
A dedicated resource within the INSDC sequence databases that is used to store raw next-generation sequencing data.

sequence similarity
See **similarity**.

sequence space
The complete collection of all protein sequences in existence.

sequence tag
Any short DNA sequence that can be used to identify a gene (*see* **expressed sequence tag**).

serial analysis of gene expression (SAGE)
A large-scale sequencing method in which short DNA tags are joined together in a concatamer for sequencing.

shielding
The influence of nearby electrons on the resonance frequency of nuclei in NMR experiments, resulting in chemical shifts that can be used to identify particular structures.

short interfering RNAs (siRNAs)
Double-stranded RNA duplexes, usually about 21 base pairs in length with short 3′ overhangs, which interact with the nuclease Dicer to initiate RNA interference.

short linear motifs (SLiMs/MiniMotifs)
Regulatory protein modules comprising short and variable sequences that are not necessarily evolutionarily related.

shotgun proteomics
The unbiased and nonselective detection and identification of all proteins in a sample.

significance analysis of interactome (SAINT) scores
A socioaffinity scoring method that also incorporates peptide counts from mass spectrometry data.

silver staining
A sensitive method for staining proteins in gels, usually based on the reduction of silver nitrate.

SIMAC (sequential elution from IMAC)
A sensitive chromatography method that allows the separation of phosphorylated proteins based on the number of phosphate groups.

similarity
In protein sequence comparisons, the percentage of correctly aligned identical and related amino acids, determined using a substitution score matrix.

SIRAS (single isomorphous replacement with anomalous scattering)
An X-ray crystallography method used to solve the phase problem, in which a protein crystal incorporates a heavy metal atom that allows anomalous scattering.

size exclusion chromatography
See **gel filtration chromatography**.

Smith–Waterman algorithm
A dynamic programming algorithm that aligns sequences by searching for local similarity.

socioaffinity scoring methods
Any method for validating protein interaction data by examining the biology of known interaction networks.

sodium dodecylsulfate (SDS)
A strongly anionic detergent that binds proteins stoichiometrically and imparts a uniform negative charge.

soft ionization methods
Ionization methods in protein mass spectrometry that do not fragment the peptide backbone, and thus generate intact peptide ions.

solid-state NMR
NMR spectroscopy used for the analysis of molecules attached to surfaces.

solution arrays
Protein microarrays in a bead or particle format in solution.

SORI-CID (sustained off-resonance irradiation collision-induced dissociation)
A form of collision-induced dissociation used in Fourier transform ion cyclotron resonance mass spectrometry in which ions circling in an ion trap are exposed to increasing pressure to induce collisions.

SOS-recruitment system (SRS)
A method for detecting interactions among membrane proteins by making cell survival dependent on interactions with the critical membrane-bound regulator SOS (CDC25).

spare parts algorithm
An algorithm that searches through a database of loop structures in order to find matches for particular protein structures.

speciation
The origin of new species.

spectral counting
A label-free quantitative mass spectrometry technique in which the number of recorded spectra corresponding to a particular peptide correlates with the abundance of that peptide.

spectral dictionaries
Spectral libraries that also contain sequence information, thus bridging the gap between *de novo* sequencing and spectral database searching.

spectral fingerprint
A unique code generated by multiple fluorescent beads.

spectral library
Collections of deposited MS/MS spectra that allow direct comparison with experimental data.

spectral reflectance imaging biosensing (SRIB)
A form of interferometry that can be used to characterize protein interactions on microarrays.

split β-galactosidase assay
A protein complementation assay that detects protein interactions by reassembling a functional molecule of β-galactosidase.

split luciferase assay
A protein complementation assay that detects protein interactions by reassembling a functional molecule of luciferase.

split TEV assay
A protein complementation assay that detects protein interactions by reassembling a functional molecule of tobacco etch virus protease, which then releases a reporter molecule by proteolysis.

split ubiquitin system
A protein complementation assay that detects protein interactions by reassembling a functional ubiquitin molecule, which cleaves off a fused transcription factor and allows it to enter the nucleus and activate a reporter gene.

split-Trp assay
A protein complementation assay that detects protein interactions by reassembling a functional molecule of tryptophan synthase, allowing yeast to survive on media lacking tryptophan.

spoke interactions
Direct reciprocal interactions between proteins within a complex.

spot detection
Any mathematical method used to detect regions in a digital image resembling a spot or blob.

spot excision robots
An automated device for the removal of spots from protein gels.

SPOT method
A method for the fabrication of protein microarrays involving the sequential addition of activated amino acids.

SPR imaging
A modification of standard surface plasmon resonance spectroscopy using a broad beam of monochromatic polarized light that allows changes across the entire surface of a chip to be recorded simultaneously and in real time.

stable isotopes
Chemical isotopes that do not undergo radioactive decay.

stable-isotope labeling with amino acids in cell culture (SILAC)
A quantitative mass spectrometry technique in which one cell culture is fed with isotopically labeled amino acids and a comparative cell culture is fed on normal medium, thus introducing a small mass difference between the peptides from each sample.

standard MS mode
A mass spectrometry mode used with triple quadrupole mass spectrometers in which only the first quadrupole is used as a mass analyzer and the others operate in RF-only mode.

static light scattering
A method to detect the formation of protein complexes by reporting changes in the hydrodynamic radius of molecules in solution.

stationary phase
The fixed solid phase in a chromatography column, usually comprising resin beads with functionalized surfaces.

sticky bait
Bait proteins that interact nonspecifically with many prey.

sticky prey
Prey proteins that interact nonspecifically with many baits.

strand propensity
The tendency for an amino acid to be found in β-strands.

strip gels
Narrow gels typically used for isoelectric focusing.

strong anion exchange (SAX) chromatography
Anion exchange chromatography with highly charged cationic resins.

structural genomics
The branch of proteomics that seeks to solve representative protein structures for each protein fold in existence.

structural proteomics
The branch of proteomics that deals with protein structures, including solving structures (structural genomics) and the analysis of protein interactions with other molecules.

structure factor
A complete description of a reflection generated by X-ray diffraction, namely, the wavelength, amplitude, and phase of the incident X-rays.

structure space
The complete collection of all protein folds in existence.

sub-proteome
Any subset of the proteome, for example, in terms of location (for example, membrane, organelle) or post-translational modification (for example, phosphorylation, glycosylation).

sub-proteomics
Any branch of proteomics that deals with a subset of the proteome (for example, membrane proteomics, plastid proteomics).

substitution score matrix
A matrix used to calculate alignment scores based on the weighting of different amino acid substitutions, usually according to the substitutions that occur in nature.

superfolds
Protein domains found in many proteins with diverse tertiary structures and functions.

support vector machines
Supervised learning models with associated learning algorithms that analyze data and recognize patterns.

suppressor mutants
Second mutations that compensate for a primary mutation and restore the original phenotype.

surface entropy reduction
The replacement of high-entropy residues with smaller residues that support protein crystallization.

surface plasmon resonance (SPR)
An optical effect caused by the reflection of monochromatic polarized light from thin metal films.

surface plasmon resonance spectroscopy
A technique that detects protein interactions on the surface of thin metal films by measuring changes in surface plasmon resonance.

surface-enhanced laser desorption/ionization (SELDI)
The direct ionization of proteins bound to the surface of a chip or microarray by MALDI, followed by analysis by mass spectrometry.

surface-enhanced neat desorption (SEND)
A variation of SELDI in which the MALDI matrix compound is incorporated into the surface of a protein chip.

SWISS-MODEL
A structural bioinformatics WebServer used for homology modeling.

Swiss-Prot
One of the forerunners of the UniProt protein sequence database.

synchrotron radiation CD (SRCD)
See **circular dichroism**.

synchrotron radiation sources
A cyclic particle accelerator that produces high-energy monochromatic radiation.

synthetic carrier ampholytes
Collections of small amphoteric molecules covering a range of pI values allowing the formation of a pH gradient in an applied electric field.

synthetic genetic array (SGA)
A systematic high-throughput platform for synthetic lethal screens in yeast.

synthetic lethal screen
The detection of enhancer mutations by crossing haploid yeast cells and screening for lethal combinations.

systematic affinity purification–mass spectrometry
The use of affinity pulldown techniques to systematically screen for protein complexes.

systems biology
The holistic analysis of complex interactions within biological systems.

tag
(a) Sequence tag—any short DNA sequence that can be used to identify a larger sequence, such as a gene (*see* **expressed sequence tag**). (b) Protein tag—any short sequence of amino acids added to a recombinant protein to facilitate its identification or purification, or to control its activity in the cell (for example, by targeting to specific compartments). (c) Peptide tag—a short *de novo* peptide sequence derived by mass spectrometry that can be used for database searching. (d) Mass tag (*see also* **mass label**)—any specific adduct conjugated to a peptide or protein to create a defined mass difference that can be detected by mass spectrometry.

tandem affinity purification (TAP)
A systematic affinity purification–mass spectrometry technique involving two protein tags for increased specificity and sensitivity.

tandem mass tag system
An isobaric quantitative mass spectrometry technique similar to iTRAQ.

tandem mass spectrometry (MS/MS)
A form of mass spectrometry in which two mass analyzers operate in series, separated by a collision cell, allowing the analysis of fragment ions.

target discovery
The identification of a potential drug target.

target validation
The gathering of information about a potential drug target to confirm it is "druggable" (suitable for drug development).

Taylor's Venn diagram
A Venn diagram showing the overlapping properties of different amino acids.

temperature factor
A measure of certainty in an electron density map generated by X-ray diffraction.

terminal amine isotopic labeling of substrates (TAILS)
A quantitative mass spectrometry technique used for the analysis of N-terminal peptides.

threading
A method for structural prediction involving the recognition of folds that can be used for structural modeling without homology at the sequence level, achieved by searching through a fold library.

three-state predictions
The assignment of amino acid residues to one of three secondary structures: helix, strand, or unstructured coil.

time of flight (TOF)
A mass analyzer that exploits the fact that heavy and light ions with the same charge will take different times to travel down a field-free flight tube.

tissue microarray
A microarray comprising small tissue samples, often as serial sections.

TOCSY (total correlation spectroscopy)
A form of NMR spectroscopy that detects groups of protons interacting through a coupled network, not necessarily adjacent bonded pairs.

top-down proteomics
A form of proteomics that begins with the fragmentation of intact proteins.

Trace Archive
A dedicated resource within the INSDC sequence databases that is used to store raw capillary sequencing data.

transcriptome
The complete complement of messenger RNAs in a cell or organism.

transcriptomics
The global analysis of gene expression involving the highly parallel detection and quantitation of messenger RNAs.

transition pairs
A matched pair of precursor (parent) and product (daughter) ions.

transmembrane barrel
A protein structure composed of β-sheets, which is inserted into lipid membranes, often found as part of a transmembrane protein.

transmembrane helix
A protein structure composed of α-helices, which is inserted into lipid membranes, often found as part of a transmembrane protein.

transposons
Mobile DNA sequences that can move to new positions within a gene.

transverse relaxation optimized spectroscopy (TROSY)
Variant of NMR spectroscopy that allows the direct analysis of large proteins and protein complexes.

TrEMBL
One of the forerunners of the UniProt protein sequence database.

triple quadrupole
A mass spectrometer comprising three quadrupoles in series.

trypsin
A serine protease that is widely used in proteomics to digest proteins into defined peptides.

two-dimensional gel electrophoresis (2DGE)
Gel electrophoresis in which different separative principles are applied in orthogonal dimensions.

two-dimensional peptide mapping
A method for phosphorylation site analysis based on the two-dimensional separation of phosphopeptides by thin-layer chromatography.

two-hybrid systems
A system for detecting protein–protein interactions by expressing a bait protein and potential prey as fusions with two fragments of a transcription factor that becomes functional when assembled by interaction between the bait and prey.

two-step elution
Elution from a chromatography column caused by a sudden rather than gradual change in the buffer composition.

ubiquitin
A small regulatory protein that controls protein localization and recycling.

ubiquitin-based split protein sensor, USPS
See **split ubiquitin**.

unconfirmed
A putative gene whose existence is suggested by the sequence but whose expression and function has not been verified.

uninterpreted MS/MS spectrum
An MS/MS spectrum that has not been interpreted to derive sequence data.

UniProt
The principal database of protein sequences.

watershed transformation method
A automated method for detecting protein spots on gel images by dividing images into catchment basins and watershed lines.

western blot
See **immunoblot**.

whole-genome shotgun
A strategy for genome sequencing in which short sequence reads are generated and assembled without a preexisting map or scaffold.

whole-proteome microarray
A protein microarray containing every protein in the proteome.

word methods
Heuristic methods for sequence alignment, which are faster than dynamic programming algorithms.

word
In sequence alignment, a short ungapped sequence of identical or near-identical letters.

Worldwide Protein Data Bank (wwPDB)
The global consortium which provides access to databases of protein structures.

X-ray diffraction
A technique for the determination of protein structures by analyzing the reflections of X-rays scattered by a protein crystal.

X-ray mini-beams
Narrow X-ray beams that can generate diffraction data from small protein crystals.

Yeast ProtoArray
A yeast whole-proteome microarray.

zoom gels
Isoelectric focusing gels with a narrow pH range.

INDEX

Note. The index covers the main text but not the color plates section or the Glossary. The suffixes B, F, and T indicate that a topic is treated only in a box, figure or table on a page separated from any relevant text discussion.

Acronyms and initialisms are listed only in their compact form, unless their expansions appear equally or more often in the text. The prefixes α (alpha) and β (beta) and numbers are sorted as though spelled out.

A

B

C

Q

R

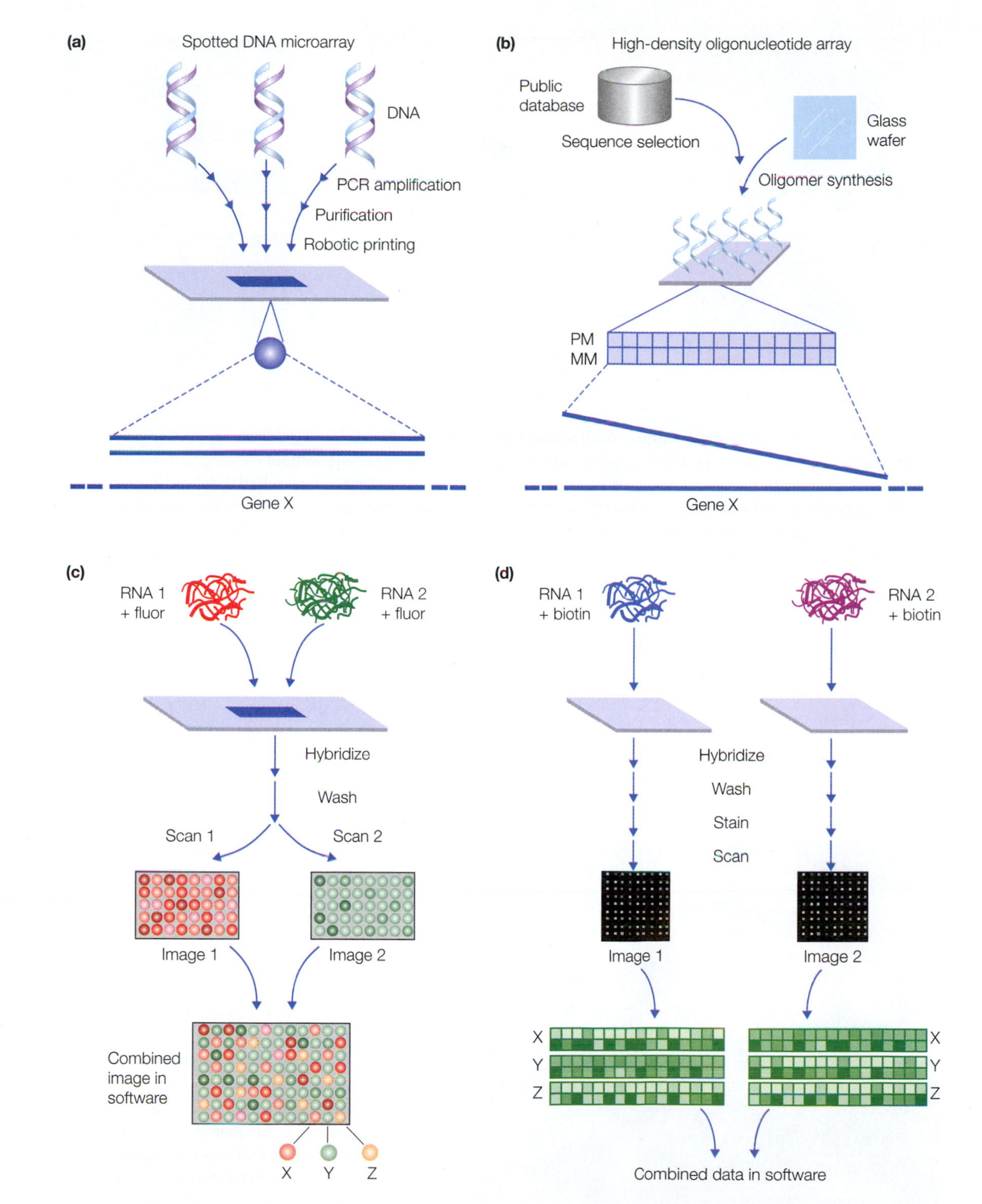

FIGURE 1.4 Expression analysis with DNA microarrays. (a) Spotted microarrays are produced by the robotic printing of amplified cDNA molecules onto glass slides. Each spot or feature corresponds to a contiguous gene fragment of several hundred base pairs or more. (b) High-density oligonucleotide chips are manufactured using a process of light-directed combinatorial chemical synthesis to produce thousands of different sequences in a highly ordered array on a small glass chip. Genes are represented by 15–20 different oligonucleotide pairs (PM, perfectly matched; MM, mismatched) on the array. (c) On spotted arrays, comparative expression assays are usually carried out by differentially labeling two mRNA or cDNA samples with different fluorophores. These are hybridized to features on the glass slide and then scanned to detect both fluorophores independently. Colored dots labeled X, Y, and Z at the bottom of the image correspond to transcripts present at increased levels in sample 1 (X), increased levels in sample 2 (Y), and similar levels in samples 1 and 2 (Z). (d) On Affymetrix GeneChips, biotinylated cRNA is hybridized to the array and stained with a fluorophore conjugated to avidin. The signal is detected by laser scanning. Sets of paired oligonucleotides for hypothetical genes present at increased levels in sample 1 (X), increased levels in sample 2 (Y), and similar levels in samples 1 and 2 (Z) are shown. (From Harrington CA, Rosenow C & Retief J (2000) *Curr. Opin. Microbiol.* 3, 285. With permission from Elsevier.)

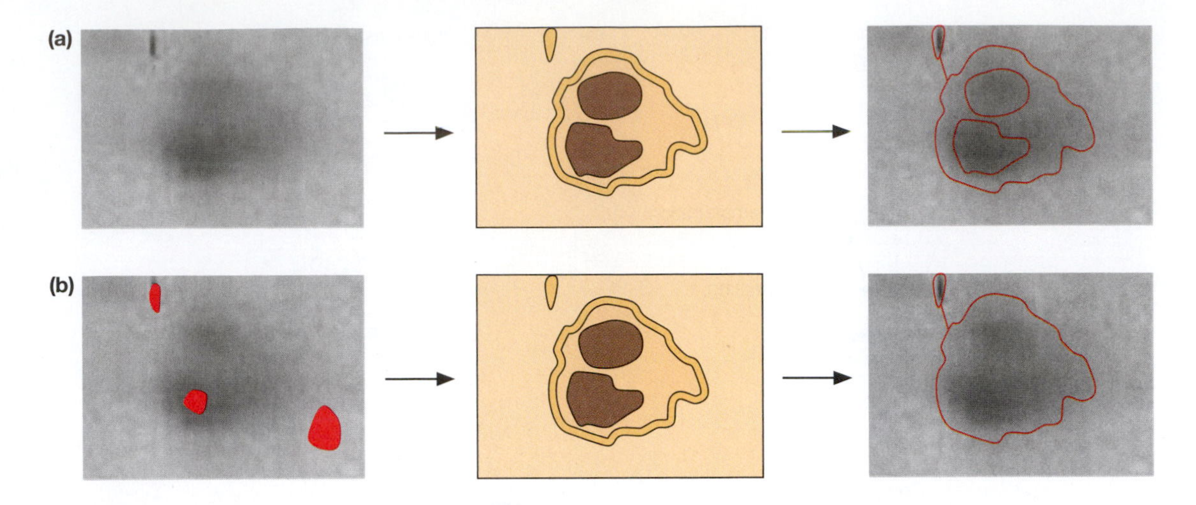

FIGURE 4.1 The watershed method for contour finding on two-dimensional gel images. (a) Any grayscale image can be considered as a topographic surface. If flooded from its minima without allowing water from different sources to merge, the image is partitioned into catchment basins and watershed lines, but in practice this leads to over-segmentation. (b) Therefore, markers (*red shapes*) are used to initiate flooding, and this reduces over-segmentation considerably. (Adapted from images by Serge Beucher, CMM/École Nationale Supérieure des Mines de Paris.)

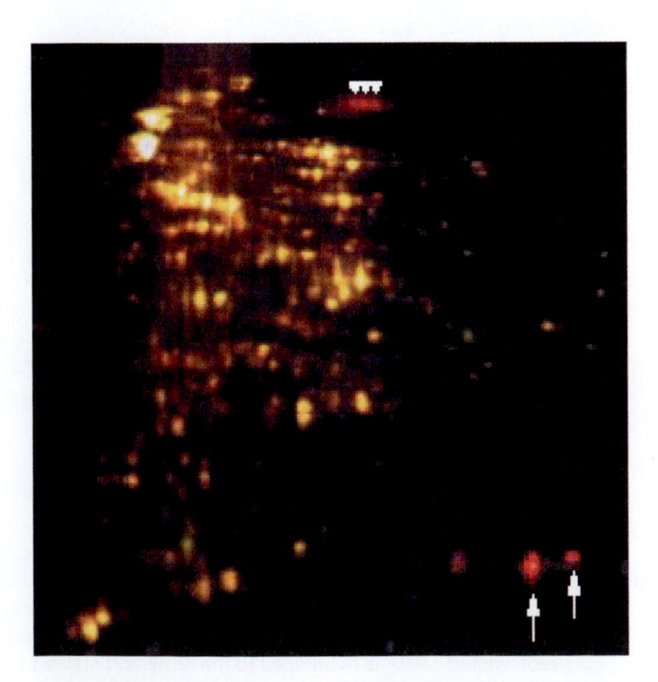

FIGURE 4.4 Two-dimensional DIGE. Overlay image of Cy3- (*green*) and Cy5- (*red*) labeled test-spiked *Erwinia carotovora* proteins. The protein test spikes were three conalbumin isoforms (*arrowheads*) and two myoglobin isoforms (*arrows*). Spots that are of equal intensity between the two channels appear *yellow* in the overlay image. As spike proteins were eight times more abundant in the Cy5 channel, they appear as *red* spots in the overlay. The gel is oriented with the acidic end to the left. (From Lilley KS, Razzaq A & Dupree P (2002) *Curr. Opin. Chem. Biol.* 6, 46. With permission from Elsevier.)

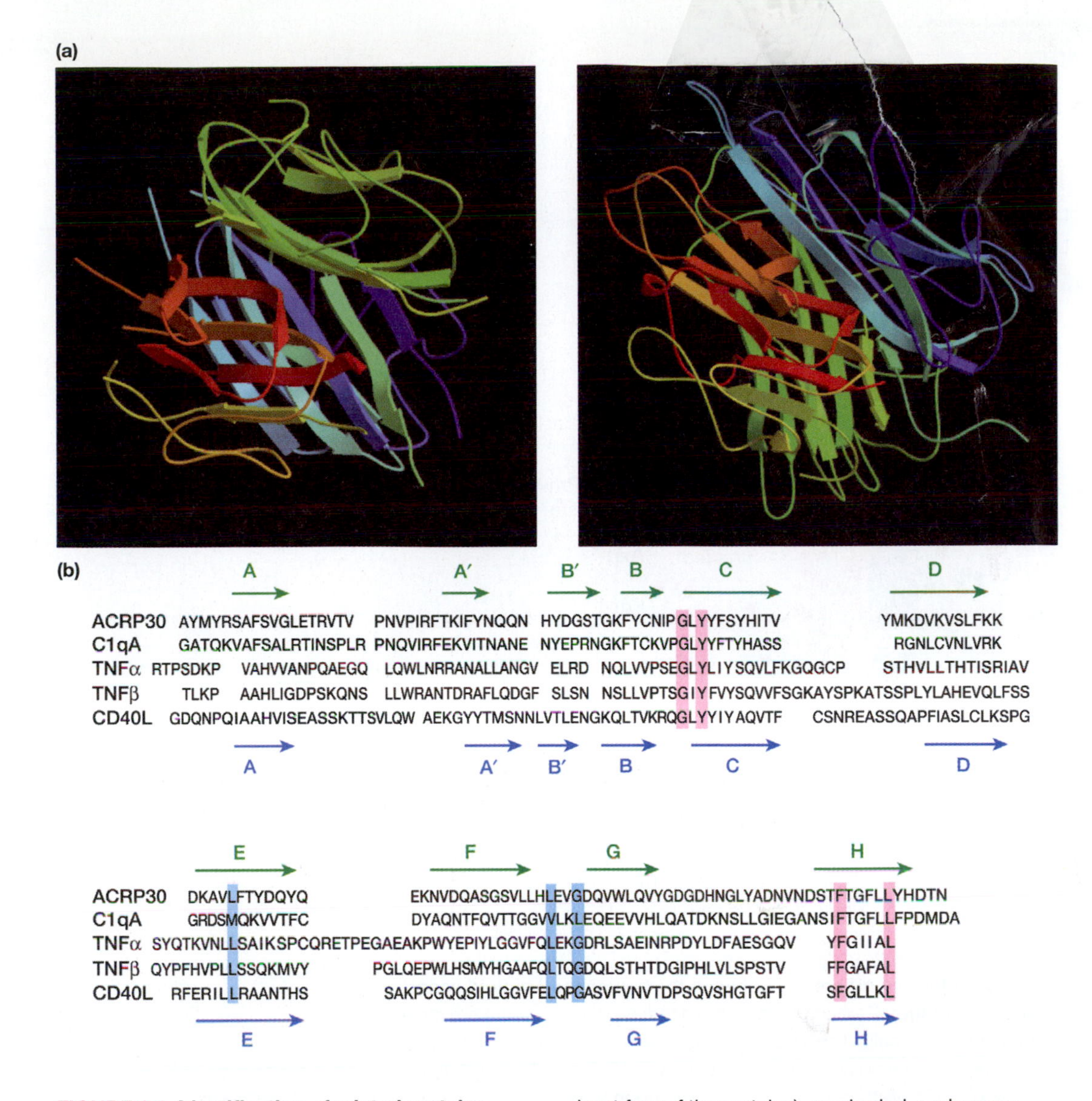

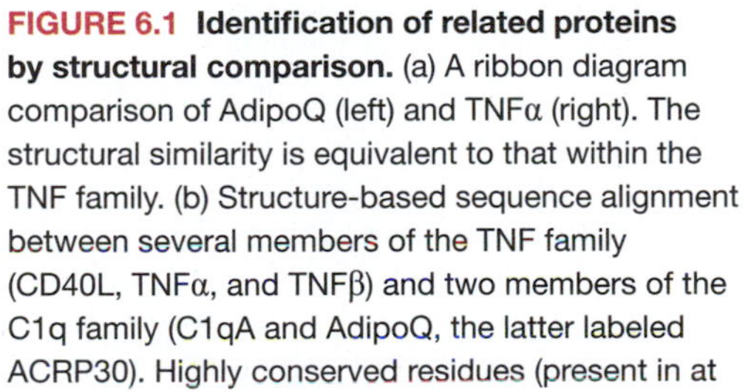

FIGURE 6.1 Identification of related proteins by structural comparison. (a) A ribbon diagram comparison of AdipoQ (left) and TNFα (right). The structural similarity is equivalent to that within the TNF family. (b) Structure-based sequence alignment between several members of the TNF family (CD40L, TNFα, and TNFβ) and two members of the C1q family (C1qA and AdipoQ, the latter labeled ACRP30). Highly conserved residues (present in at least four of the proteins) are shaded, and arrows indicate β-strand regions in the proteins. There is little sequence similarity between AdipoQ and the TNF proteins (for example, 9% identity between AdipoQ and TNFα), so BLAST searches would not identify a relationship. (Adapted from Shapiro L & Harris T (2000) *Curr. Opin. Biotechnol.* 11, 31. With permission from Elsevier. Images courtesy of Protein Data Bank.)

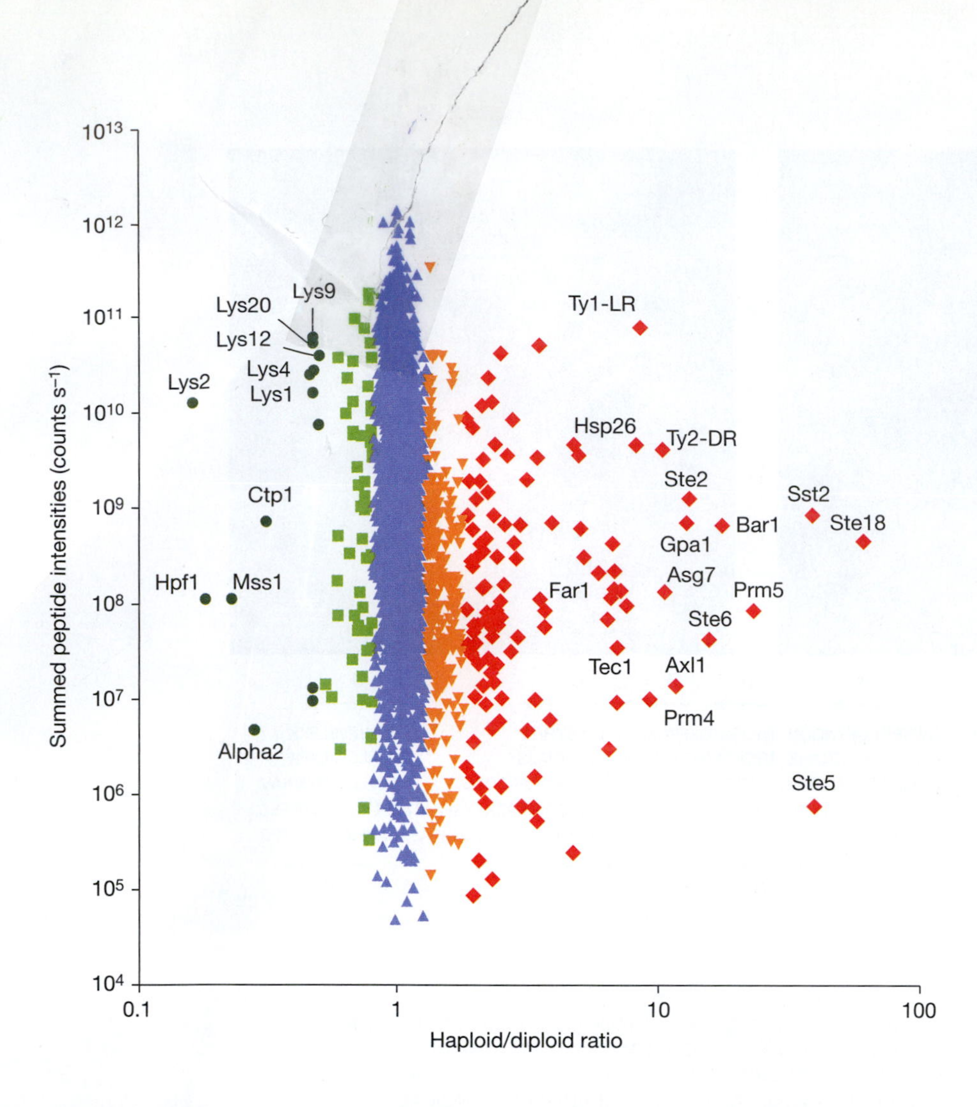

BOX 4.5 FIGURE 2 Quantitative difference between the haploid and diploid yeast proteome (overall fold change). Proteins to the left (becoming deeper *green*) are more strongly represented in haploid cells. Proteins to the right (becoming deeper *red*) are more strongly represented in diploid cells. (From de Godoy LMF, Olsen JV, Cox J et al. (2008) *Nature* 455, 1251–1254. With permission from Macmillan Publishers Ltd.)

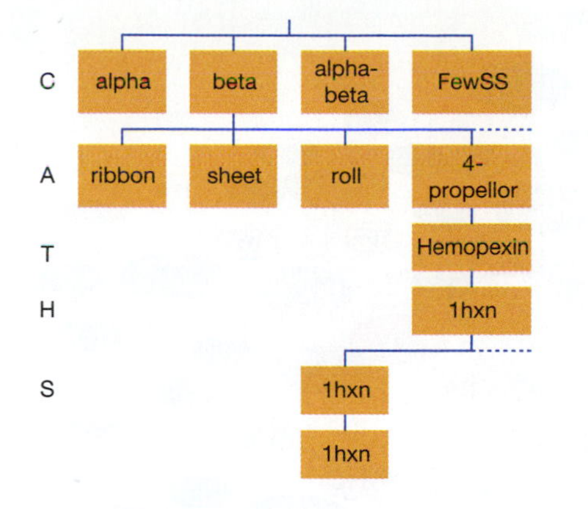

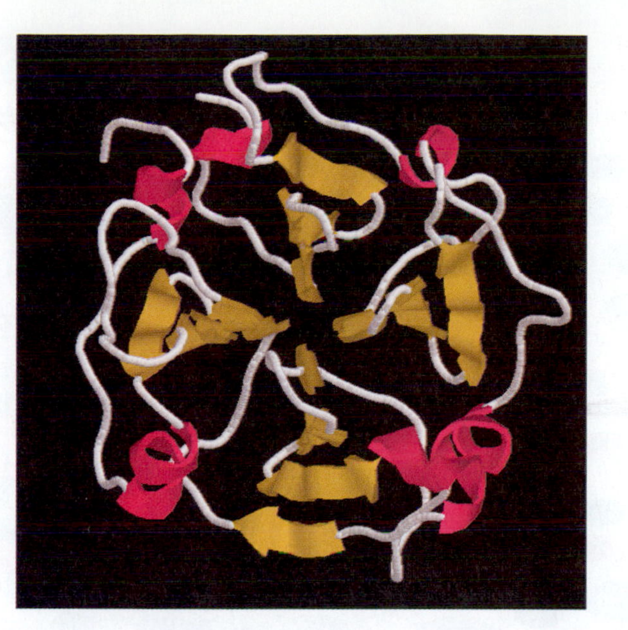

FIGURE 6.8 Structural classification of proteins using the CATH database. The protein shown is hemopexin, a protein rich in β-sheets with few α-helices. (Courtesy of Christine Orengo.)

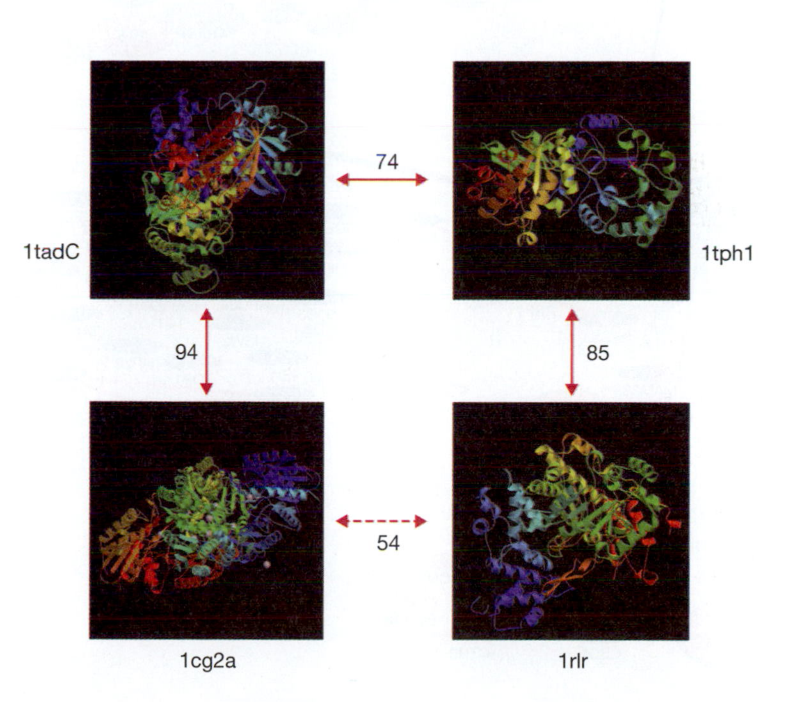

FIGURE 6.9 The Russian doll effect. Four proteins are illustrated that show continuous structural variation over fold space. Each of the proteins shares at least 74 structurally equivalent residues with its nearest neighbor, but the two extreme proteins show only 54 structurally equivalent residues when compared directly. Key: 1cg2a, carboxypeptidase G2; 1tadC, transducin-K; 1tph1, triose phosphate isomerase; 1rlr, ribonucleotide reductase protein R1. (From Domingues FS, Koppensteiner WA & Sippl MJ (2000) *FEBS Lett.* 476, 98. With permission from Elsevier. Images courtesy of Protein Data Bank.)

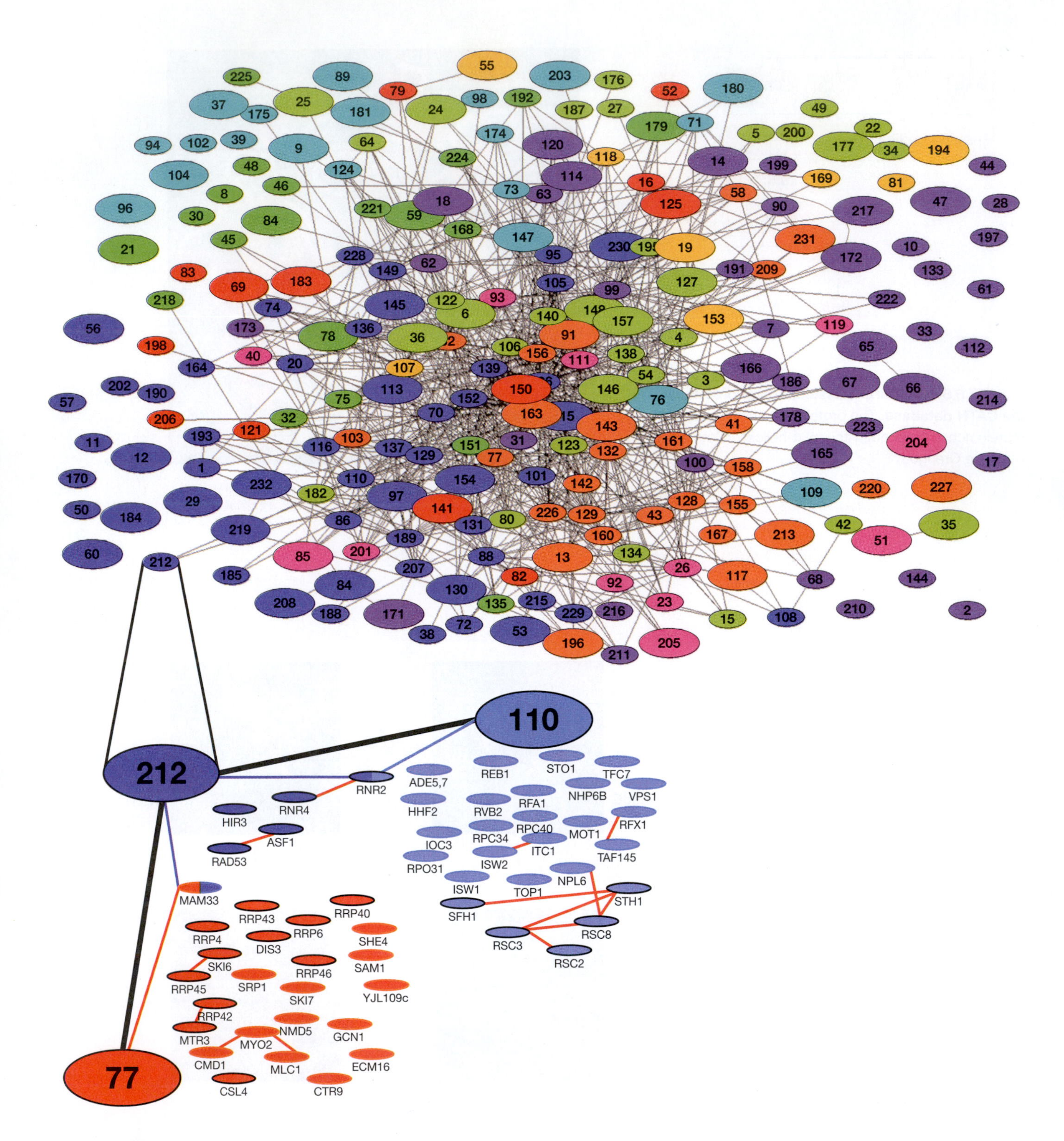

FIGURE 7.19 The protein complex network, and grouping of connected complexes. Links were established between complexes sharing at least one protein. For clarity, proteins found in more than nine complexes were omitted. The graphs were generated automatically by a relaxation algorithm that finds a local minimum in the distribution of nodes by minimizing the distance of connected nodes and maximizing the distance of unconnected nodes. In the upper panel, cellular roles of the individual complexes are color-coded: *red*, cell cycle; *dark green*, signaling; *dark blue*, transcription, DNA maintenance, chromatin structure; *pink*, protein and RNA transport; *orange*, RNA metabolism; *light green*, protein synthesis and turnover; *brown*, cell polarity and structure; *violet*, intermediate and energy metabolism; *light blue*, membrane biogenesis and traffic. The lower panel is an example of a complex (TAP-C212) linked to two other complexes (TAP-C77 and TAP-C110) by shared components. It illustrates the connection between the protein and complex levels of organization. *Red* lines indicate physical interactions as listed in the Yeast Proteome Database. (From Gavin AC, Bösche M, Krause et al. (2002) *Nature* 415, 141. With permission from Macmillan Publishers Ltd.)

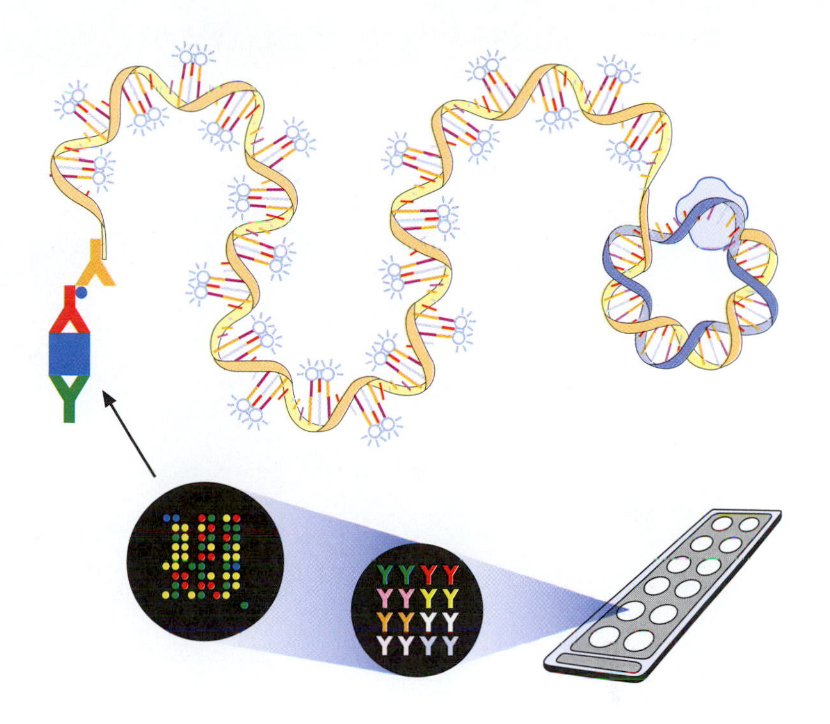

FIGURE 9.5 Sensitive protein detection using the RCA antibody chip. The chip is divided into 16 Teflon wells, each containing an array of 256 antibodies as probes. When a protein, represented by the *blue* square, is captured by one of the probes, it can be recognized using a second, biotinylated antibody (*red*), which is subsequently detected by a tertiary universal antibody connected to a circular oligonucleotide. A strand-displacing DNA polymerase can use this circular template, generating a long concatemer. (From Kingsmore SF & Patel DD (2003) *Curr. Opin. Biotechnol.* 14, 74. With permission from Elsevier.)